FLUID POWER

Third Edition

HARRY L. STEWART

and

JOHN M. STORER

Bobbs-Merrill Educational Publishing
Indianapolis

Copyright © 1980, 1973, and 1968 by The Bobbs-Merrill
Company, Inc.
Printed in the United States of America

The Bobbs-Merrill Company, Inc.
4300 W. 62nd Street
Indianapolis, Indiana 46268

Third Edition
First Printing 1980

Library of Congress Cataloging in Publication Data
Stewart Harry L
    Fluid power.

    Includes index.
    1. Fluid power technology. I. Storer, John M.,
joint author. II. Title.
TJ840.S758   1980   621.2   79-9123
ISBN 0-672-97224-7

Contents

CHAPTER 44

Preface

In this day of mechanization, automation, and mechanical expansion of the muscles of man—whether his efforts are in the agricultural, automotive, industrial production, marine, or aviation fields or in the more sophisticated space exploration program—fluid power touches the daily lives of a continually increasing number of our population. If an individual is to master these mechanical muscles, he must grasp and keep pace with the basic fundamental principles which are the foundation for this vast technological explosion.

The purpose of this text is to present a fundamental understanding of the physical principles of fluid power in a logical, building-block manner, along with a practical working knowledge of the components normally utilized in designing, installing, operating, and maintaining fluid power systems—whether they are simple or complex. Sufficient basic data and principles are provided to orient the totally inexperienced quickly; even those who have been exposed to the subject previously will find many suggestions and aids which can increase their efficiency.

This text is aimed at the vocational or engineering student who is preparing himself for industry, the man in the shop who is responsible for keeping these mechanical marvels in operation, and the tool or machine designer who is striving to increase production. Also, it enables the salesman of fluid power components to achieve a more thorough understanding of the total field in which his products are used.

Chapter 1

Introduction to Fluid Power

As a student of fluid power, you are in one of two classes—either you have had practically no formal contact with fluid power and fluid power components. Or you have had some degree of experience with the design, application, and maintenance of one or more devices of fluid power equipment. This course is, of necessity, pointed toward the first class, with the additional intention of pursuing the subject in sufficient depth and breadth to project us into and beyond the experience of as many students in the second class as is practical. Participation by students with a variety of experience in the latter class can be quite helpful. Experience is one of our greatest teachers. However, this can, in turn, be a source of great frustration if only questions and problems without practical answers are produced. One of the objectives of this course is to provide as many answers as possible. Since the application of fluid power has become world-wide in recent years, some students have, without doubt, experienced problems and answers that are normally not dealt with in this type of course. Active participation by these students is invited and encouraged.

1

MODERN CONCEPT

The term *fluid power* is used to denote the technology that deals with the transmission and control of energy by means of pressurized fluids. The simple transfer of fluids, as with water in either a well or an irrigation system, does not fall within this concept of fluid power, because the fluid in these systems is not used for the transmission of energy.

The modern fluid power concept has appeared only within the last four hundred years, with only an extremely slow expansion of its applications and uses until after the turn of the present century. For example, the advent of the Industrial Revolution with the technical capacity for producing interchangeable machined parts made possible the production in quantity of the snugly fitted parts that are required in fluid power components. World War I brought a pronounced expansion of the use of fluid power, particularly by the manufacturers of naval equipment. Since World War I and especially near the middle of the 20th Century, the expansion in fluid power as an industry is in line with the explosive increase in the number of manufacturers that are totally or partially involved in the production of fluid power components. This growth has increased many times during the past 50 years (Fig. 1-1).

Joint Industry Conference

Probably the most common users of fluid power components have been in the field of production machinery. As a result of the industry's concern for uniformity of quality and performance specifications, representatives of fluid power equipment manufacturers held a series of meetings in Detroit in 1951. From those meetings (since referred to as the *Joint Industry Conference*, or *JIC*), a set of standards, fluid power symbols, and specifications that have become widely accepted and used throughout industry was evolved. The symbols, in particular, have equipped the fluid power designer and user with a common written language that transcends design variations, trade names, and even the normal language barriers between nations.

National Fluid Power Association

By 1954, the fluid power component manufacturers accepted this obvious challenge and formed the *National Fluid Power Associa-*

Fig. 1-1. Farm tractors use a number of hydraulically operated attachments. Here is a loader attachment which eliminates many backbreaking hours on the farm. Note the hydraulic cylinders that power the lift arms. These cylinders utilize *Teflon* rings and seals. The operator can actuate the loader with little effort through the hydraulic control valve. (Courtesy International Harvester Company, Pay Line Group)

tion, or *NFPA*, for the purpose of further strengthening the professional stature and service value of the fluid power industry. The development of standards for the fluid power industry and fluid power technology was a foremost objective of the *NFPA* at the time of its establishment—and this area is the most prominent single area of activity in the association today.

NFPA is an association of manufacturers of fluid power systems and components with a membership of over one hundred seventy companies throughout the United States. In addition to its interests in standardization, the association provides its membership with a variety of marketing, statistical, management, public relations, and educational services. In all these areas it represents the fluid power industry of the United States to other industries, governmental agencies, and the general public.

Until 1960, the development of fluid power standards was also a primary interest and purpose of the *Joint Industry Conference* (JIC), sponsored primarily by the automotive industry, but participated in by other consumers, manufacturers of fluid power equipment, and general interests. *JIC* was responsible for developing the first generally accepted standards for fluid power symbols and drafting practices and for the widely used *JIC Hydraulic and Pneumatic Standards for Industrial Equipment*. In 1958, the *JIC* fluid power symbols were adopted by the *American Standards Association* as American Standards.

Later Developments

In 1960, the automotive industry ceased sponsoring the *Joint Industry Conference*. Recognizing the necessity for providing further continuity to the excellent work of *JIC*, the *National Fluid Power Association* proposed that the development of future fluid power standards become a joint activity of *NFPA* and the *American Standards Association* (*ASA*). At a general conference called by *ASA* on September 28, 1961, *NFPA* was officially named the sponsor of the newly organized *ASA* Sectional Committee, B93, Fluid Power Systems and Components, within the structure of *ASA*.

Under this administrative arrangement, *NFPA* is responsible for developing and reviewing recommended industry standards and for processing fluid power industry standards through *ASA* Sectional Committee B93, for adoption as American standards. Current standards activity of *NFPA* includes not only the updating and revision of portions of the *JIC* recommendations but also the development of recommended standards in many additional areas.

One of the first and most notable efforts of the *NFPA* was to prepare and offer to the industry a "Glossary of Terms" which is felt by many to have done away with much of the confusion that existed due to nonstandard terms.

The *NFPA*, in turn, fostered the *Fluid Power Society*, a technical organization devoted to furthering the growing technology of fluid power—the exchange of ideas, techniques, and information, and the updating and upgrading of the members' knowledge of fluid power.

Fluid Power Society

It was realized at the time of the formation of the *Fluid Power Society* that there was a need for an organization to serve all individuals interested in any phase of fluid power—research, development, design, application, installation, operation, maintenance, and education—and serving all fields of utilization—including industrial, aerospace, marine, mobile, materials handling, farm implements, etc.

The *Fluid Power Society* has made outstanding contributions to fluid power education in the past few years. It is a founding member of the *Council on Fluid Power Education* which is made up of representatives from six international educational and engineering organizations. It cosponsored, with the *National Fluid Power Association* and Wayne State University, the first summer institute on fluid power education for college teachers. In 1965, the association secured a government contract for $234,000 to sponsor seven summer institutes for vocational and technical teachers.

The *Fluid Power Society* is a cosponsor of the *National Conference on Fluid Power*, a cosponsor of the *Fluid Power Seminars* conducted by the Milwaukee School of Engineering, and a cosponsor of seminars offered in cooperation with the *National Fluid Power Association*. It also cooperates with other societies on annual fluid power meetings, including the *American Society for Engineering Education*, and the *American Vocational Association*.

ADVANTAGES

The rapid endorsement and use of fluid power has been due to a number of highly favorable characteristics. Some special-purpose applications are quite complex. However, most components and systems in common use are simple in construction. The wide availability of both rotating and linear force components permits the elimination of complicated systems of cams, gears, and levers. Motion can be transmitted without the slack and the backlash that are so difficult to eliminate in the use of solid machine parts. The ease with which corners can be turned with fluid transmission lines permits a minimum of linkage and a maximum of force directions (Fig. 1-2). A larger force can be controlled by a much smaller force. This permits sensitive, but comparatively weak, manpower to be multiplied many times (Fig. 1-3). The hydraulic braking system on a modern vehicle is a typical example. Fluid systems conveniently can be protected from overloading or overpowering by using simple pressure-limiting or relief devices. These systems can be made to duplicate the discerning touch of the human hand, but with a power force that far exceeds it (Fig. 1-4). The fluids themselves are not

Fig. 1-2. A not uncommon street sight is this hydraulically powered mobile unit used for fire fighting, tree trimming, and many aerial maintenance operations. Both azimuth and elevation can be controlled either from the personnel cage or from the base location. (Photo, courtesy Mobile Aerial Towers, Inc.)

subjected to wear or breakage, and the fluid systems are economical to operate.

The extreme flexibility of fluid power systems does, however, give rise to a number of problems. Since fluids have no characteristic shape, they must be confined positively throughout the entire system, and they must be prevented from flowing to places where they are not supposed to flow. Special consideration must be given the structural design and the relation of the various components in a fluid power system. Leakages must be prevented or anticipated, since they can be expensive and can cause malfunctions within a system. Proper sizing and strength of fluid conductors and the intelligent selection of working components are essential for the proper operation of a fluid system.   The pressures set up in a fluid system must be controlled, along with the movement of the fluids in the system. This movement causes friction, both within the liquid and against its containing surfaces. If this friction is excessive, accumulations of heat and serious losses in efficiency can occur.

Dirt particles cannot be permitted to enter or to accumulate in the fluid system, since they can clog functional orifices (openings or ports) and create excessive wear of the component parts. Chemical action may cause corrosion of the metal parts and breakdown of the seal compounds.

Fig. 1-3. Two joystick hand levers actuate boom, stick, bucket, and swing on this heavy-duty excavator. Powerful, dependable hydraulic components deliver the high flows for rapid swing, lift, and dump—or high pressure for maximum digging forces. (Courtesy Caterpillar Tractor Co.)

It is most important and necessary to understand how a fluid power system works, both in terms of the general principles common to all physical mechanisms and of the peculiarities to be expected in specific situations. Although solids differ in various ways from liquids and gases and although all substances—lead, tin, water, oil, oxygen, etc.—have their own special properties, all forms of matter have basic fundamental patterns of action and reaction (Fig. 1-5). When

Fig. 1-4. This low-profile shuttle car, driven by two 20 hp traction motors, features a four-wheel drive that is guided by *Orbitrol* hydraulic steering for ease of handling and good maneuverability. (Courtesy Joy Manufacturing Company)

Fig. 1-5. This liquid-filling machine controls the flow of liquids by an air signal given when the desired height is reached, providing extreme filling accuracy. (Courtesy Pneumatic Scale Corporation)

the principles of the physical sciences became clarified in early modern times, new potentialities for understanding and action were opened for everyone. Therefore, when a person thoroughly grasps these principles, he is in position to increase his power to understand, to control, and to use the latent forces of natural phenomena.

Although solids are rigid and fluids are nonrigid, the student can apply his knowledge of levers to hydraulic situations. In the same manner that a child weighing 50 pounds (22.7 kg) can be made to balance another child weighing 100 pounds (45.4 kg) by sitting at twice the distance from the fulcrum or balancing point, a force of 50 pounds (222.4N) can be applied over a given area in a hydraulic system to balance a force of 100 pounds (444.8N) applied over twice that surface area.

Each machine for accomplishing mechanical movement has its advantages and disadvantages, as well as its special capabilities and limitations. Ropes and chains, for example, can pull, but they cannot push. Solid rods can both push and pull, but they cannot be extended around corners in the same manner that chains, ropes, or hydraulic elements can be extended. Gears operate continuously in a rotary motion, whereas a lever usually operates in a back-and-forth motion. Although each machine obeys the same general laws of physics, it also possesses special properties that must be considered.

The inventor or engineer who attempts to move an object, therefore, is faced with several possibilities. The object that is to be moved may be either pushed with a stick or pulled with a chain. Either a jet of water or a stream of air may be played on it, or another object may be thrown against it. If a weight is to be lifted, the engineer may choose among levers, gears, ropes and pulleys, chain hoists, screw jacks, hydraulic jacks, or the force of flotation. Part of the system may make use of electricity, heat, or even light, as with doors that open automatically when people approach.

PRINCIPLES

The most important single fact or determinant that must be considered in developing an understanding of fluid power and its applications is stated in Pascal's law, which dates back to 1653 A.D. This law simply states *that pressure set up in a liquid or gas exerts a force equally in all directions.* The pressure vector *is perpendicular to all the containing surfaces.*

When a blow is struck on one end of a solid bar, the main force of the blow is transmitted in a straight line through the bar to the opposite end (Fig. 1-6). This occurs because the bar is rigid. The direction of the blow is nearly the sole determiner of the direction of the transmitted force. The more rigid the bar becomes, the less force is lost inside the bar or transmitted outward at right angles to the direction of the blow.

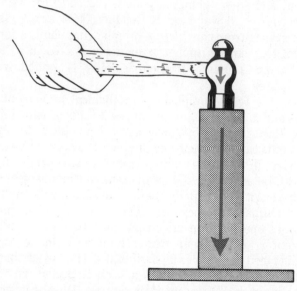

Fig. 1-6. The main force resulting from a blow struck on one end of a solid bar is transmitted in a straight line through the bar to the opposite end of the bar.

When a blow is applied to the upper end of a column containing a confined liquid, however, the force not only is transmitted in a straight line through the liquid to the opposite end but also is transmitted equally and undiminished in all directions throughout the column—forward, backward, and sidewise—so that the containing vessel is literally "filled" with pressure (Fig. 1-7).

For the same reason, a flat hose becomes circular in its cross section when it is filled with water under pressure. The outward push of the water is equal in all directions (Fig. 1-8). Water leaves the hose at the same velocity through different leaks, regardless of the side of the hose on which they might occur.

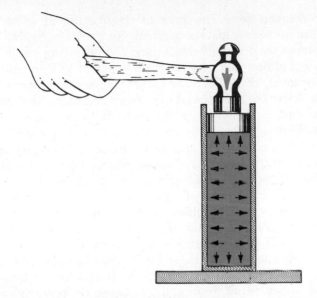

Fig. 1-7. When a blow is struck at the upper end of a column of confined liquid, the force is transmitted not only in a straight line through the column but also equally and undiminished in all directions throughout the column.

Another determinant was discovered experimentally by the British physicist Robert Boyle (1627-1691). It is known as Boyle's law. This law states *that a given number of molecules of a gas confined in a specific volume will increase in pressure as the volume is reduced; if the temperature of the gas remains constant, the pressure will vary inversely with the volume.*

Compression of a gas while maintaining a constant temperature is called *isothermic compression.* This is not practical to achieve, because it assumes 100 percent mechanical and volumetric efficiency—a Utopian accomplishment with no friction losses. Conversely, compression of a gas with no loss of the resulting heat is called

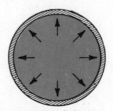

Fig. 1-8. A flat hose becomes circular in cross section when it is filled with water under pressure. The outward "push" of the water is equal in all directions.

adiabatic compression. The practical results are at some point between the isothermic and the adiabatic compression, depending on the equipment used, its efficiency, and the operating conditions.

The effects of flow and flow rates are explaind by Bernoulli's law (Daniel Bernoulli, 1700-1782): *The higher the speed of a flowing fluid or gas, the lower the pressure. As the speed decreases, the pressure increases, and, conversely, as the speed increases, the pressure decreases.* The significance of this law is most important in the selection of fluid conductors and in the sizing of flow components.

The temperature of a gas directly affects both pressure and volume, as defined in Charles' law (Jacques Charles). With each degree that the temperature of a gas rises, expansion occurs by the same fraction of its original volume, if the pressure remains constant. If the temperature is in Celsius (C), this fraction, which is called a constant, is 1/273 of the volume at 0° Celsius (C).

Charles' law indicates the reason why it is important to maintan reasonably low temperatures within an air receiver to avoid the energy losses that result from cooling between the receiver and the point of air usage. These losses may be read directly as a loss in pressure without a loss of air.

Although other laws of physics bear on the function of fluid and mechanical applications, the four laws mentioned above are the most basic and fundamental to the fluid power concept. The practical application of one or more of these laws is shown in every application of fluid power throughout industry.

Chapter 2

Areas, Forces, Pressures, and Volumes

The laws and theories presented in the first chapter are seldom used directly at the design, application, or maintenance levels, although they comprise the fundamental determinants that make the performances of fluid power components predictable. Pascal's law is the basis for calculating the actual forces that can be expected from a given fluid power component.

AREAS AND FORCES

If the force vectors within an air cylinder are diagrammed, it can be seen that the only usable vectors are those directed against the piston or force member. The force vectors shown in the lateral cross section (Fig. 2-1) are completely balanced and completely contained by the rigid cylinder wall. Therefore, these force vectors are static. They perform no work. The same force vectors are also shown in the longitudinal cross section (Fig. 2-2), along with the dynamic longitudinal force vectors which are directed against the movable force member. The latter force vectors are the only ones of concern in

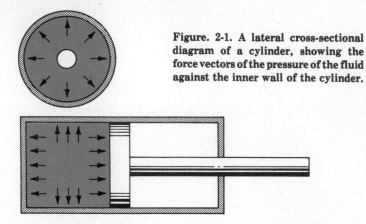

Figure. 2-1. A lateral cross-sectional diagram of a cylinder, showing the force vectors of the pressure of the fluid against the inner wall of the cylinder.

Fig. 2-2. Diagram of the lateral and longitudinal force vectors produced within the chamber of a cylinder.

determining the deliverable force of the cylinder component. The deliverable force is determined readily from two factors: (1) fluid pressure within the component, and (2) the effective area of the force member measured in a plane perpendicular to the direction of movement.

The most often used mathematical formula in the fluid power field is the formula for circular areas: $A = \pi \times r^2$, or 3.1416 times the radius squared. Since the force to be delivered by a cylinder is the product of the fluid pressure and the moving force area against which it is applied within the cylinder, the cylinder force formula is $F = A \times pressure$; with the force F, in pounds [newtons (N)]; the piston area A, in square inches (square centimeters); and the fluid pressure, in pounds per square inch [pascals (Pa) or kilopascals (kPa)].

The force formula is also used to calculate the resultant forces of fluid pressure against a square or rectangular area, such as the vane of a fluid motor or gear tooth, or the vane of a vane-type torque generator. The area of a vane is calculated simply by multiplying the vane length and the vane width, all in inches (centimeters).

Areas other than square, rectangular, or circular areas are commonly encountered in fluid power components, but it is rarely necessary to calculate them. These areas are common in vane- and gear-type pumps and motors, but the manufacturer's specifications usually give this information in terms of displacement per revolution.

Pressure is defined as *force divided by the area over which it is distributed*. As shown in Fig. 2-3, if the input force A is 100 pounds (444.8 N) and the area of the piston A is 10 square inches (64.5163 cm²), then the pressure in the liquid is [(100 ÷ 10) or (444.8 ÷ 64.5163)], or 10 pounds per square inch (10 psi), or 68.9 kPa. The same pressure also acts on piston B, so that each square inch of its area is pushed upward with a force B of 10 pounds per square inch (68.9 kPa). In this instance, a column of liquid having uniform cross section is considered, so the area of piston B is the same as that of piston A, or 10 square inches (64.5163 cm²). Therefore, the upward or output force B on piston B is 100 pounds (444.8 N), which is the same as the input force A applied to piston A. In other words, all that is accomplished, in this instance, is to move the 100 pounds (444.8 N) of force around a bend, but the basic principle that is diagrammed in Fig. 2-3 applies to nearly all mechanical uses of fluid power.

At this point, it should be noted that, since Pascal's law is independent of the shape of the container, it is not essential that the cross-sectional area of the tube connecting the two pistons is equal to the cross-sectional area of the pistons throughout its length. A connection of any size, shape, or length can be used, if an

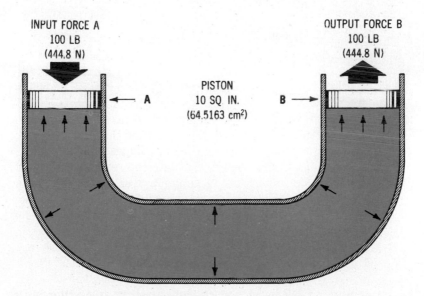

Fig. 2-3. Pressure in a fluid system.

unobstructed passage is provided. Therefore, the fluid system in Fig. 2-4, with a relatively small bent pipe connecting the two cylinders, performs in exactly the same manner that the system diagrammed in Fig. 2-3 performs.

In Figs. 2-3 and 2-4, hydraulic systems having pistons equal in area were considered, wherein the output force is equal to the input force. In Fig. 2-5, a fluid power system is shown in which the input piston A is much smaller than the output piston B. If it is assumed that the surface area of piston A is 2 square inches (12.9 cm²) and that the surface area of piston B is 20 square inches (129.0 cm²), an input force of 100 pounds (444.8 N) applied to piston A creates a pressure of 50 psi (344.7 kPa) within the fluid. The upward force on piston B, therefore, is 50 pounds (222.4 N) on each square inch (6.45 cm²) of its surface area, or 1000 pounds (4448.2 N). In this example, the original or input force is multiplied tenfold.

As shown in the example, this multiplication of force is determined by the relative sizes of the output piston B area and the input piston A area. The ratio of the output force to the input force remains at ten to one (10:1), regardless of the input force—as long as the piston area ratio is ten to one (10:1).

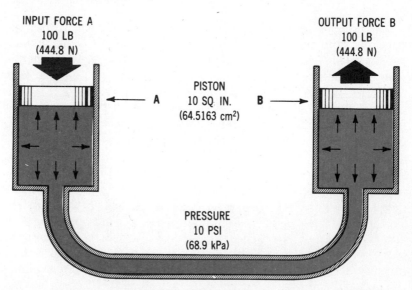

Fig. 2-4. Pressure in a fluid system in which the diameter of the connecting tube is smaller than the diameter of the pistons.

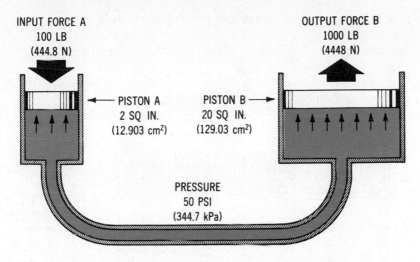

Fig. 2-5. Pressure in a fluid system in which the diameter of the input piston is much smaller than the diameter of the output piston.

The fluid system, of course, operates in the same manner when the output and the input pistons are reversed. If the input piston and the output piston are reversed (see Fig. 2-5), the output force then is one-tenth the input force. In some instances, reversal of the pistons is desirable. If two pistons are used in a fluid power system with a common fluid pressure, the force acting on each piston is directly proportional to its surface area, and the magnitude of the force on each piston is the product of the pressure and its surface area.

A different situation exists in the example diagrammed in Fig. 2-6. A cylinder A contains a single piston B with a piston rod C attached to one side of the piston and extending from the cylinder at one end. The liquid, under equal pressure, is admitted to both ends of the cylinder. The opposed faces of the piston B act as two pistons opposing each other. The surface area of the left-hand face is equal to the entire area of the cylinder, say 6 square inches (38.7 cm²). However, the surface area of the right-hand face is equal to the area of the cylinder minus the area of the piston rod, which is 2 square inches (12.9 cm²), leaving an effective area of $(6-2)$ $(38.7-12.9)$, or 4 square inches (25.8 cm²) on the right-hand face of the piston. The pressure on both faces of the

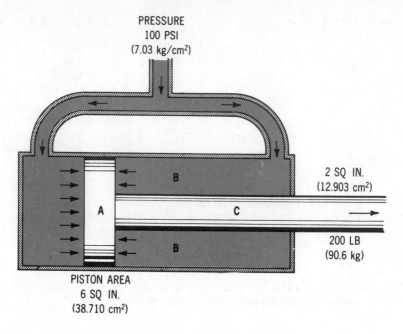

PRESSURE
100 PSI
(7.03 kg/cm²)

2 SQ IN.
(12.903 cm²)

200 LB
(90.6 kg)

PISTON AREA
6 SQ IN.
(38.710 cm²)

Fig. 2-6. A fluid system in which the fluid is applied at equal pressures to both sides of the piston.

piston is, of course, the same, say 100 psi* (7.03 kg/cm²). Applying the rule stated previously, the force pushing the piston to the right-hand side is equal to its area times the pressure (6 × 100), or 600 pounds (38.7 × 7.03, or 272.1kg). Likewise, the force pushing the piston to the left-hand side is equal to its area times the pressure (4 × 100), or 400 pounds (25.807 × 7.03, or 181.4kg). Therefore, a net unbalanced force of 200 pounds (90.6 kg) moves the piston toward the right-hand side. Since all the other forces are in balance, the net effect is the same as if the piston and the cylinder were equal only to the size of the piston rod.

In the fluid systems shown in Figs. 2-3 and 2-4, the surface area of each piston is 10 square inches (64.5163 cm²). Therefore, if one piston is pushed downward a distance of 1 inch (2.54 cm), 10 cubic inches

*The recommended units of force and pressure are newtons (N) and pascals (Pa) or kilopascals (kPa). Since these units are not directly interchangeable with the units used in these calculations, force is stated in kilograms (kg) and pressure is stated in kilograms per square centimeter (kg/cm²).

(163.87 cm³) of fluid are displaced. Since hydraulic oil, for example, is a nearly incompressible fluid, this volume of liquid is moved to another area; that is, it moves the other piston. Since the area of the second piston is also equal to 10 square inches (64.5163 cm²), the piston moves a distance of 1 inch (2.54 cm) to accommodate the 10 cubic inches (163.87 cm³) of fluid. Since the pistons are equal in area, they are moved equal distances but in opposite directions.

Applying the same principle to the fluid system diagrammed in Fig. 2-5, only 2 cubic inches (32.774 cm³) of liquid are displaced if the piston A is pushed downward a distance of 1 inch (2.54 cm). To accommodate these 2 cubic inches (32.774 cm³) of fluid, piston B is forced to move only 1/10 inch (0.254 cm), because its area is 10 times that of piston A. Therefore, the second basic rule for two pistons in the same fluid system can be stated as: *the distances that the two pistons are moved are inversely proportional to their surface areas.*

Further study of the fluid systems reveals that in the relation of each set of pistons, *force times distance moved of the input piston is equal to force times distance moved of the output piston.* Therefore, if a fluid system provides for an increase in force, a decrease in length of travel results. Conversely, a decrease in force provides an increase in length of travel.

A basic statement can be made that, neglecting friction, in any hydraulic system (or any other mechanical system) *the input force multiplied by the distance through which it moves is equal to the output force multiplied by the distance through which it moves.* This relation of force to distance moved is a fundamental key to the extreme flexibility of fluid systems.

In the preceding paragraph, friction was neglected for the sake of simplicity. However, it should be remembered that some friction is always present in either fluid or mechanical movements. Also, heat is produced whenever work is performed against friction. Therefore, heat is a form of energy because it can be produced from work. It is known in actual practice that friction represents a loss of efficiency, but this does not mean destroying energy itself. This means only that some of the energy in the system is changed to another form of energy which cannot be used for the problem at hand. The latter form of energy is neither usable nor available, but it exists in the form of lost heat. The presence of heat in a fluid system, then, can be interpreted logically as an indication of energy loss or inefficiency in the same degree that heat is present.

PRESSURE AND VOLUME

The relation of pressure and volume, as stated in Boyle's law, is significant both in the compression of air and in the expansion of air as an energy transmission medium. Air exists as free air in the atmosphere surrounding us, and it has weight. The weight of the air results in a pressurizing of the atmosphere at a given point, caused by the column of air immediately above that point bearing downward on it. The standard atmosphere at sea level is weight-pressurized at 14.7 pounds per square inch (1.0333 kg/cm²). This pressure value does not register on the pressure gauges in an air system, because the gauges indicate compression that is over and above the standard atmospheric pressure. To find the actual atmospheric pressure at a given time and at a given location, a barometer is used. Barometer readings are in *inches of mercury*, rather than in *pounds per square inch*, and they are not pertinent in this discussion. The average value of the atmospheric pressure or the barometer reading at sea level is 29.92 inches of mercury (1.0331376 kg/cm²), which is approximately equivalent to 14.7 pounds per square inch (1.0333 kg/cm²). Pressure readings that take into account the standard atmospheric pressure are referred to as *absolute pressure*, or psia. Pressure readings from a standard pressure gauge are referred to as *gauge pressure*, or psig.

As already mentioned in Boyle's law, *volume varies in inverse proportion to pressure*. By reducing the volume of an enclosed gas at constant temperature to one-half its original volume, its absolute pressure is doubled. If a volume of free air is compressed to 100 psig (114.7 psia) or 7.03 kg/cm² (8.06 kg/cm²a), it then occupies a volume of only 1/7.8 its original volume. This is calculated by dividing 114.7 by 14.7 (8.06 kg/cm² ÷ 1.03 kg/cm²) to determine the number of "atmospheres" that are contained within the compressed volume—in this example, 7.8 atmospheres. The number of atmospheres at a given pressure can be found by the simple formula:

$$A = \frac{psig + 14.7}{14.7} \ or \ A = \frac{psig + 1.0333}{1.0333}$$

The volume of an air cylinder with a bore of 2 inches (5.08 cm) and a stroke of 10 inches (25.4 cm) is found by multiplying the area of the bore by the length of stroke (3.1416 × 1 × 1 × 10) or (3.1416 × 2.54 × 2.54 × 25.4), or 31.4 cubic inches (514.8 cm³) of compressed air. To calculate the air compressor capacity required to operate the air cylinder at 20 strokes per minute at 60 psig (4.218 kg/cm²g), the total volume represented must be determined. For the sake of simplicity, the surface areas on each side of the piston can be considered equal, ignoring the rod area. On a 2-inch (5.08 cm) bore cylinder with a 5/8-inch (1.5875 cm) rod diameter, the rod area is practically negligible. In a double-acting cylinder (one that both pushes and pulls by air pressure), 31.4 cubic inches (514.8 cm³) of compressed air are used for each stroke direction, or a total volume of 62.8 cubic inches (1029.6 cm³) per cycle. By cycling 20 times per minute, (62.8 × 20) (1029.6 × 20) or 1256 cubic inches (20,592.0 cm³) of compressed air are consumed per minute.

Since air compressors are rated in cubic feet per minute (cfm), it is necessary to convert the 1256 cubic inches into cubic feet. One cubic foot is equal to (12 × 12 × 12), or 1728 cubic inches. By dividing 1256 by 1728 (1256 ÷ 1728), it is found that approximately 0.73 cubic foot (20,671.4 cm³) of compressed air is used.

A further step is necessary, however, because air compressor ratings are stated in terms of free air, rather than in terms of compressed air. Then, the problem is to determine how many cubic feet of free air (air volume at atmospheric pressure) are represented by 0.73 cubic feet (20,671 cm³) of air at 60 psig. By using the above formula for the number of atmospheres:

$$\frac{60 + 14.7}{14.7} = 5.1 \text{ atmospheres (approx.)}$$

The volume of free air is then (5.1 × 0.73 cubic foot of compressed air), or approximately 3.7 cubic feet of free air per minute (1746.2 cm³/sec). By employing the rule of thumb, *a 1-horsepower compressor input delivers approximately 4 cubic feet of free air per minute (1887.8 cm³/sec)*; the 2-inch (5.1 cm) bore by 10-inch (25.4 cm) stroke air cylinder, cycling at 20 times per minute, requires approximately 1 horsepower of compressor capacity to sustain its operation at 60 psig.

By applying these calculations to each air-operated component, total air consumption can be determined. Some allowance should be made for inevitable losses that result from leakages. If additional applications are being considered for an existing air system, these calculations can be most helpful in determining whether the capacity of an existing system will be exceeded by such an addition.

Two industrial production machines that use fluid power to maintain efficient production rates and consistent quality of product can be seen in Figs. 2-7 and 2-8. These machines are typical of the use of fluid power to meet the ever-increasing demands of high-production schedules in industry.

Fig. 2-7. Part of a system consisting of 13 machines that completely manufacture a six-cylinder diesel engine block at a rate of 16 blocks per hour. The system performs all the machining operations and assembles the bearing caps and cylinder liners. (Courtesy The Cross Company, Fraser, Michigan)

Fig. 2-8. Cast aluminum converter housings are milled, bored, faced, and grooved on this three-position hydraulically operated shuttle machine. (Courtesy The Cross Company, Fraser, Michigan)

Chapter 3

Horsepower, Torques, and Levers

James Watt, Scottish inventor of steam engine fame, first used the term *horsepower* to compare the power of his engines with the power of the horse. The horsepower unit deals with two factors—the amount of work done and the amount of time required to do it. These two factors can be broken down further into force, distance, and time.

HORSEPOWER CALCULATIONS*

The *horsepower* unit is standardized and recognized universally as: *550 foot-pounds of work accomplished in 1 second of time, or 33,000*

---

*Since the horsepower unit is an English unit, it is described in English rather than in metric units. The basic metric unit of power, ironically, is the watt, which makes a direct conversion from English units to metric units cumbersome and impractical. Simple substitution of metric units in the following calculations is not recommended, since the results will not match or agree with metric calculations based on the watt as the basic unit. Since most authorities seem to agree that the metric equivalent to a horsepower is 745.7 watts, the reader may make the conversion if he so desires.

*foot-pounds per minute. One foot-pound is the amount of work
required to lift a weight of 1 pound to a height of 1 foot.*
The basic horsepower formula is:

$$horsepower = \frac{force \times distance}{550 \times time \ (in \ seconds)}$$

The horsepower required to perform an amount of work that is
equivalent to lifting a 10-ton weight to a height of 10 feet, in 20
seconds (Fig. 3-1), with a freight elevator can be determined by
converting to the factors in the horsepower formula, as follows:

1. Ten tons (10 × 2000), or 20,000 pounds of force are required.

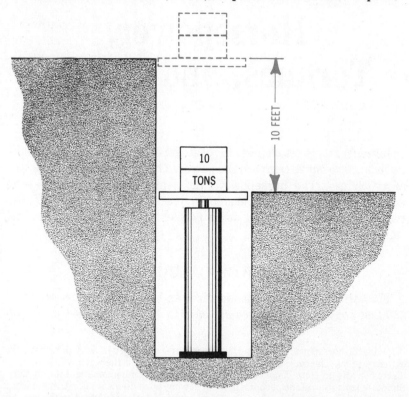

**Fig. 3-1. Work required to lift a load of 10 tons (10.2 metric tons) to a height of 10
feet (3.2 meters).**

2. Dividing the pounds of force required by the number of foot-pounds per horsepower unit (20,000 ÷ 550), or 36.36 horsepower units are required.
3. The horsepower units multiplied by the height in feet (36.36 × 10), or 363.6 foot-pounds of work are required.
4. Then, by dividing 363.6 by the time, in seconds (363.6 ÷ 20), or 18.18 horsepower are required.

The above factors can be substituted in the basic horsepower formula, as follows:

$$hp = \frac{20,000 \times 10}{550 \times 20} = 18.18 \text{ horsepower}$$

The horsepower value above is *theoretical horsepower* required. It does not consider the horsepower required other than merely to balance the load. Neither does it account for friction nor loss of efficiency in the mechanism used to apply the lifting motion to the freight elevator. Available motor sizes indicate that no less than a 20-horsepower motor is needed. If losses in efficiency and safety factors are considered, a 25- to 30-horsepower motor is the practical minimum size.

A study of the mathematics involved in the horsepower formula reveals that a shortage of input horsepower results in a workable freight elevator. However, a sacrifice of either the weight-lifting capacity or the time required to perform the lift is required. This same principle of sacrifice applies to fluid power applications, as will be pointed out later.

A hydraulic cylinder application can be analyzed in the same way that the foregoing freight elevator application was. In the hydraulic cylinder application, assume that a force of 6 tons is needed through a distance of 18 inches and that the 18-inch stroke is accomplished in 6 seconds. Also, assume that the cylinder is operating within a pressure range of 1000 pounds per square inch (psi). The following calculations are necessary to find the horsepower required:

1. The piston area is found by dividing the 6 tons of force by 1000 psi (12,000 ÷ 1000), or an area of 12 square inches.
2. A hydraulic cylinder with a 4-inch bore has a piston area of 12.57 square inches. Since this cylinder provides no safety factor for

force, experience dictates the selection of a larger bore, say 5 inches, with a piston area of 19.6 square inches. Therefore, the pressure against a 5-inch diameter piston is (12,000 ÷ 19.6), or 612 psi.

3. Thus, the theoretical pressure has been determined. However, it is good practice at this stage to allow for pressure losses in the lines, fittings, etc. Therefore, the pressure value can be rounded off at 700 psi.

4. The volume of oil needed to fill the cylinder for the full 18-inch stroke can be determined by the formula: *volume = area × length*. Thus, a volume of (19.6 × 18), or 352.8 cubic inches of oil is required.

5. Since the required pump volume is expressed in gallons per minute, enough fluid to complete an 18-inch stroke in 6 seconds is the equivalent of (18 × 10), or 180 inches of stroke in (6 × 10), or 60 seconds. Since both distance and time were multiplied by 10, the volume is multiplied by 10 (352.8 cubic inches × 10), or 3528 cubic inches per minute. By dividing 3528 cubic inches per minute by 231 (cubic inches in 1 gallon), the required pump capacity (3528 ÷ 231), or 15.3 gallons per minute (gpm) is indicated.

These calculations show that the application requires a 5-inch bore by 18-inch stroke hydraulic cylinder and a fluid supply of 15.3 gpm at 700 psi. To convert this information to hydraulic horsepower, the following formula can be used:

$$hydraulic\ hp\ =\ gpm\ \times\ psi\ \times\ 0.00058$$

$$(kw\ =\ kg/cm^2\ \times\ liters\ per\ min.\ \times\ 0.001634)$$

Since the formula determines theoretical horsepower, allowances should be made for both mechanical efficiency and volumetric efficiency for any pump-electric-motor installation. Hydraulic pump manufacturers usually provide the required input-horsepower values for their pumps at different pressures. Therefore, it is wise to be guided by these values in selecting these components.

A better understanding of the hydraulic horsepower formula and the tools for evaluating the mechanical and volumetric efficiencies of the pumps can be gained by an analysis of the preceding horsepower calculations. Substituting in the horsepower formula above, the theoretical horsepower is:

$$hp = 15.3 \times 700 \times 0.00058$$
$$= 6.2$$

The theoretical horsepower formula can be compared with the basic horsepower formula, as follows:

1. The force against the piston surface area is (700 × 19.6), or 13,720 pounds.
2. The number of horsepower units is (13,720 ÷ 550), or 24.95.
3. The force is applied through a distance of 18 inches, or 1½ feet; therefore, (24.95 × 1½), or 37.42 foot-pounds of work is required.
4. Since the time (6 seconds) is only one-sixth as fast as the time required in the basic horsepower formula, (37.42 ÷ 6), or 6.2 hp is required, which is the same value arrived at in the hydraulic horsepower formula.

If, as the calculations indicate, 6.2 hp is the theoretical horsepower requirement, the factor (0.00058) in the hydraulic horsepower formula can be validated, as follows:

1. The gpm × psi (15.3 × 700) multiplied by the unknown factor is equal to 6.2 hp.
2. Then, the factor can be determined from the equation:

$$factor = \frac{6.2}{15.3 \times 700} = 0.00058$$

The hydraulic horsepower formula has several practical applications, as follows:

1. A typical pump manufacturer's performance curve, for example, may indicate that the pump capacity is 12.5 gpm at 600 psi, requiring an input of 5 hp. Applying the hydraulic horsepower formula, the calculations are:

   theoretical hp input = 12.5 × 600 × 0.00058 = 4.35 hp

then, theoretical horsepower (4.35) divided by actual horsepower (5.0) equals mechanical efficiency, as shown by the equation:

$$efficiency = \frac{4.35 \times 100}{5.0} = 87 \; percent$$

2. If the same performance curve indicates 13 gpm at "zero" pressure and 12.4 gpm at 1000 psi, a volumetric efficiency of approximately 95.4 percent at 1000 psi is indicated. This 4.6-percent loss of volume is attributed to internal leakage back to the suction side, and is called "slippage."

3. The performance curve may also indicate the need for ⅓ horsepower to pump 13 gpm at "zero" pressure. This value is known as "idle horsepower." This is usually an average value for the pump model. As a result of manufacturing tolerances, this average value can vary considerably among the different pumps in a production run. A true volumetric efficiency percentage should be based on the percentage of actual volume delivered against the volumetric displacement of the pumping element.

TORQUES AND LEVERS

Torque is a term that is used to describe a rotational force. It is a unit of force only. Unlike horsepower, it includes neither time nor distance factors. Torque is a turning or twisting moment of force, such as that produced in a rotating shaft or other device rotating around a central pivot point (Fig. 3-2).

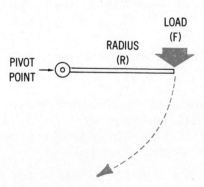

Fig. 3-2. Torque diagram.

In Fig. 3-2, the radius R represents the radius of the arc of movement, or the distance from the pivot to the point of load application. The load F represents the amount of force applied in the form of load. The torque or rotational force at the shaft or pivot point is equal to the load force times the distance from pivot center ($F \times R$). If the load F is in pounds and the distance R is in inches, the torque value is in terms of *inch-pounds*. If the distance R is in feet, the resulting torque is stated in *foot-pounds*.

Torque calculations can be illustrated by assigning values to the load F and the distance R factors as they appear in the diagram in Fig. 3-2. If the load F is assumed to be 200 pounds (90.72 kg) and the distance R is 9 inches (22.9 cm), the torque produced at the pivot point is (200×9) (90.72×22.9) or 1800 inch-pounds (20.74 kg.m).

Although the speed of rotation and the distance moved are not involved in the torque calculation from the practical standpoint, acceleration of the load F by gravity alone results in a gradual increase in torque, ranging from zero with the load stationary to maximum torque as its maximum velocity is reached. Deceleration of the load F from maximum velocity to zero or near zero results in an increase in torque. This is due to the momentum of the load F. This acceleration and deceleration principle can become an important factor, as will be pointed out later in a discussion of the application of the rotary actuator component. Levers and leverages involve mathematical calculations that can be either simple or complex, depending on the degrees of arc through which the lever moves.

Application of levers may either multiply or reduce the effective force of a cylinder. (See Fig. 16-1 in Chapter 16 for lever types.) The first-class lever may either increase or decrease the cylinder force, depending on the relative lengths of the lever arm on each side of the fulcrum or pivot axis. This is also true for the second-class lever; but, in the third-class lever, the resulting lever force is always less than the cylinder force. If the lever force is greater than the cylinder force, the cylinder stroke is greater than the lever stroke, and vice versa. In other words, *force times distance is equal to force times distance*. This principle can often be helpful in matching the cylinder bore and stroke to the work that is to be performed.

Chapter 4

Motions and Graphics

The hydraulic circuit designer is faced with a twofold problem: (1) to determine the type and size of force component required to perform the desired work, and (2) to select the type and capacity of the pressure-generating components that can supply the fluid energy to the force components. The first part of the task involves the type of motion required (straightline or rotational) and the magnitude of the force required (pounds of thrust or unit-pounds of torque). The second part of the task involves the volumetric requirements of the first portion and is influenced by the time or speed factor.

MOTIONS

For example, a hydraulic cylinder with a 4-inch (10.2 cm) bore by 10-inch (25.4 cm) stroke and a piston rod diameter of 1 3/4 inches (4.44 cm) requires 125.8 cubic inches (2061.5 cm³) of fluid to complete the extension stroke and 101.8 cubic inches (1668.2 cm³) of fluid to complete the retraction stroke. This is a combined or total volume of 227.6 cubic inches (3729.7 cm³) of fluid for the complete cycle of the cylinder, or approximately 1 gallon (3.8 liters). The size or volumetric capacity of the pump cannot be determined until the cylinder volume is modified by the time allowed for the stroke cycle of the cylinder.

A two-minute stroke cycle can be powered adequately by a pump having only ½ gpm (1.9 liters/min) volumetric capacity. If the application calls for the stroke cycle to be completed in 2 seconds, the fluid must be delivered at a 30-gpm (113.6 liters/min) rate for 1 gallon (3.8 liters) of fluid to be supplied in 2 seconds. Also, a given machine motion may require more than a single speed and resulting volume rate for a specific cylinder. The overall task may be complicated further by including multiple cylinders with varied speeds and cycles within the total cycle of the complete hydraulic circuit (Chart 4-1). The need for a practical method of charting the individual cylinder demands, as well as the total cycle volumes, then arises (Charts 4-2 and 4-3).

GRAPHICS

A practical method of charting is shown in Charts 4-2 and 4-3. These charts are relatively simple to prepare, after the force components have been chosen to fit the work demands. This type of chart, however, can serve several purposes in the detailed design of a circuit. Graphically, it can represent the sequence of motions of the force components, the timing of each individual component and its relation to other components and the total time cycle, the total volume requirement, the maximum and minimum volumetric demands for each unit of time, and a profile of the varying volume requirements throughout the entire cycle. If the cycle sequence is flexible enough to permit, individual component speeds and sequences can be altered to modulate the volume of fluid that is supplied the circuit.

The total cylinder volume required for the complete cycle is 695.0 cubic inches (11,388.96 cm³) of fluid (see Chart 4-1). Since this is an automatic-cycling-machine, completing a cycle each 10 seconds, (695.0 × 6), or 4170 cubic inches (68,333.79 cm³) of fluid per minute, is required. This value is then converted to gallons (liters) per minute by dividing by the number of cubic inches (cm³) in 1 gallon (3.8 liters). Thus, (4170 ÷ 231), or 18.0 gpm (68.14 liters per minute) is required.

Chart 4-1 is based on the assumption that a pump of 18.0-gpm (68.14 liters per minute) capacity is available for the application. Complete flexibility of cylinder speeds is also assumed, with the

Chart 4-1. Time Required for Extension and Retraction Strokes

(Total Cycle Time: 10 Seconds)

Sequence	1	2	3	4	5	6	7	8	9	10
Cyl. 1	0.27 sec									
Cyl. 2	1.8 sec		0.7 sec							
Cyl. 3			0.7 sec		1.94 sec				2.83 sec	
Cyl. 4					1.94 sec				2.83 sec	
Cyl. 5						2.44 sec			2.83 sec	
Total Volume 695 in³ (11,388.96 cm³)	69.5 (1138.896)	69.5 (1138.896)	69.5 (1138.896)	69.5 (1138.896)	69.5 (1138.896)	69.5 (1138.896)	69.5 (1138.896)	69.5 (1138.896)	69.5 (1138.896)	69.5 (1138.896)

Cyl. 1 = 2-in. bore, 6-in. stroke, w/1-in. dia. piston rod (50.8 mm bore, 152.4 mm stroke, w/25.4 mm rod)
Cyl. 2 = 4-in. bore, 10-in. stroke, w/1¾-in. dia. piston rod (101.5 mm bore, 254 mm stroke, w/44.45 mm rod)
Cyl. 3 = 1½-in. bore, 20-in. stroke, w/⅝-in. dia. piston rod (38.1 mm bore, 508 mm stroke, w/15.875 mm rod)
Cyl. 4 = 3¼-in. bore, 4-in. stroke, w/1⅜-in. dia. piston rod (82.55 mm bore, 101.6 mm stroke, w/34.925 mm rod)
Cyl. 5 = 6-in. bore, 6-in. stroke, w/2½-in. dia. piston rod (152.4 mm bore, 152.4 mm stroke, w/63.5 mm rod)

extension stroke ————— retraction stroke ————▶

Chart 4-2. Fluid Demand [in³ (cm³)] During Extension and Retraction Strokes

(Total Cycle Time: 10 Seconds)

Sequence	1	2	3	4	5	6	7	8	9	10	11
Cyl. 1	18.85 (308.89) ↑			14.14 (231.7) ↓							
Cyl. 2		62.83 (1029.59)	62.83 (1029.59)		25.4 (416.25) ↓	25.4 (416.25)	50.8 (832.5)				
Cyl. 3				23.56 (386)	11.78 (193) ↑					14.6 (239.25)	14.6 (239.25)
Cyl. 4						11.06 (181.26)	22.12 (362.48) ↑			13.62 (223.19)	13.62 (223.19)
Cyl. 5								84.83 (1390.1)	84.83 (1390.1)	70.1 (1148.7)	70.1 (1148.7)
Total Volume 695 (11,388.96)	18.85 (308.89)	62.83 (1029.59)	62.83 (1029.59)	37.7 (617.8)	37.18 (609.27)	36.46 (597.47)	72.92 (1194.94)	84.83 (1390.1)	84.83 (1390.1)	98.32 (1611.17)	98.32 (1611.17)

Cyl. 1 = 2-in. bore, 6-in. stroke, w/1-in. dia. piston rod (50.8 mm bore, 152.4 mm stroke, w/25.4 mm rod)
Cyl. 2 = 4-in. bore, 10-in. stroke, w/1¾-in. dia. piston rod (101.6 mm bore, 254 mm stroke, w/44.45 mm rod)
Cyl. 3 = 1½-in. bore, 20-in. stroke, w/⅞-in. dia. piston rod (38.1 mm bore, 508 mm stroke, w/15.875 mm rod)
Cyl. 4 = 3¾-in. bore, 4-in. stroke, w/1⅜-in. dia. piston rod (82.55 mm bore, 101.6 mm stroke, w/34.925 mm rod)
Cyl. 5 = 6-in. bore, 6-in. stroke, w/2½-in. dia. piston rod (152.4 mm bore, 152.4 mm stroke, w/63.5 mm rod)

extension stroke ⟶   retraction stroke ⟶

Chart 4-3. Fluid Demand [in³ (cm³)] During Extension and Retraction Strokes (With Standard Oversize Rod Diameters)

Total Cycle Time: 10 Seconds

Sequence	1	2	3	4	5	6	7	8	9	10
Cyl. 1	18.85 (308.9)			9.94 (162.89)						
Cyl. 2		62.83 (1029.6)	62.83 (1029.6)		19.14 (313.6)	38.29 (627.5)				
Cyl. 3				19.47 (319)	9.73 (159.4)				9.82 (160.9)	9.82 (160.9)
Cyl. 4					11.06 (181.2)	22.12 (362.15)			10.3 (168.8)	10.3 (168.8)
Cyl. 5							84.83 (1390)	84.83 (1390)	47.1 (771.8)	47.1 (771.8)
Total in³ Volume (cm³)	18.85 (308.9)	62.83 (1029.6)	62.83 (1029.6)	29.41 (481.9)	28.87 (473.1)	60.41 (989.9)	84.83 (1390)	84.83 (1390)	67.22 (1101.5)	67.22 (1101.5)
Accumulator in³ Factor (cm³)	+40.9 (+670.2)	−3.08 (−50.5)	−3.08 (−50.5)	+30.34 (+497.2)	+0.68 (+11.14)	+0.66 (+10.82)	−25.08 (−411)	−25.08 (−411)	−7.41 (−122.4)	−7.47 (−122.4)

extension stroke ⟶⟵ retraction stroke

Cyl. 1 = 2-in. bore, 6-in. stroke, w/1⅜-in. dia. piston rod. (50.8-mm bore, 152.4-mm stroke, w/34.93-mm rod)
Cyl. 2 = 4-in. bore, 10-in. stroke, w/2½-in. dia. piston rod. (101.6-mm bore, 254-mm stroke, w/63.5-mm rod)
Cyl. 3 = 1½-in. bore, 20-in. stroke, w/1-in. dia. piston rod. (38.1-mm bore, 508-mm stroke, w/25.4-mm rod)
Cyl. 4 = 3¾-in. bore, 4-in. stroke, w/2-in. dia. piston rod. (82.55-mm bore, 101.6-mm stroke, w/50.8-mm rod)
Cyl. 5 = 6-in. bore, 6-in. stroke, w/4-in. dia. piston rod. (152.4-mm bore, 152.4-mm stroke, w/101.6-mm rod)

cylinder speeds to be determined entirely by the constant-volume fluid supply. The sequential relation of the cylinders is the same in all three charts. In Chart 4-1, a constant volume of fluid supplied and consumed is represented with cylinder speeds matching the constant supply of fluid available.

In the circuit represented by Charts 4-2 and 4-3, however, it is assumed that fairly rigid requirements of time and sequence exist. Cylinder No. 1 is to extend completely in the first second of the 10-second cycle. Cylinder No. 2 then extends fully during the second and third seconds. Cylinder No. 1 retracts in the fourth second, and Cylinder No. 3 extends in 1½ seconds, carrying halfway into the fifth second. Cylinder No. 4 and Cylinder No. 5 are also charted within the 10-second cycle in the order of their extension strokes. Cylinders No. 3, 4, and 5 return simultaneously in the ninth and tenth seconds to complete the cycle. Each cylinder is charted for a constant speed throughout. If a cylinder should be required to traverse a part of its stroke at one speed and the remainder at another speed, its required volume of fluid for each speed should be allocated properly to the time segments of the chart.

In Chart 4-2, the greatest demand for fluid is indicated in the ninth and tenth seconds of each cycle. When this demand is projected to a full-minute rate by multiplying by 60, (98.32 × 60), or 5899.2 cubic inches (96,670.19 cm³) of fluid per minute, is required. When converted to flow rate, (5899.2 ÷ 231), or 25.54 gpm (96.68 liters per minute), is required. Assuming that the sequence of time must be maintained, 25.54 gpm is the capacity that must be supplied, rather than the 18.0 gpm (68.14 liters per minute) shown in Chart 4-1.

At this point in the analysis, some consideration should be given to building economies that are practical, without detracting from the desired cycle and time sequence. One possibility is the use of oversize piston rods. When the push stroke of the cylinder is used to perform the work, the diameter of the piston rod neither adds to nor subtracts from the push force of the cylinder. In Chart 4-3, standard oversize rod diameters are utilized, with a resultant reduction in the volume of fluid required to accomplish the return or pull strokes of the cylinders.

As can be noted in Chart 4-3, the volume requirement in the ninth second of the cycle has been reduced to 67.22 cubic inches (1101.5 cm³) of fluid. The greatest demand for fluid occurs in the seventh and eighth seconds. This demand, which remains unchanged, is for the cylinder push stroke and is unaffected by the rod diameter. When this demand is projected to a full-minute rate by multiplying by 60, (84.83 × 60), or 5089.8 cubic inches (83,406.5 cm³) per minute, is required. When converted to flow rate, (5089.8 ÷ 231) or 22.0 gpm (83.28 liters/minute) is required; or a savings of 3.54 gpm (13.4 liters per minute) is effected by using oversize rods in Cylinders No. 3, 4, and 5.

The total volume demand for the 10-second cycle is now 597.5 cubic inches (9791.2 cm³) of fluid. When this volume is projected to a full-minute value by multiplying by 6, (597.5 × 6), or 3585 cubic inches (58,747.4 cm³) per minute, is the total demand rate. In turn, when converted to flow rate, (3585 ÷ 231), or 15.5 gpm (58.67 liters/minute), is the total volume required to complete the cycle. If this system were powered, however, with a pump having a capacity of 15.5 gpm (58.67 liters/minute), less than adequate speeds of cylinders during the second, third, seventh, eighth, ninth, and tenth seconds would result. This fact is revealed in Chart 4-3 by the set of values opposite "Accumulator Factor." These values also indicate a surplus of volume during the first, fourth, fifth, and sixth seconds of the cycle.

The next logical step in effecting economies is to apply the surplus of fluid in the surplus portions of the cycle to those portions in which the chart shows an inadequate supply. To accomplish this step, some method must be found for storing the surplus fluid during periods of surplus delivery and for using it to augment the pump capacity during the periods of demands in excess of pump capacity.

The *accumulator*, a fluid power component that is to be discussed later, is the device used for this purpose. In conjunction with the use of oversize piston rods, the accumulator has reduced the calculated volume requirement from 25.54 gpm (96.68 liters/minute) to 15.5 gpm (58.67 liters/minute), which, in turn, represents an appreciable savings in horsepower. Sound engineering practice at this point suggests the selection of a pump capacity in excess of the 15.5 gpm (58.67 liters/minute) value, allowing for pilot losses, for clearance flows within the components, and for loss in efficiency due to anticipated wear. The selection of the proper accumulator size is discussed in a later chapter.

Chapter 5

The Nature of Fluids

The choice of a hydraulic fluid is at least as important as the selection of the hydraulic components themselves. The efficient operation of hydraulic equipment depends on the viscosity and quality of the hydraulic fluid, as well as on the design and construction of the mechanical parts. A properly refined petroleum oil is the best hydraulic fluid for most applications.

TYPES OF FLUIDS

If large volumes of hydraulic fluid are needed, water is sometimes used, such as in hydraulic elevators, large forging presses, etc. The use of water in hydraulic systems has certain disadvantages, as follows: (1) it is limited to temperatures above its freezing point and below its boiling point; (2) it promotes rusting of steel and other metal parts; and (3) it provides limited lubrication for moving parts. One advantage is that water is noninflammable.

In some hydraulic applications, a fire-resistant fluid is required. Die-casting equipment, forging equipment, hydroelectric welders, heat-treating, foundry, and steel mill equipment, industrial trucks, and mining equipment are only a few of the applications where protection against fire hazards is desirable. Several different types of hydraulic fluids have been used with varying success. The hydraulic fluids most widely used are:

1. *Water-Oil Emulsions.* These fluids are mixtures of water, petroleum oil, and an emulsifying agent (soluble-oil type). They afford good fire protection, because the steam released when the fluid hits a hot surface blankets the fire. The lubricating properties, corrosion resistance, and stability are fair to good, and they have little or no effect on seals—except for the natural rubber seals. The emulsions should be carefully and correctly maintained if they are to remain effective (Fig. 5-1).

2. *Glycol-Water Fluids.* These hydraulic fluids are mixtures of

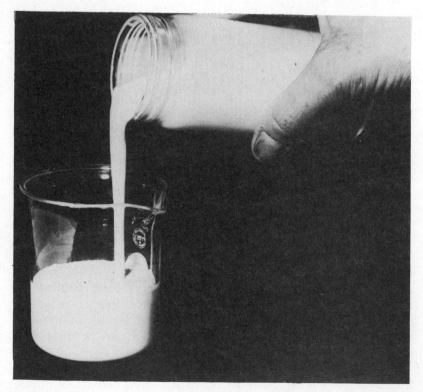

Fig. 5-1. This fire-resistant emulsion fluid passed stability test at 200°F (38°C). Note lack of separation. (Courtesy E. F. Houghton & Co.)

water, glycol, thickener, lubricant, and certain additives which enhance anticorrosion, antiwear, and lubrication properties. Their fire resistance is excellent—if their water content is maintained. Corrosion resistance is fair—except in radial piston-type pumps—where rusting may be a problem. The mixtures are also slightly corrosive to zinc or cadmium. Wear resistance is good at the lower pressures; however, at the higher pressures, some gear-type and some piston-type pumps show excessive wear. The high density of these fluids causes some problems in pump starvation. Most types of seals are not affected by these fluids (Fig. 5-2).

3. *Phosphate Esters.* These esters are used either unmixed or mixed with certain other substances. They provide excellent fire resistance, and their lubricating quality is good. The phosphate esters are noncorrosive, except for the fact that they soften many types of paint or pipe compounds. They are stable in continuous operation up to about 300° F. These fluids do not protect against rusting caused by water contamination, unless they are fortified with rust inhibitors. The effect on seals must

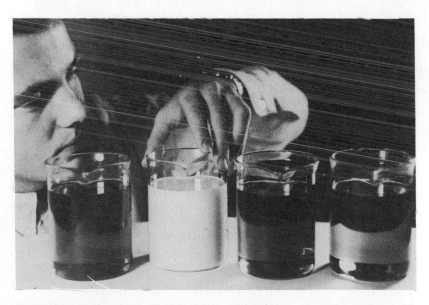

Fig. 5-2. Water-glycol, emulsion fluid, phosphate ester, and petroleum oil fluid (left to right). (Courtesy E. F. Houghton & Co.)

be analyzed carefully—seal swelling can occur with B432 and *Neoprene* types of rubber (Table 5-1). (See Fig. 5-2.)

PROPERTIES OF FLUIDS

Hydraulic fluids serve several purposes other than the primary function of transmitting or controlling power. A suitable hydraulic fluid retains its characteristics over a relatively long period of time (stability), protects the hydraulic system against rust and corrosion, provides lubrication for the moving parts, separates readily from moisture and other contaminants, and is readily available and reasonably economical (Table 5-2). Properly refined petroleum oils satisfy these requirements.

Table 5-1.
Recommended Packing Materials for Fire-Resistant Fluids

Fluid	Packing Material
Water-oil emulsions	Nitrile rubber (*NBR, Buna-N, Hycar*) (packing should be selected by testing various rubber materials). Braid impregnated with *Teflon* suspensoid.
Water-glycol	Natural rubber. Leather (treated for water resistance).
Phosphate ester	Butyl rubber. Silicon rubber. Leather impregnated with *Thiadol. Byram.* Braid with soap-glycerine lubricant which will not wash out. Braid lubricated by impregnation with *Teflon* suspensoid.
Silicate esters	Nitrile rubber. Chloroprene rubber. Polyacrylate rubber. Leather impregnated with *Thiokol.* Braid impregnated with *Teflon* suspensoid.
Silicon base	Nitrile rubber. *Byram.* 1F4 Rubber (formerly *Poly FBA*). Braid impregnated with *Teflon* suspensoid.

Table 5-2. Comparison of Hydraulic Fluids

	Petroleum Oils (including wear resistant types)	Chemical Water Additives	Water Emulsions	Water Glycols	Oil Synthetic Blends	Phosphate Esters
Flame resistance Bare flames Hot surface	Poor Poor	Very Good Good	Fair Fair	Very good Good	Fair Fair	Good Very good
Cost (times petroleum)	1	1	1	2½	3	4
Stability	Excellent	Excellent	Good	Excellent	Good	Excellent
Lubricity in pump Balanced vane Gear w/bushings	Excellent Excellent	Poor Poor	Good Excellent	Very good Excellent	Excellent Excellent	Excellent Excellent
Temperature limits, °F (°C)	0-130 ($-17.8 - 54.4$)	50-120 ($10 - 48.9$)	50-120 ($10 - 48.9$)	Below 0-120 ($-17.8 - 48.9$)	0-130 ($-17.8 - 54.4$)	Below 0-130 (must select right viscosity grade)($-17.8 - 54.4$)
Corrosion protection	Very good	Good	Good	Good	Very good	Very good
Compatibility	Excellent	Good (consider water hardness)	Very good (except paint)	Very good (except paint)	Fair (except paint, rubber, plastic)	Good (except paint, rubber, plastic)

There is no fire-resistant fluid that is an exact replacement for oil. Each type of fluid has specific characteristics. A simple comparison in Table 5-2 brings out the main features of each type of fluid. *Note:* Keep in mind that this type of "simple" comparison chart is always incomplete. It is intended only to be a loose general guide. Always consult your local reliable fluid representative for specific information and advice when making your fluid selection.

Since petroleum oils are used in most hydraulic applications, their properties are of utmost importance in the performance of hydraulic systems. The following properties are directly concerned with the performance of hydraulic oils in hydraulic systems (See Fig. 5-2.):

1. *Viscosity or Body.* This is one of the most important properties. Viscosity measures the resistance to flow offered by the oil. Hydraulic pumps, motors, and valves depend on close fits for creating and maintaining oil pressure. Leakage through these clearances results in loss of pump efficiency, loss of pressure and precision control, and increase in temperature. Leakage losses are greater with the lighter oils (low viscosity). On the other hand, if the viscosity of the hydraulic oil is too high, the resistance of closely fitted moving parts increases; the flow resistance through valves, lines, and passages increases (dropping the system pressure); and the equipment is sluggish in operation.

   Since the pump is, perhaps, the most critical hydraulic component with respect to viscosity, the basic viscosity recommendations established by the pump manufacturer should be followed. Viscosity is determined by means of a viscometer, which measures the time of flow of a given quantity of oil through a capillary tube at a specified temperature. The results are expressed in *seconds, Saybolt Universal (SSU or SUS).* Viscosity of hydraulic oils is usually measured at 100° F (37.8° C).

   Petroleum oils characteristically become thinner as the temperature increases, and they become thicker as the temperature decreases. The viscosity index (V.I.) of an oil is a measure of this property. It is a value that is calculated from viscosities determined at 100° F (37.8° C) and at 210° F (98.9° C). A high viscosity index indicates a relatively low rate of change in viscosity with respect to the temperature of the oil.

   Petroleum oils, in general, possess a sufficiently high V.I. for most hydraulic applications. In some instances, where the operating temperature is considerably higher than the starting temperature, oil having a V.I. that is too low can be too thick at the starting temperature and too thin at the operating temperature. For this reason, oil-cooling devices may be required to prevent an increase in operating temperatures.

Petroleum oils may vary widely in their V.I. characteristics. The origin of the crude oil, the method of refining and blending and other processes influence the V.I. of the finished hydraulic oil. Various substances (additives) are added to hydraulic oils to improve their V.I. characteristics. These additives are often used in hydraulic oils for aircraft service where temperature extremes are encountered.

2. *Chemical Stability.* Although properly refined petroleum oils possess inherent resistance to deterioration, oxidation of hydraulic oils may take place under certain conditions. The products of oxidation in an oil can be extremely harmful to hydraulic equipment. Some of the results of excessive oil oxidation are: (1) increased viscosity of oil; (2) depositing of gummy oxidation products on pumps, motors, or valves; (3) formation of heavy sludge which settles in low points in the system; and (4) formation of acid products which corrode metal parts.

Several factors influence the oxidation of hydraulic oils, such as high temperatures or the presence of air, water, dirt, and other contaminants. The temperature effect on oil oxidation is quite pronounced. For each 20° rise in temperature, the oxidation rate is approximately doubled. An obvious solution to the problem of high temperature is to provide an oil cooler. Another solution is to check the operation of the hydraulic system for those factors which contribute to overheating.

The presence of air entrapped in the hydraulic fluid provides the oxygen that contributes to oil oxidation. Oil under pressure absorbs air which has leaked into the system. This air, subjected to alternate compression and decompression, heats rapidly and exposes the oil to further oxidation. As the pressure is released, the air no longer in solution tends to foam.

The presence of water is difficult to control. The water may enter the hydraulic system either from a defective water cooler or from the condensation of air moisture in the reservoir. Water not only influences the oxidation of hydraulic oils but also mixes with the oil (emulsification) and the products of oil oxidation (which have a high affinity for water) to form heavy sludge. The ability of the hydraulic oil to separate from the water (demulsibility) is an important property.

The presence of other contaminants, such as metallic particles, dust, dirt, etc., has a pronounced effect on oil oxidation. These materials act as catalysts by assisting the oxidation reaction. The use of filters, the proper handling of hydraulic oils, and the periodic cleansing of the hydraulic systems can reduce oil contamination.

The additives are used to impart additional or special properties to the oils. Inhibited or fortified hydraulic oils contain oxidation inhibitors which minimize the formation of injurious oxidation products. Properly refined base oils also minimize the tendency for oil oxidation.

3. *Corrosion Resistance.* Petroleum oils usually provide adequate protection against rusting of iron and steel parts. Under certain conditions, this protection may be inadequate. The corrosion effects encountered with hydraulic oils are of two types: (1) corrosion of iron or steel influenced by the presence of water and air (rusting), and (2) corrosion of metals from the acid by-products of oil oxidation. Rusting destroys the surface finish of finely machined parts, reduces clearances, increases friction and wear, and clogs small orifices with rust particles. Acid corrosion destroys surfaces, pits and etches metal parts (which leads to eventual failure of the part), and increases surface roughness and wear. The oxidation-inhibitor additives help to control acid corrosion. The rust-preventive additives control rusting by adhering tenaciously to the metal surfaces to prevent contact by moisture.

4. *Demulsibility.* The ability of an oil to separate from water is an important property. When water and oil are subjected to the violent agitation, churning, and continual recirculation that are typical in hydraulic equipment, emulsions may form. These emulsions increase friction and wear of moving parts, and they increase oil oxidation, rusting, and acid corrosion.

5. *Pour Point.* "Pour" refers to the ability of an oil to flow at low temperatures. As the temperature decreases, oil thickens as a result of viscosity changes. The small amount of wax still

present congeals (changes from a liquid to a solid) the oil at a given low temperature. Five degrees above this temperature is known as the pour point. The heavier the "body" of the oil, the more difficult it is to distinguish, at low temperatures, between an impeded oil flow due to high viscosity and an impeded oil flow due to pour point. At or near the pour point, the wax separates from the oil, and it may clog the small orifices. At or below the pour point, the oil does not flow into the intake port of the pump.

Hydraulic systems, in general, are not exposed to temperatures low enough to reach the pour point of the hydraulic oil used. Exceptions are mobile or outdoor equipment. In these applications, oils that contain pour-point—improving additives may be necessary. The pump manufacturers usually recommend a pour point that is 15° to 20°F below the lowest operating temperature.

6. *Defoaming.* Hydraulic oil under pressure absorbs air which enters the hydraulic system through defective packings and leaky lines. Or the air may be sucked through the pump intake if the reservoir's oil level is too low. When the pressure is released, the air may leave the solution and form a foam on the surface of the oil. Although most well-refined oils do not foam readily, foam-depressant additives are added to the hydraulic oils, in some instances, to enable them to break the foam and separate quickly from the entrapped air.

Failure of the oil to separate from the entrapped air results in irregular actions of pumps, valves, and motors and in excessive heating of the system as the air pockets are compressed. Some cylinders are provided with an air-bleed provision to relieve the trapped air. The air usually separates either in the cylinder or other high point in the hydraulic system or in the reservoir as the fluid is returned.

7. *Aniline Point.* This point is a measure of the hydraulic oil's affinity for some types of organic compounds. It is the temperature at which the oil and the aniline dye added to it become miscible (capable of being mixed). The chief effect of an undesirable aniline point in a hydraulic oil is on the rubber compounds used in the dynamic seals of the fluid-power components. Most standard seals are compounded for compatibility with hydraulic oils that possess an aniline point ranging from 180° to 200°F (82.2° to 93.3°C). An oil with a high aniline

point causes the seals to shrink, and an oil with a low aniline point causes them to expand. Either expansion or shrinkage may alter the compression fit of the seals, causing them to seize or to fail to seal. Since most commercially available hydraulic oils fall within the aforementioned aniline-point range, the seal problem seldom arises. However, the problem can become acute when aromatic or nonhydraulic oils are employed indiscriminately in a system.

Some of the properties or tests applied to hydraulic oils are of little or no significance to the users of hydraulic oils, and they are listed here merely as a matter of information, as follows:

1. *Gravity.* The determination of the gravity of an oil has little significance as an indication of quality or performance. This test was formerly used to determine the type of crude oil. Modern refinery techniques have rendered this test unnecessary.
2. *Color.* Color controls are used by oil refiners to maintain uniformity of their products. Color bears no relation to the quality of the oil.
3. *Flash Point.* The flash-point test determines the temperature at which a heated oil vaporizes. Hydraulic equipment is not usually operated at high temperatures; therefore, this property is of little importance. (See Figs. 5-3 and 5-4.)
4. *Carbon Residue.* This test indicates the amount of carbon deposit remaining when oils are subjected to destructive distillation. This property is related neither to quality nor to performance of hydraulic oils.
5. *Neutralization Number.* The acidity of hydraulic oils is determined by measuring the quantity of standard alkaline solution required to neutralize the acid. This quantity is called the "neutralization number." It is of no value in rating new hydraulic oils. The rate of increase of the neutralization number may be of some significance in rating hydraulic oils in use. However, the amount of acids present does not necessarily correlate with oil decomposition or with corrosion of metals.

To summarize the foregoing information, the user of hydraulic oils is concerned principally with viscosity. Hydraulic oils with viscosities

Fig. 5-3. A measured amount of hydraulic oil is poured onto molten aluminum at 1300°F (704°C) in pot. Oil bursts into flame immediately. (Courtesy E. F. Houghton & Co.)

that are recommended by the manufacturer of the hydraulic equipment (considering operating temperatures) should be selected.

The user also is concerned with such properties as chemical

Fig. 5-4. A measured amount of water-glycol fluid is poured onto molten aluminum at 1300°F (704°C) in pot. Water-glycol boils; glycol content burns later with slow flame. (Courtesy E. F. Houghton & Co.)

stability, corrosion resistance, demulsibility, pour point, and defoaming ability, if the hydraulic application involves unusual requirements with respect to these properties. A reputable supplier of hydraulic fluids can assist in making suitable recommendations. Several other requirements, which are sometimes specified for petroleum oils, are not applicable to oils for hydraulic use. These properties are gravity, color, flash point, carbon residue, and neutralization number, and they need not be specified when purchasing hydraulic oil.

Top-quality hydraulic oils are made from carefully fractionated, center-cut selected crude oils which are refined by modern processes to remove the undesirable constituents. The additives discussed in this chapter are added to improve these properties. Perhaps the best indication of quality in a hydraulic oil is its performance in actual service. Freedom from gummy deposits, heavy sludge, corrosion, and other problems traceable to the hydraulic fluid are the best indication that the choice of a hydraulic oil has been made wisely (Fig. 5-5).

Fig. 5-5. Pump rings and vanes were run for more than three years with water glycol before normal maintenance. There was a minimum of wear and no rust. (Courtesy E. F. Houghton & Co.)

Chapter 6

Fluid Preparation Devices—Design

The greatest hazard in any fluid-power system is, without doubt, *foreign matter* in the fluid media. "Foreign matter" is an all-inclusive term that includes any material, ranging from moisture to nuts and bolts. This may be lint, dust, sand, metal particles, or the "glass" from a diamond ring. Foreign matter can enter the system from an outside source, it can be generated within the system due to wear of the component parts, or it can be "built-in" at the original assembly of the system. A new system is usually the "dirtiest" hydraulic system. Therefore, a new system should contain provisions for purifying itself from the time it is placed in service.

Each metal-to-metal contact that is in motion within a system is a potential source of fine metal particles that can become entrained in the fluid medium. Engaging a threaded pipe or fitting into a mating port can create metal particles that are either microscopic in size or barely visible to the naked eye. Scale and other matter may become detached from the walls of the fluid conductors and become a part of the fluid stream. The fluid itself may hold particles in suspension that are too small to be easily visible. Or they may be as large as or larger than many of the orifices and clearances through which the fluid is to be transmitted.

A fluid breakdown can result in the formation of sludge and acids. This results from chemical reactions within the fluid, and may be caused by water, air, heat, and pressure, as well as incompatible fluids. Sludge is not always abrasive; however, it is recognized as the source of resinous and gummy coatings on moving parts, and it can clog the passages. Acids can be the cause of pitting and corrosion of critical moving parts, thereby causing malfunctions.

The *degree of filtration*, which is the maximum size of the particles that a filter is designed to pass, is expressed in microns. A *micron* is a unit of length equal to one-millionth of a meter, or 0.000039 inch. Particles smaller than approximately 40 microns are microscopic in size, and they cannot be seen with the naked eye. Fog consists of 50-micron particles, and a pollen grain is 60 microns in size. A grain of table salt is approximately 100 microns, and a dressmaker's pin has a diameter of about 600 microns. The diameter of a human hair is approximately 70 microns, depending on whether it is from the head of a blonde, brunette, or redhead.

Some fluid power components have manufactured clearances in the 0.0001 to 0.0004-inch (0.0025 to 0.01 mm) range. Hydraulic components are usually lubricated by the operating fluid. The mating metal surfaces that are fitted to 0.0004-inch (0.01 mm) clearances between members can act as a filter for the particles that are larger than approximately 100 microns. This results in a collection and build-up of contaminants immediately ahead of the clearance gaps. These contaminants may interfere with and possibly prevent proper lubrication of the mating surfaces. Premature wear and early failure of many components can be traced to this type of contaminant action. If the contaminants that squeeze through the clearances are abrasive, their wearing action on the mating surfaces can be quite rapid and extensive (Fig. 6-1).

HYDRAULIC SYSTEMS

The use of filters is an important step in keeping dirt out of the system and in reducing hydraulic maintenance. The ideal filter removes all foreign matter in the oil. However, it permits the full volume of oil to flow as demanded or pumped by the pump without any restriction over a reasonable service period. In actual operation,

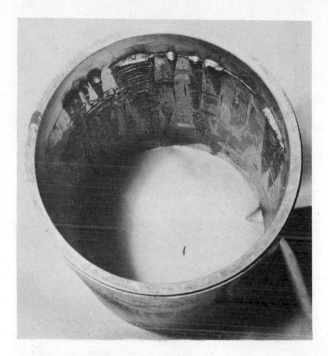

Fig. 6-1. A scored hydraulic cylinder tube. (Courtesy Logansport Machine Co., Inc.)

all the foreign matter cannot, practically, be removed. The capacity of a filter element to pass fluid at minimum pressure drop across the element declines as more of the minute openings become loaded with the contaminants that are being removed. As a result of this loading, a filter element either suffers a loss of flow capacity or operates with an increasing pressure drop.

Oil filters are available in three basic types: (1) size, or mechanical; (2) absorbent; and (3) adsorbent.

Photos of filter media taken at similar magnifications using a scanning electron microscope are shown in Fig. 6-2.

1. *Mechanical Filters.* The mechanical filter is the most widely used type of filter in industrial hydraulic systems. The fluid is forced by pressure through an area of minute openings, pores,

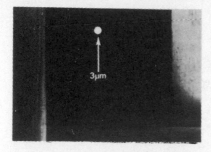

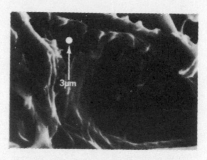

Suction strainer (200 Mesh). Typical 74+ μm demonstrates primary function—to stop larger particles and prevent catastrophic pump failure.

Typical commercial "10μm nominal" filter medium. Our tests show an absolute rating between 25 to 50 μm.

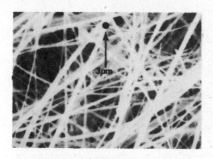

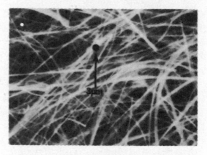

Pall "UN" grade silt control filter medium rated at 6μm absolute and 2μm mean pore size. Nominal rating of 1.5 μm.

Pall "UP" grade Ultipor filter medium rated at 3.0μm absolute and 0.9μm mean pore size. Nominal rating of 0.45μm.

Fig. 6-2. A comparison of filter materials from a 200-mesh suction strainer element to a filter material rated at 3.0 μm absolute. (Courtesy Pall Industrial Hydraulics Corp., Subs. Pall Corp.)

or tortuous passages that retard and trap solid particles. In its most common form, the surface-type filter, the oil flows through the openings in closely woven fabric or treated paper.

The resin-impregnated-cellulose pleated element in Fig. 6-3 is available in 3-, 10-, or 25-micron particle selection. This pleated element design provides a large filtering area in a relatively small space. The cellulose elements require care in the choice of proper paper, as well as the type and degree of

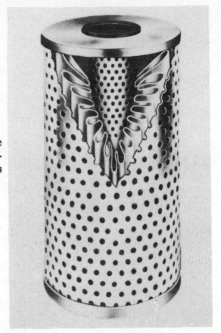

Fig. 6-3. A resin-impregnated-cellulose pleated throwaway type of filter element. (Courtesy Schroeder Brothers Corporation an Alco Standard Co.)

impregnation. The amount of "cure" is also important. The proper balance between these two factors will provide a medium that has the desired resistance to normal system temperatures and the mechanical strength to hold its rating and resist collapse. A typical temperature for this medium would be 250°F (121°C).

When a fire-resistant fluid is to be used in a hydraulic system, the filter medium must be compatible with the fluid, and the size of the filter element should be larger than that used for petroleum-base fluids.

Filter elements may also be made of pleated stainless steel or monel wire cloth. Some have a micron rating as low as five microns, but are generally available in the larger pore size. The metal elements are usually of the reusable type, whereas the resin-impregnated-cellulose pleated elements are usually of the disposable type. Cutaways of filter elements in line-type filters suitable for pressures to 5000 psi (34,473.8 kPa) can be seen in Fig. 6-4.

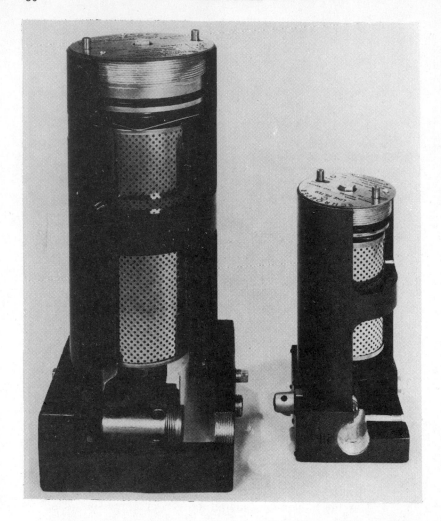

Fig. 6-4. Cutaways of 5000 psi (34,473.8 kPa) hydraulic filters with resin-impreg-nated-cellulose pleated throwaway-type filter elements. These filters are also available with stainless steel wire cloth reusable elements. (Courtesy Schroeder Brothers Corporation an Alco Standard Co.)

The edge-type filter is another form of mechanical or size filter. This type of filter provides for the oil to flow radially through the space between paper or metal discs, or ribbon. It consists of a stack of wheel-shaped metal discs, each separated from the next disc by a thin metal spacer. The fineness of filtration is determined by the thickness of these spacers. The stack is closed at the bottom and is open to the filter outlet at the top. Dirty inlet oil flows into the space around the element and is forced through the small openings between the discs. All solids larger than the openings are held back. Clean oil flows upward through the stack of discs to the outlet. Metal edge-type filter elements are commonly spaced to hold back particles of approximately 0.0035 inch (0.089 mm), but some elements are spaced to hold back smaller particles.

One edge-type element is made of a treated-paper ribbon wound in a spiral to form a tube. The dirty oil is channeled to the outside, and passes through the small openings between the layers of ribbon.

The *depth-type* filter is designed to pass the dirty oil through a relatively thick layer of waste, felt, or fiber, such as cellulose, with an increase in compactness of the filter material in the direction of oil flow.

2. *Absorbent Filters*. These filters are closely related to the depth-type unit described in the preceding paragraph. They are similar in construction and are made of porous materials, such as cotton waste, paper, wood pulp, cloth, asbestos, etc. The coarser particles, as well as the fine insolubles, are filtered out by mechanical absorption. These filters do not remove oil-soluble oxidation products.

3. *Adsorption Filters*. The adsorption filters contain materials, such as fuller's earth, boneblack, charcoal, activated clay, or chemically treated paper or waste which remove impurities by chemical action. These materials may be in the form of a bed or in the form of a renewable cartridge.

They present a large surface area to the oil flowing through the element. The insoluble oxidation products and the solid contaminants of extremely small micron size are removed by size filtration. These filter materials also tend to remove certain additives that are used in many modern fluids. Therefore, the use of adsorbent filters is not generally recommended where inhibited hydraulic oils are in service. Filters of this type can be provided with inert filter elements for use with additive-type oils.

PNEUMATIC SYSTEMS

The pneumatic system is no less subject to contamination than the hydraulic system. It is not a closed system, because the air, after it has been used, is exhausted to the atmosphere, and the supply is constantly replenished from the atmosphere. The air we breathe has been estimated to contain approximately one million particles of dust per cubic inch. It may be recalled from Chapter 1 that the volume of air in a 100-psi (689.5 kPa) pneumatic system represents nearly eight times the volume of air taken from the atmosphere in the free state. This indicates that, in the compressed air system, the concentration of particles in suspension is nearly eight times the concentration in the atmosphere. Along with pipe scale, solidified compressor oil, and pipe compound, these particles can be extremely harmful to a compressed air system. Moreover, when the air is drawn into a compressor, depending on the weather conditions, the humidity can range upward to 100 percent. As the air is compressed and cooled, the water vapor condenses as free water. Even when an aftercooler is used, some moisture in vapor form is swept downstream from the receiver. The moisture that is in suspension condenses into large droplets, because the air is cooled as it moves farther from the compressor and the high temperatures that were reached in compression. This condensate is a deterrent to the proper operation of the components. It can cause rusting, pitting, and clogging of the grooves, orifices, and closely fitted parts. Other undesirable results from this moisture are the destruction of the lubricants and the freezing of the air lines. The first step in preparing the compressed air is to be used in a pneumatic system is to remove the contaminants.

The two main types of devices used to remove these contaminants are the: (1) air line filter, and (2) "air dryer." The air line filter is the more common device. One design of air line filter (Fig. 6-5) removes the impurities or contaminants in two operations: (1) by centrifugal force, in which the heavy contaminants are thrown out, and (2) through a porous medium in which the lighter contaminants are filtered out. The inlet passage directs the incoming air through a louvered splash shield, throwing it downward and outward in a whirling pattern. The larger impurities and the condensates are "thrown out" by centrifugal action. Then, they collect on the sides of the bowl and spiral downward past the baffle into the lower quiet chamber for draining off through the drain petcock. Therefore, only

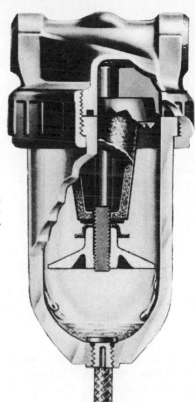

Fig. 6-5. An air line filter which employs a porous-bronze filter element. (Courtesy Watts Regulator Company)

the dry, centrifugally cleaned air reaches the porous-bronze filter element, where the finer particles are filtered out. The louvered splash shield keeps the condensate and the larger particles from lodging on the element. The porous-bronze filter elements are available for removing particles as small as 5 microns, with 40-micron filtration being the standard size in this unit.

Another filter design is shown in Fig. 6-6. In this filter, which has a metal bowl, the direction of flow is through the main element where mist particles are coalesced (unite to form new particles) and submicron solid particles are removed from the air stream. The removed liquids are collected by the foam sock and fall into the bottom of the bowl where they are manually or automatically expelled from the filter. The clean, oil-free air then passes onward and downstream to the point of service.

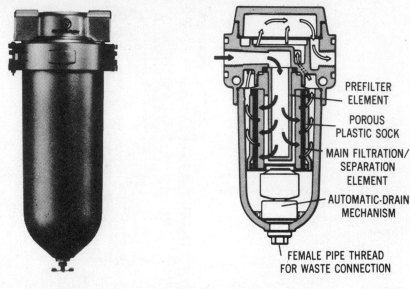

PREFILTER
ELEMENT

POROUS
PLASTIC SOCK

MAIN FILTRATION/
SEPARATION
ELEMENT

AUTOMATIC-DRAIN
MECHANISM

FEMALE PIPE THREAD
FOR WASTE CONNECTION

Fig. 6-6. An air line filter designed for use where exceptionally clean, dry air free from oil is required (left) and cutaway (right). (Photos courtesy of C. A. Norgren Co.)

Another design of filter is shown in Fig. 6-7. This filter provides three-stage filtration in a very compact form. A 5-micron prefilter element removes dirt. The main filtration/separation element in the bottom converts the oil and water mist to a liquid form and removes the submicron particles. Liquids coalesced (united to form new particles) inside the element are eventually forced through the outer wall. A porous plastic sock surrounds the main filter element, preventing retransmission of droplets as they drain downward along the outer surface of the main element and fall into the "quiet zone" at the bottom of the bowl. The top indicator element turns red, if the lower element fails, if there is leakage past the O-ring seal, or if there is a drain mechanism malfunction. In normal circumstances, the indicator acts as an absorbent to help remove traces of hydrocarbons.

Other designs are available in air filters. However, the principle of contaminant removal is quite similar to that discussed above. Filter units are available with transparent plastic bowls, metal bowls, or plastic bowls with metal bowl guards. An optional feature on some units is the automatic drain or dumping of the collected contaminants.

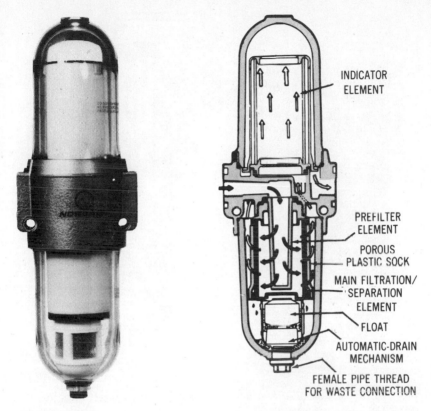

INDICATOR
ELEMENT

PREFILTER
ELEMENT

POROUS
PLASTIC SOCK

MAIN FILTRATION/
SEPARATION
ELEMENT

FLOAT

AUTOMATIC-DRAIN
MECHANISM

FEMALE PIPE THREAD
FOR WASTE CONNECTION

Fig. 6-7. A filter which provides three-stage filtration in a very compact form (left) and cutaway (right). (Photos courtesy of C. W. Norgren Co.)

Air line filters are manufactured in sizes from 1/8 in. NPT to 2 in. NPT and for pressures up to 150 psig (10.5 kg/cm²). Some air filters with metal bowls are available for 250 psig (17.5 kg/cm²). A filter and a lubricator equipped with metal bowl guards are shown in Fig. 6-8.

The second chief means of contaminant removal is commonly referred to as the "air dryer." Several types of air dryers are available. Among them are the refrigeration type, the twin-tower desiccant type, and the deliquescent (dissolve gradually and become liquid by attracting and absorbing moisture from the air) desiccant type. Dryers are designed to deliver air that is "dried" to an extremely low dew point. A refrigeration-type air dryer is diagrammed in Fig. 6-9.

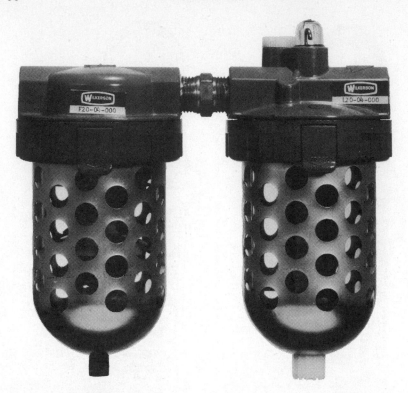

Fig. 6-8. An air line filter and lubricator equipped with metal bowl guards. (Courtesy Wilkerson Corp.)

The most practical method of lubricating pneumatic components is to supply a lubricating fluid to them by introducing the fluid into the air stream after the undesirable contaminants have been removed. This is done with air line *lubricators*.

Many lubricators make use of a venturi, an orifice (opening or port) or a flow sensor to develop a differential pressure, which lifts or siphons the oil from the bowl. A needle valve can be used in the oil tube to adjust the flow of oil into the air stream. In a wick-type lubricator, the oil is transferred from the bowl to the air stream through a wick. The action of a wick-type lubricator can be adjusted either by increasing or by decreasing the length of the wick that extends into the air stream. Wicks are made of various porous materials.

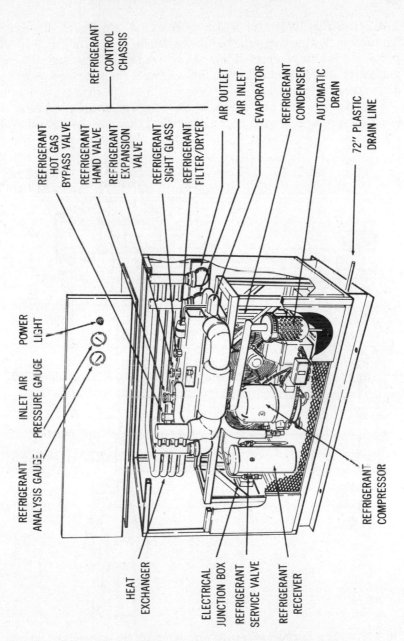

Fig. 6-9. A refrigeration-type air dryer. (Courtesy Wilkerson Corp.)

A *Micro-Fog* lubricator which delivers oil into the air line only when air is flowing through the lubricator can be seen in Fig. 6-10. A portion of the air is directed into the oil reservoir *B* through the fog generator *A*. The remaining air flows around the flow sensor to the downstream system.

Velocity of the air through the fog generator creates a low-pressure area sensed in the sight-feed dome *1*. As a result, oil

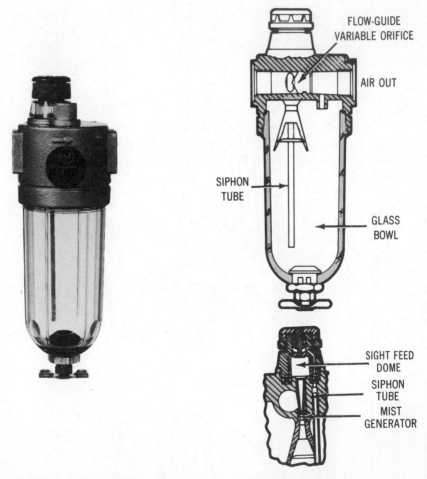

Fig. 6-10. A *Micro-Fog* lubricator (left) with automatic flow sensing and cutaway (right). (Photos courtesy of C. A Norgren Co.)

flows up the siphon tube *12* to the sight-feed dome and drips into the fog generator *A*.

The oil meets the high-velocity air in the fog generator and generates oil fog in the upper portion of the reservoir. Most oil particles larger than 2-micron particle size settle back into the reservoir. The smaller particles remain airborne and travel through the passageway *C* into the downstream airline and onward to its points of lubrication. These small oil particles are the *Micro-Fog* (see Fig. 6-10).

The flow sensor deflects in proportion to the airflow demand and governs the *Micro-Fog* generation. This results in the lubricant being injected into the air line in proportion to the airflow, regardless of flow variations. An oil-feed adjusting knob is located on top of the lubricator.

Another lubricator design is shown in Fig. 6-11. This lubricator inserts only one of the thirty-three drops of lubricant in the sight glass into the downstream air. This lubricator can be adjusted for very small quantities of oil, and this works well where only minute quantities of oil are required. The mist generator breaks about 3/11ths of the total oil seen at the sight glass dome into 2-micron particle size or smaller. These small particles of oil remain suspended in the air line for hundreds of feet. The mist particles are created by passing a portion of the incoming air into the oil-mist generator. This air and oil mist is sprayed onto a baffle plate located in the bowl area. Here particles of oil larger than 2-micron size are coalesced (form new particles) and fall into the bowl to be reused. Since the bowl area is always under pressure which cannot be relieved, these lubricators cannot be filled with oil, unless the pressure is turned off and the air pressure is vented from the bowl. The height of the oil level in the bowl is critical and cannot be filled above the baffle plate. Note that the lubricator in Fig. 6-11 is equipped with a bowl guard.

To assure peak performance with a lubricator, it is usually recommended that an oil that is not more viscous than SAE 10 oil be used. A nondetergent oil that will have no adverse effects on the polycarbonate transparent lubricator bowls or on the seals and gaskets on any of the downstream components should be selected. It is important that the oil selected can readily be broken down into a mist.

Lubricators are available with connections from 1/8 in. NPT to 2 in. NPT. They are also available with plastic bowls, plastic bowls with

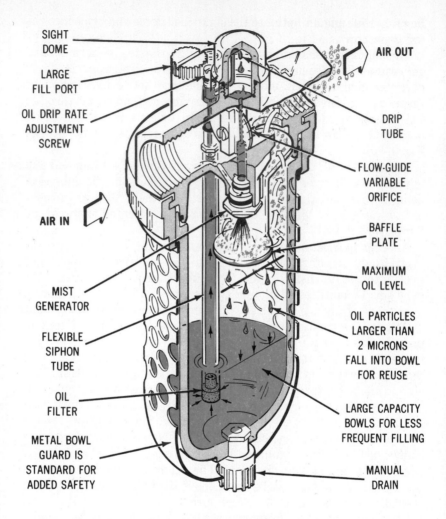

Fig. 6-11. An air line lubricator that produces an oil mist having 2-micron particle size or smaller. Note the metal bowl guard. (Courtesy Wilkerson Corp.)

metal guards, or metal housings with sight gauges. Some lubricators are available with oil-level alarms. Oil reservoirs are available in many sizes.

Both the filter and the lubricator can be obtained as separate units, or they are available as a unit assembly combined with a pressure regulator having an optional gauge. A combination of the separate units is shown in Fig. 6-12. An assembly unit combining all these component functions into a single housing is shown in Fig. 6-13. The importance of these units cannot be overemphasized. Their presence in a pneumatic system extends the service life of the pneumatic components; therefore, their value far exceeds the cost of installing them.

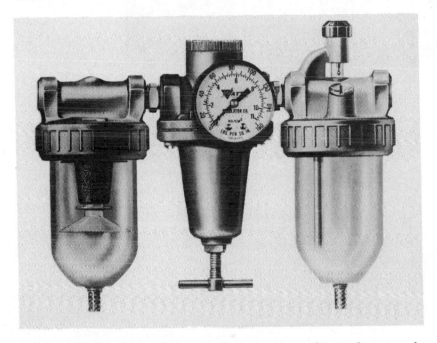

Fig. 6-12. Unit assembly combining the filter, pressure regulator and gauge, and lubricator in a pneumatic system. (Courtesy Watts Regulator Company)

Fig. 6-13. A combination unit in which a filter, pressure regulator, pressure gauge, and a lubricator are built into a single housing. These units are often found on machine tools. (Courtesy Logansport Machine Co., Inc.)

Chapter 7

Fluid Preparation Devices—Application

The two basic concepts of hydraulic-system filtration are: (1) full-flow filtration, with all of the oil in the system passing through a filter each time it circulates through the system, and (2) proportional filtration, with only a portion (5-15 percent) of the fluid passing through a filter element on a given passage through the system. There are advantages and disadvantages for either concept, depending on the type and degree of filtration required and on the location of the filter unit in the system.

HYDRAULIC APPLICATIONS

The suction-type filter is known as a sump strainer, and it is the most universally used filtration unit. It is usually a surface-type filter (Fig. 7-1). The filter is submerged in the hydraulic reservoir, and it is attached to the inlet end of the suction line of the pump. The sump-type filter is a full-flow type, and is generally considered to be the minimum provision for preventing contamination of the system from the reservoir. These filters are less expensive in terms of

Fig. 7-1. Cutaway view of a take-apart sump-type filter, showing the position of the magnetic rods. (Courtesy Marvel Engineering Company)

Courtesy Marvel Engineering Company

original cost. However, they are more inconvenient, since they are usually relatively inaccessible (hard to get at). The inaccessibility factor often leads to neglect by the machine operators and the maintenance personnel.

The inconvenience or inaccessibility problem has been alleviated by removing the filter element from the reservoir and enclosing it in a housing that is a part of the external suction piping to the pump. In this design, an external plunger or lever can be incorporated into the unit to provide a visible indication of the relative loading of the element (Figs. 7-2 and 7-3). These indicators, which are optional accessories, can be obtained with an electric switch either attached or built-in. The switch can be wired to a warning light or a warning bell. Or it can be wired to the motor control of the machine, so that the machine is shut off when the filter becomes loaded to the point of becoming nonfunctional. This type of indicator operates on a predetermined pressure drop across the element. Discretion should be used in permitting a switch-type unit to shut down the machine, since some operations cannot tolerate a failure in hydraulic pressure during the operating cycle.

Both the submerged-type and the external-type suction filters are

Courtesy Marvel Engineering Company

Fig. 7-2. Vacuum indicator with conduit electrical connector provides a visual indication of the relative loading of the filter element. (Courtesy Marvel Engineering Company)

available with a bypass feature which limits the amount of pressure drop or vacuum that can be transmitted to the pump inlet. When the filter element becomes so loaded with contaminants that its flow capacity is reduced to the point of "starving" the pump, the built-in relief valve opens, permitting the oil to bypass the filter element.

It is extremely important that a suction filter be sized adequately in relation to the pump. Also, it should be remembered that the only force available for forcing the fluid through a suction filter is the atmospheric pressure (14.7 psi, or 101.3 kPa), pushing downward on the oil in the reservoir. The reservoir, in some unusual designs, is pressurized to increase the inlet pressure or to "supercharge" the pump. This practice may create other problems, however, since pressurizing of the reservoir is reflected in higher pressures in the return and drain lines.

The line-type pressure filter (Figs. 7-4, 7-5, and 7-6) is a second

Fig. 7-3. Dirt Alarm indicates when it is time to replace the filter element. (Courtesy Schroeder Brothers Corporation an Alco Standard Co.)

type of full-flow filter unit. As the name indicates, this unit is placed directly in the pressure line, and it filters all the oil immediately on delivery from the pump. This unit is more expensive than the suction-type filter, since it is encased in an envelope capable of withstanding full system pressures. In this position, the filter does not protect the pump from contaminants that may enter the reservoir from outside the system or from contaminants that may be generated within the components downstream from the filter. The line-type pressure filter can withstand errors in sizing and negligence better than the suction-type filter, since much more pressure is available to force the oil through the element. If more pressure is required,

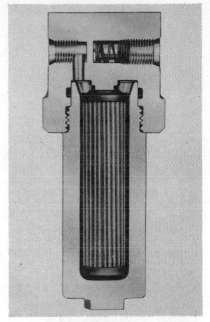

Fig. 7-4. Cutaway of a high-pressure line type of filter. (Courtesy Marvel Engineering Company)

Courtesy Marvel Engineering Company

however, more pressure is wasted and unavailable to perform work. If the pressure drop across the element is too large, a complete collapse of the element may occur. These units are also available with the same type of indicator, built-in bypass, and/or switch features. They are convenient and accessible for servicing, and they provide extremely small micron filtration, since pressure is available for forcing the oil through the finer openings. Like the suction-type filters, elements with magnets for attracting the ferrous particles are available as an optional feature.

The line-type return filter (Fig. 7-7) is a third type of full-flow filter unit. This unit is installed in the line returning the fluid to the reservoir, and it filters all the oil in the system, except the oil returned through the drain lines. The return filter is lower in cost than the pressure filter, but it is more expensive than the suction-type units. These return line filters are usually built to withstand a pressure of 100 to 300 psi (689.5 to 2068.4 kPa). Their chief advantage is that they return only clean oil to the reservoir, thereby protecting the pump from the contaminants generated within the other components of the system.

Fig. 7-5. High-pressure filters and filter elements are built in a variety of sizes and configurations. Various micron sizes are available. (Courtesy Pall Industrial Hydraulics Corp., Subs. Pall Corp.)

The sizing of the return line filter should be analyzed carefully, regarding the components within the system. The flow rate through the return line can far exceed the output volume of the pump. For example, the pull stroke of a 2:1 ratio rod-type cylinder can result in a return line flow that is 200 percent of the pump capacity. The existence of accumulators in the system may cause the value to double again, if they are used to accelerate further the return stroke of the 2:1 ratio rod-type cylinder. The return line filter, then, should always be sized to fit the return flow, regardless of pump capacity.

A combination of filter types, with respect to location and fineness of filtration, is often the most practical means of meeting the cleanliness requirements of a fluid system. In a system that employs a servo valve, as well as other less demanding components, it may be more practical to provide for 5-micron filtration immediately ahead of the servo valve, while 40-, 60-, or even 100-micron filtration may be quite adequate for the other components. In this instance, however,

Fig. 7-6. High-pressure hydraulic filter with *Dirt Alarm.* Elements are available to provide 3- to 150-micron filtration. (Courtesy Schroeder Brothers Corporation an Alco Standard Co.)

it may be wise to provide for some type of fine filtration in the line ahead of the 5-micron filter, to avoid shortening the life of the 5-micron element by too-rapid loading (Fig. 7-8).

Proportional-flow or bypass filtration is preferred in some systems, because clogging of the filter with contaminants does not cut off the oil flow in the system. These filters can be connected into the hydraulic circuit in several ways: (1) supply line bypass circuit; (2) return line bypass circuit; and (3) relief valve bypass circuit (Fig. 7-9).

The supply line bypass circuit is recommended where the bleeding of oil from the main supply line does not affect circuit efficiency (when large pump capacity is available). The return line bypass circuit is used where back pressure is not objectionable and where full pump capacity is needed. The relief valve bypass circuit may be used where back pressure in the return line cannot be tolerated and where the

Courtesy Marvel Engineering Company

Fig. 7-7. Line-type return filters with standard element (left) and extended length, "long-life" element (right). (Courtesy Marvel Engineering Company)

volume of oil discharged through the relief valve is sufficient to provide adequate filtering. The various installations of full-flow and proportional-flow filter units within a system are shown in Fig. 7-10.

A schematic in which high-pressure filters are used in the power loop of a hydraulic transmission is shown in Fig. 7-11. These are silt-control filters containing reverse-flow valves (RFV) which allow filtered flow in one direction and free flow in the opposite direction. This provides optimum protection of the power loop. It prevents a pump failure from immediately causing a motor failure, and vice versa. Also, if the loop filters are located as near the pump/motor as possible, it protects both the pump and the motor from chips released from lines and fittings. A cutaway view of a filter assembly with

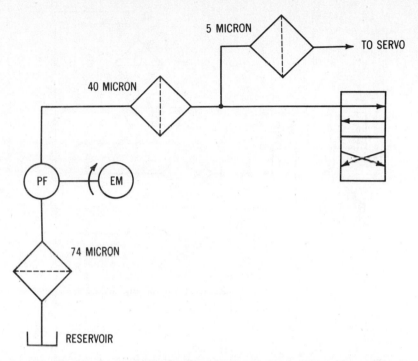

Fig. 7-8. Diagram in which fine filtration is placed ahead of the 5-micron filter to avoid shortening its usefulness by too rapid loading.

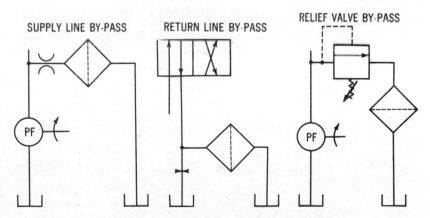

Fig. 7-9. Diagrams of three types of bypass filtration, showing supply line bypass (left); return line bypass (center); and relief valve bypass (right).

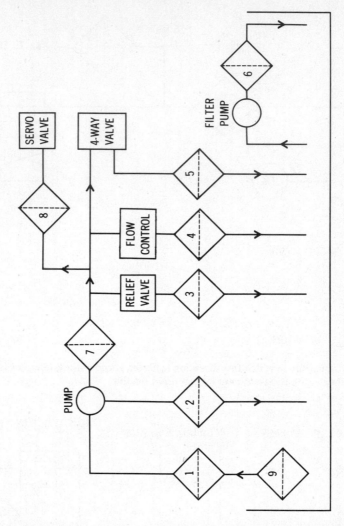

Fig. 7-10. Schematic of a composite full-flow and partial-flow filter installation in a system.

reverse-flow valve for high-pressure loop of a hydraulic transmission can be seen in Fig. 7-12.

PNEUMATIC APPLICATIONS

The pneumatic or compressed air system, like the hydraulic system, should always be considered a "dirty" system. It can be

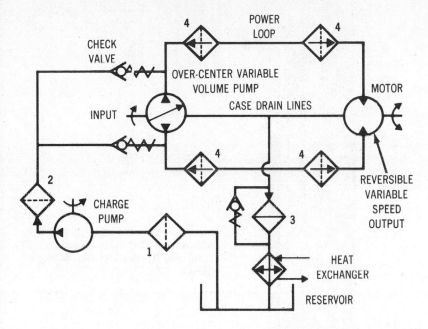

1. Classic suction chip control filter. Often specified by hydrostatic transmission manufacturers to be non-by-pass 10 μm nominal.
2. This filter should be silt control, but can be used where charge pump is separate. Ideal location where loop filters (4) cannot be installed.
3. Possible location of silt control filter where loop filters (4) cannot be installed. Protects the heat exchanger in the event of a failure.
4. High pressure silt control filters containing Reverse Flow Valve (RFV) which allows filtered flow in one direction (arrow) and free flow in the other.

Fig. 7-11. Schematic of hydrostatic transmission circuit showing use of high-pressure filters with reverse-flow valves. (Courtesy Pall Industrial Hydraulics Corp., Subs. Pall Corp.)

assumed that contamination and foreign matter are present at all times and that it is necessary to protect the delicate equipment from these contaminants. In addition to contaminants in the atmosphere that is charged into the pneumatic system, carbon deposits and decomposed compressor oil may be added at the compressor. Scale and corrosion deposits may break loose in the air receiver and the conduit lines to be carried through the air stream to the point where the air is used.

In an air system, the first point that requires attention is the intake

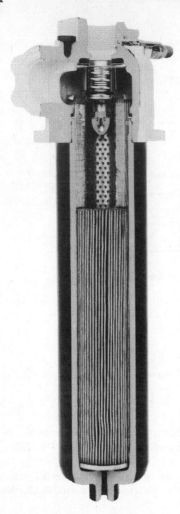

Fig. 7-12. Cutaway of hydraulic filter
with reverse-flow valve. (Courtesy Pall
Industrial Hydraulics Corp., Subs. Pall
Corp.)

line of the air compressor. Although nearly all compressor
manufacturers offer compressor intake filters with their machines,
the installation conditions at the actual plant site sometimes dictate
variations from the standard filter. Since conditions may vary widely
among different plants and industrial locations, the specific
conditions should be analyzed carefully.

The location of the compressor intake is one of the more important
decisions to be made. If the compressor is located in a portion of the
plant where dust, dirt, chemical fumes, or smoke is present, the

compressor intake should be located at a distance from the area. A common means of avoiding the polluted atmospheres in some plant areas is to locate the compressor intake outside the building. The compressor inlet should be located where exhaust fans or exhaust vents cannot carry polluted air to the intake line. If nonpaved or dusty parking lots are nearby, it may be wise to elevate the compressor intake line—perhaps, to install it on the roof of the plant. If a roof-type installation is made, care should be taken to protect the intake filter from weather. When the filter is located at an appreciable distance from the compressor, a larger-than-recommended size may be advisable to permit use of the larger intake lines that may be required to compensate for the added distance.

Air compressor intake filters are available in a variety of designs and types. Probably the most common types of intake filters are the oil bath and the dry filters. The type to be used on a given installation should be determined by the type of atmosphere, the compressor manufacturer's recommendations, and the availability of maintenance for the unit. In choosing the location of the intake filter, availability of maintenance should be an important consideration.

The second important point that should be considered with regard to contaminant removal in the system is the portion immediately after the air leaves the compressor and before it reaches the air receiver. It is a general, and nearly universal, recommendation that all stationary industrial compressors should be provided with a water-type cooling device (aftercoolcr), to lower the air temperature following the temperature rise caused by compression. By cooling the air, condensation of entrained moisture within the receiver is encouraged. The receiver, then, is the third important point to be considered in the system. Since it is comparatively large in volume and since the air is provided with an opportunity to deposit its condensed moisture, it is wise to take advantage of this contaminant collecting characteristic of the air receiver. Either a manual or an automatic means of draining these contaminants from the air receiver should be provided.

Another important item to consider is drying the air after it leaves the compressor (Fig. 7-13). When the dryer is installed between the compressor/aftercooler and the storage tank, it must be sized for maximum compressor capacity, since the dryer handles either full compressor flow or virtually no flow at all. The advantage of this installation position *A* is that the storage tank often can act as a

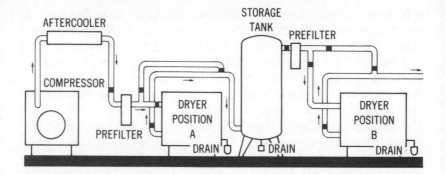

Fig. 7-13. Layout of pneumatic system showing optional location of dryers. (Courtesy Wilkerson Corp.)

"buffer" against momentary flow requirements beyond the capacity of the dryer. In position A, one can be assured of dry air always flowing into the storage tank and system, even under temporary high demands. Install flexible piping around the dryer to isolate it from vibration.

The installation of the dryer in position B is often selected. The advantage of installing the dryer after the storage tank is that the unit often can be sized for average airflow, rather than for strict compressor capacity. The potential problem with this installation is that under full flow, the dew point of the air flowing out of the dryer will rise, and there will be a greater pressure loss (Δp) across the dryer. These temporary conditions can be calculated prior to selecting a dryer to be installed after the storage tank.

The last important point to be considered in a pneumatic system is the point of actual air consumption, immediately ahead of the air-operated tools or devices. Generally, it is recommended that not more than two tools or devices should be operated from the air delivered by a single air line filter. If several tools or devices are used, the size of the air-filter unit should be selected carefully on the basis of maximum air demand. Normally, there is a pressure loss or pressure drop through an air filter, resulting from the turbulent nature of the flow of air through it. This pressure drop may be abnormally high, and it may be detrimental to the proper functioning of the work units if the filter unit is undersize.

Unlike the air cleaning devices that have been discussed, air line lubrication usually is accomplished best at the point of air consumption. Selection of a lubricator by line size may be acceptable for average-demand applications. However, for critical applications, the maximum and minimum airflow requirements should be determined first. Minimum airflow tables and pressure-drop curves for the various flow rates should be checked carefully in the manufacturer's specification sheets. The pressure-drop curves, in particular, should be examined carefully to prevent the occurrence of a pressure drop that is not in excess of that which the system or operation can tolerate.

Generally speaking, lubrication efficiency is reduced when the volume of the air in the air line between the lubricator and the cylinder exceeds the air-displacement volume of the cylinder. For this reason, it is wise to locate the lubricator as near as practical to the cylinder or air tool. Where small-bore short-stroke cylinders using a small volume of air are encountered, the reverse-flow type of lubricators should be used. These reverse-flow lubricators are placed as near as possible to the cylinder and in the air line leading to the rod-end port of the cylinder. The lubricators are installed between the control valve and the cylinder or tool.

For extremely difficult and problem-creating cylinder lubrication requirements, an injection-type lubricator may be necessary. This type of lubricator delivers a full flow of oil by means of a capillary tube directly to the cylinder inlet or inlets. One lubricator of this type provides a number of outlets for servicing several troublesome points.

As mentioned previously, lubrication must be added to the compressed air lines to protect and lubricate the precision air instruments properly. However, the application engineer should be cautious, since too much lubrication can be detrimental to the proper functioning of finely machined air tools; it can clog the small orifices (openings or ports), making the components sluggish in their operation.

Chapter 8

Fluid Preparation Devices—Maintenance

A new hydraulic system may not be a clean system. This may be caused by the relatively simple process of handling and assembling the component parts of the system. The act of turning the threaded pipes or fittings into their mating parts often shears off the thread crests, freeing them to be picked up by the fluid stream. Small pieces of welding slag may flake off from the remote crevasses (cracks), and a few grains of foundry or core sand may wash out of the cast-iron components. Foreign matter may be transferred from the hand tools to find their way into the plumbing or fittings, and the lint from the rags that are used to keep the assembly process clean may contribute to the problem they are intended to prevent. For these and many other reasons, the first few hours of operation for a new fluid system may find a damaging amount of contaminants circulating through the fluid stream.

HYDRAULIC FILTERS

Since there are so many sources of contaminants involved, it is not unusual for the filter element (Fig. 8-1) in a new system to become

Fig. 8-1. Hydraulic filter protecting the hydraulic system on a welding machine in the plant of a major automobile manufacturer. (Courtesy of Pall Industrial Hydraulics Corp., Subs. Pall Corp.)

loaded dangerously within the first few hours of operation. An alert maintenance crew anticipates an apparently premature saturation of the filter elements, and either cleans or exchanges them after the fluid has been circulated thoroughly through the system. Once the filters have been cleaned or exchanged, they can be expected to remain serviceable for a longer operating period. The regular maintenance cycle, however, should not be established until at least two, and preferably three, cleanings have been performed. Thereafter, the maintenance cycle can be determined by inspection and experience, as each system establishes its own maintenance schedule. This schedule should be adapted to each system, depending on the rate of wear of the system components, the capacity of the filter units, the degree of filtration, the amount of contaminants in the system area, and how well the system is designed to repel contamination.

The malfunction of one component often leads to failure of another component, unless proper maintenance procedures are practiced. If one of the components in a hydraulic system fails, the system should not be operated until it has been cleaned thoroughly. If the component that has failed is merely replaced and the system is then returned to operation, a chain reaction of component failures may have been started. The problem is contamination—bits and pieces of the component that failed may be transported by the fluid throughout the system to damage the other components seriously.

If a pump fails, for example, the oil may have carried particles of hard metal to all parts of the system. Metal particles moving at high velocity are extremely abrasive. Therefore, their rifling action may scratch and cut the precision parts in a new replacement pump, thereby speeding the failure of other components that otherwise continue to work satisfactorily. These unseen foreign particles of hard metal may be extremely small (microns) in diameter, but they can cut the inside of rubber hose lines to shreds and damage the packings in the cylinders.

Rubber and metal particles in the oil can speed up oxidation if they are permitted to remain in the oil, but mechanical damage is a more serious problem. In some instances, only 20 percent of the rated service life from the new replacement component may be realized if contaminants like these are not cleaned out. They do not belong in a hydraulic system.

After replacing a failed component, the system should be cleaned thoroughly before it is returned to operation. The following nine points should be included in any check list for cleaning a hydraulic system:

1. Drain the oil from the system.
2. Clean or replace all filters, strainers, and sumps.
3. Thoroughly clean the reservoir to remove all contaminants.
4. All the other major components in the system should be cleaned and inspected. Break all hose connections and blow out the lines and components with cleaning solution and air.
5. Charge the system with a flushing oil that is recommended by

the equipment manufacturer. Oil that is used to refill the system, if cut 50 percent with kerosene, is a satisfactory flushing agent.

6. Operate the equipment under no load. Cycle each operation at least two or three times.

7. Drain the flushing oil from the system. Break the connections located at the lowest points in the system. Be sure to drain the reservoir. Drain all the flushing fluid and *throw it away*.

8. Clean the filters and strainers again; they may contain contaminants that have been dislodged by the flushing oil.

9. Refill the system with an oil recommended by the original equipment manufacturer. Then the machine is ready to return to work.

Kerosene alone should be avoided as a flushing agent, because it is incompatible with most of the hydraulic fluids and the rubber-seal compounds. If kerosene alone is used, flush the system thoroughly to remove it completely.

Solvents and chemical cleaners should be avoided for three reasons: (1) They provide poor lubrication for the pump while they are in the system. (2) They are difficult to remove from the system. Only a trace of a commercial chlorinated solvent can nearly destroy the extra oxidation resistance provided by additive-type hydraulic oils. (3) Solvents can be highly corrosive to steel and copper parts, if they are permitted to remain in the system and become mixed with water.

Regular maintenance of the filters in a hydraulic system can indicate the general condition of the entire system. When a dirty element is removed or replaced, it is good practice to perform a microscopic examination of the particles that have been trapped by the filter. For example, a large rubber residue is an indication of possible deterioration or excessive wear of a seal or seals somewhere in the system. Rubber hoses, if used, may be breaking down or shedding internally. Also, a cylinder barrel or tube may have become scored and may be chewing into the piston seals methodically. If brass particles are found, the brass or bronze bearings in the system components should be examined. If an excess of gummy paste is present, the fluid should be examined carefully for possible breakdown. This type of fluid breakdown may be caused by excessive heat. Therefore, hot spots in the circuit should be suspected. If the breakdown is caused by oxidation of the fluid, the fluid should be examined carefully for chemical composition, since moisture or an undesirable chemical may have entered the system. A record should be kept of the general condition of the filter and the contaminants found at each cleaning for purposes of comparison. By this means, many breakdowns and component failures can be anticipated and avoided by taking preventive or corrective measures before a component becomes completely nonrepairable and inoperable.

Washable filter elements can be kept in serviceable condition for a considerable period of time, if they are handled and cleaned properly. If it is remembered that the oil-flow direction is from outside to inside, the cleaning solution should flow in the opposite direction to clean the element properly. Simple flushing, however, is not considered adequate for restoring the filter element to its original condition. In some facilities, the condition of the element is determined by testing its capacity for passing light from a bulb through the screen. This method is extremely successful if experienced personnel conduct the test. The results of the test can be evaluated by comparing the reconditioned filter element with a new or unused element.

Sonic cleaning devices provide the most effective and positive method of cleaning the wire-cloth type of filter elements. This method thoroughly dislodges even the smallest particles and removes them. The screen is not damaged, and the tiny passages between the wire strands are not altered. The filter elements should be handled with extreme care, since they are constructed from delicate materials.

Before returning a filter element to the system, the entire surface of the screen should be examined closely for weak spots, tears, or enlarged openings. A punctured screen permits the fluid to be filtered to follow the path of least resistance and to avoid the remaining filtering areas of the screen. This type of filter element should be replaced immediately and disposed of.

Good maintenance practice includes good housekeeping for the external areas of the entire system. Also, it contributes to the cleanliness of the inner working parts and components of the system. Care should be taken to eliminate all external leakage whenever it is discovered. If it is possible for the working fluid to leak from the system, it is possible for dirt and contaminants, as well as crippling air, to enter the system through the same leakage paths. The resulting accumulation of fluid also tends to trap airborne contaminants. This makes the housekeeping and maintenance tasks more difficult.

PNEUMATIC FILTERS

The air line filters (Fig. 8-2) and the filter elements should be

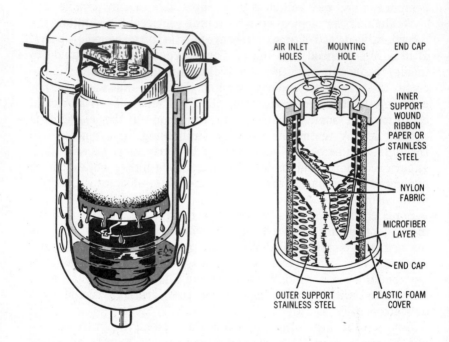

Fig. 8-2. Air line filter (left) with cutaway (right) designed to remove liquid oil, water, and solid particles from the air supply. Contaminated air enters the interior of the elements; then it flows through the inner support core where it is diffused over a larger area. The air passes through fine-woven nylon fabric that serves to support the filter media. Air then flows through layers of random inert microborosilicate fibers, where the principles of ultra-fine filtration take place; then it passes through the outer layer of fine-woven nylon fabric and the stainless steel outer support. The coalesced liquids, with entrained solids, enter the porous plastic-foam cover that surrounds the element; the foam cover is bonded to the end caps to prevent bypassing of the air, especially during the pulsating flows. The coalesced liquids and solids gravitate to the bottom section of the foam cover and drip off the end cap into the sump where they are discharged by the automatic drain or removed manually. (Courtesy Wilkerson Corp.)

cleaned with the same care and detail that are employed in cleaning the hydraulic filter elements. If plastic bowls are used, extreme care should be exercised to prevent their destruction by exposure to chemicals and solvents that can cause them to craze (develop a mesh of fine cracks) and shatter. They should be cleaned with soap and water. In some instances, materials recommended by the manufacturer should be used.

The normal whirling motion of the moisture and the heavy contaminants that are thrown against the inner wall of the filter bowl often cause the plastic bowls to become opaque. If the plastic bowls are to serve their intended purpose, they must be cleaned regularly or as the need arises.

Occasionally, moisture appears underneath the lubricant in the bottom of the air line lubricators. This is the result of excessive entrained moisture in the air supply above the capacity of the filter to remove it. The lubricators should be checked regularly, particularly in warm, humid weather, to eliminate the possibility of excess moisture building up to the point where it is reinjected into the air stream and fed as an emulsion into the precision-type air tools.

To borrow an analysis from the advertising field, proper use and maintenance of fluid preparation devices does not really cost—it pays. Proper maintenance cannot be overemphasized.

Chapter 9

Fluid Power Symbols

Since the beginning of man, symbols have been used to convey ideas, messages, etc. Until after 1950, no apparent effort was made to develop a set of symbols for the fluid power industry. The symbols that were used in some circuits had little meaning, except to those persons who were immediately involved. In other circuits, the complexity involved in attempting to explain the operation of the various components made the cost of designing a circuit exorbitant. This type of circuit diagram is shown in Fig. 9-1.

A set of fluid power symbols approved by the *American National Standards Institute, Inc.* is shown in this chapter. These symbols can be of inestimable value to designers, installation and maintenance personnel, sales engineers, etc.

RULES FOR SYMBOLS

Symbols are used to show connections, flow paths, and functions of the components represented. Conditions occurring during transition from one flow-path arrangement to another can be indicated by symbols.

The locations of ports, direction of shifting of spools, and the

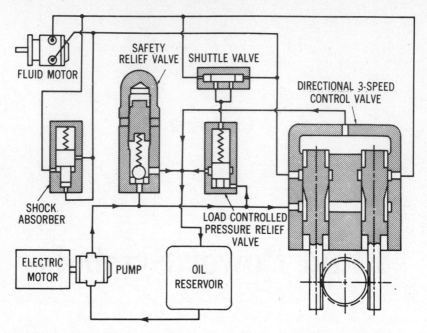

Fig. 9-1. Diagram of a fluid circuit in which the components are shown in cross section, rather than by symbols.

positions of the control elements on the actual component are not indicated by symbols. Symbols are not used to indicate construction, and they are not used to indicate values, such as pressure, flow rate, and other component settings. The symbols can be rotated or reversed without affecting their meaning—except in lines to reservoirs, a vented manifold, and an accumulator.

Lines

The width of the lines does not affect the meaning of the symbols. Line widths should be nearly equal for all symbols.

Solid line. A main line conductor, outline, or shaft.

Dash line. A *pilot* line for a control.

Dotted line. A *drain* line.

Center line. An outline for an *enclosure*.

Lines *crossing* (not necessarily at a 90° angle), but *not connected*.

Lines *joining*. Usually indicating a *pressure line* and a *connector*.

Basic Symbols

The basic symbols can be any suitable size. Sizes can be varied for emphasis or clarity; however, relative sizes should be maintained.

Circles and semicircles.

Large and small circles can be used to indicate that a component is the *main* component and the other is an *auxiliary* component.

Triangle.

Arrow.

Square and rectangle.

Letter combinations are sometimes used with basic symbols, but they are not necessarily abbreviations. In multiple-envelope symbols, the flow condition nearest an actuator symbol occurs when the control is permitted or caused to actuate. Each symbol shows the normal, at-rest, or neutral condition of the component, unless multiple diagrams are provided to show the various phases of circuit operation.

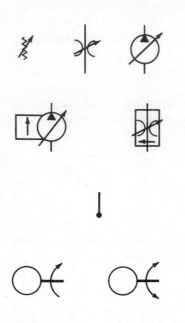

An arrow at approximately 45 degrees through a symbol indicates that the component can be adjusted or varied.

An arrow parallel to the short side of a symbol, inside the symbol, indicates a pressure-compensated component.

A line terminated in a dot represents a thermometer, and is the symbol for temperature cause or effect.

An arrow (on the near side of a shaft) indicates the direction of rotation of a rotating shaft.

FLUID CONDUCTORS

These lines can be seamless steel, aluminum, or copper tubing, or iron pipe.

Pressure Lines

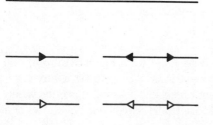

A *working* line (main) or pipe carrying either air or oil under pressure.

Hydraulic line, indicating *direction* of flow.

Pneumatic line, indicating *direction* of flow.

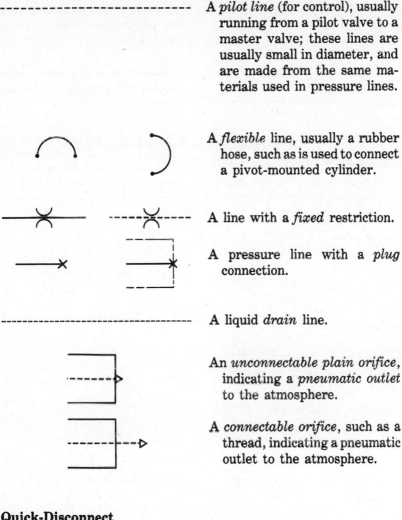

A *pilot line* (for control), usually running from a pilot valve to a master valve; these lines are usually small in diameter, and are made from the same materials used in pressure lines.

A *flexible* line, usually a rubber hose, such as is used to connect a pivot-mounted cylinder.

A line with a *fixed* restriction.

A pressure line with a *plug* connection.

A liquid *drain* line.

An *unconnectable plain orifice*, indicating a *pneumatic outlet* to the atmosphere.

A *connectable orifice*, such as a thread, indicating a pneumatic outlet to the atmosphere.

Quick-Disconnect

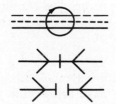

Conductor with a *rotating coupling*.

Conductor *without* checks *connected* (upper) and *disconnected* (lower).

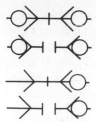

Conductor with *two* checks *connected* (upper) and *disconnected* (lower).

Conductor with one check *connected* (upper) and *disconnected* (lower).

FLUID STORAGE AND ENERGY STORAGE

Conventionally, reservoirs are drawn in the horizontal plane. All lines enter and leave the reservoir from above. A line entering or leaving below the reservoir is shown only when the bottom connection is essential to the circuit function.

Reservoir

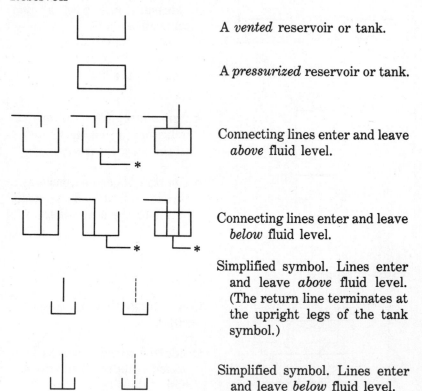

A *vented* reservoir or tank.

A *pressurized* reservoir or tank.

Connecting lines enter and leave *above* fluid level.

Connecting lines enter and leave *below* fluid level.

Simplified symbol. Lines enter and leave *above* fluid level. (The return line terminates at the upright legs of the tank symbol.)

Simplified symbol. Lines enter and leave *below* fluid level.

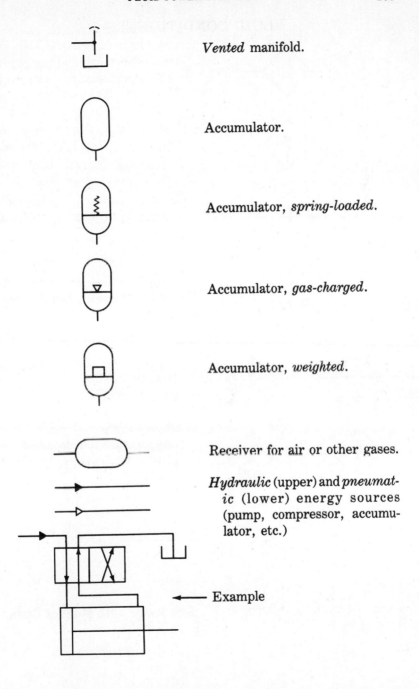

Vented manifold.

Accumulator.

Accumulator, *spring-loaded*.

Accumulator, *gas-charged*.

Accumulator, *weighted*.

Receiver for air or other gases.

Hydraulic (upper) and *pneumatic* (lower) energy sources (pump, compressor, accumulator, etc.)

Example

FLUID CONDITIONERS

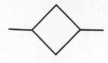

Device for controlling the physical characteristics of the fluid.

Heater. The inside triangles indicate that heat is introduced.

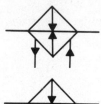

Heater. Outside triangles indicate a *liquid* heating medium.

Heater. Outside triangles indicate a *gaseous* heating medium.

Temperature controller. Outside triangles indicate a liquid or gaseous medium.

Temperature controller. The temperature is maintained between two predetermined limits.

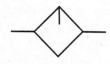

Lubricator *without* drain.

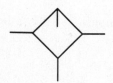

Lubricator with *manual* drain.

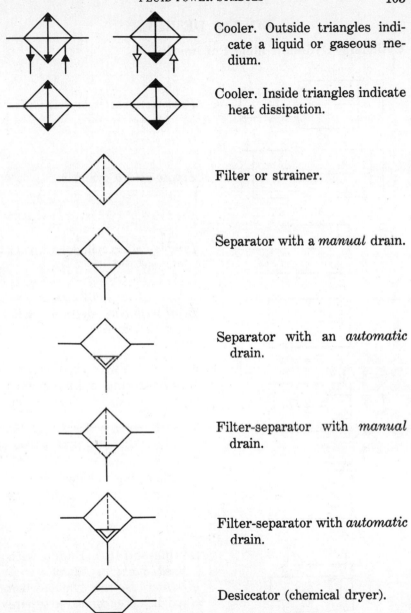

Cooler. Outside triangles indicate a liquid or gaseous medium.

Cooler. Inside triangles indicate heat dissipation.

Filter or strainer.

Separator with a *manual* drain.

Separator with an *automatic* drain.

Filter-separator with *manual* drain.

Filter-separator with *automatic* drain.

Desiccator (chemical dryer).

LINEAR DEVICES

Cylinders (Hydraulic annd Pneumatic)

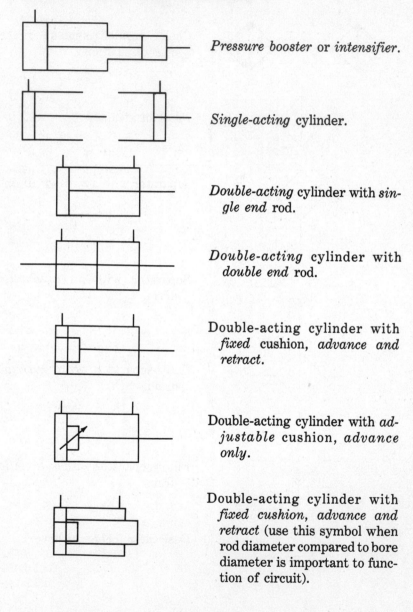

Pressure booster or *intensifier*.

Single-acting cylinder.

Double-acting cylinder with *single end* rod.

Double-acting cylinder with *double end* rod.

Double-acting cylinder with *fixed* cushion, *advance and retract*.

Double-acting cylinder with *adjustable* cushion, *advance only*.

Double-acting cylinder with *fixed cushion, advance and retract* (use this symbol when rod diameter compared to bore diameter is important to function of circuit).

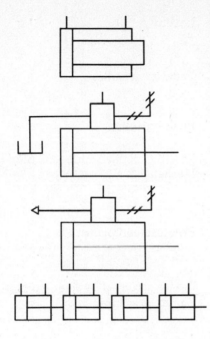

Double-acting cylinder *without cushion* (use this symbol when diameter of rod compared to diameter of bore is important to function of circuit).

Hydraulic *servo positioner* (simplified).

Pneumatic *servo positioner* (simplified).

Discrete positioner (combination of two or more basic symbols).

ACTUATORS AND CONTROLS

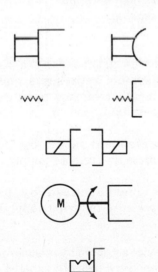

Manual (a general symbol without indicating a specific type, such as foot, hand, leg, arm, etc.).

Spring actuator or control.

Electrical actuator or *solenoid-actuated* (single-winding) control.

Electrical actuator or *reversing motor*.

Detent. A notch should indicate each detent in the actual component. A short line indicates the *detent in use*.

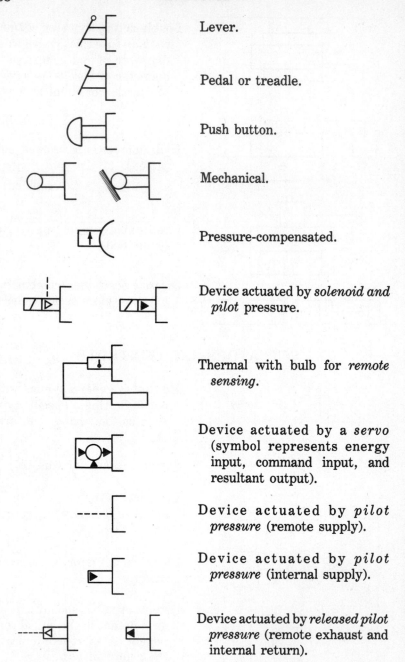

Lever.

Pedal or treadle.

Push button.

Mechanical.

Pressure-compensated.

Device actuated by *solenoid and pilot* pressure.

Thermal with bulb for *remote sensing*.

Device actuated by a *servo* (symbol represents energy input, command input, and resultant output).

Device actuated by *pilot pressure* (remote supply).

Device actuated by *pilot pressure* (internal supply).

Device actuated by *released pilot pressure* (remote exhaust and internal return).

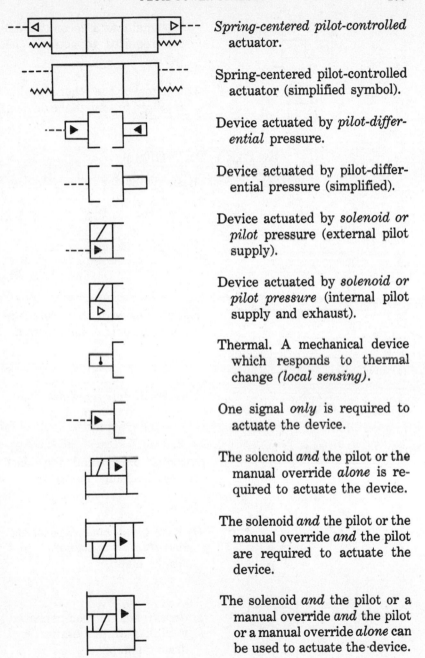

Spring-centered pilot-controlled actuator.

Spring-centered pilot-controlled actuator (simplified symbol).

Device actuated by *pilot-differential* pressure.

Device actuated by pilot-differential pressure (simplified).

Device actuated by *solenoid or pilot* pressure (external pilot supply).

Device actuated by *solenoid or pilot pressure* (internal pilot supply and exhaust).

Thermal. A mechanical device which responds to thermal change *(local sensing)*.

One signal *only* is required to actuate the device.

The solenoid *and* the pilot or the manual override *alone* is required to actuate the device.

The solenoid *and* the pilot or the manual override *and* the pilot are required to actuate the device.

The solenoid *and* the pilot or a manual override *and* the pilot or a manual override *alone* can be used to actuate the device.

One signal *and* a second signal are required to actuate the device.

One signal *or* the other signal is required to actuate the device.

ROTARY DEVICES

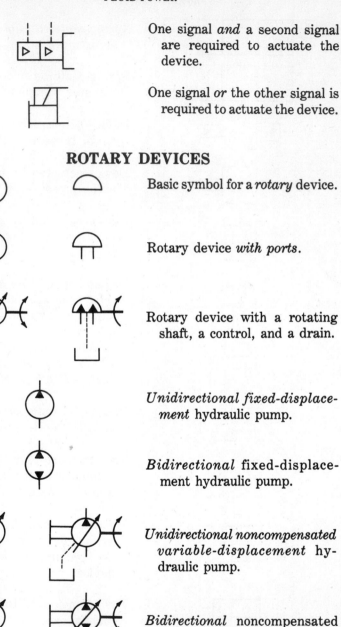

Basic symbol for a *rotary* device.

Rotary device *with ports*.

Rotary device with a rotating shaft, a control, and a drain.

Unidirectional fixed-displacement hydraulic pump.

Bidirectional fixed-displacement hydraulic pump.

Unidirectional noncompensated variable-displacement hydraulic pump.

Bidirectional noncompensated variable-displacement hydraulic pump.

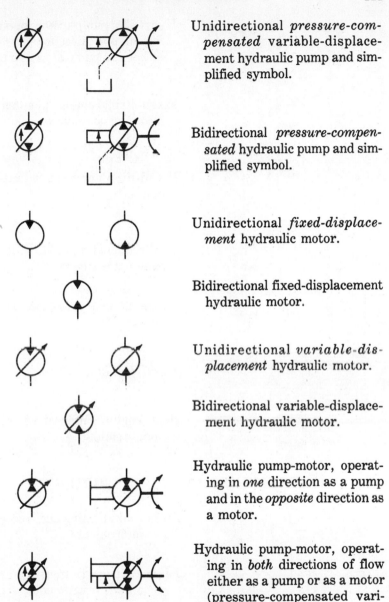

Unidirectional *pressure-compensated* variable-displacement hydraulic pump and simplified symbol.

Bidirectional *pressure-compensated* hydraulic pump and simplified symbol.

Unidirectional *fixed-displacement* hydraulic motor.

Bidirectional fixed-displacement hydraulic motor.

Unidirectional *variable-displacement* hydraulic motor.

Bidirectional variable-displacement hydraulic motor.

Hydraulic pump-motor, operating in *one* direction as a pump and in the *opposite* direction as a motor.

Hydraulic pump-motor, operating in *both* directions of flow either as a pump or as a motor (pressure-compensated variable-displacement).

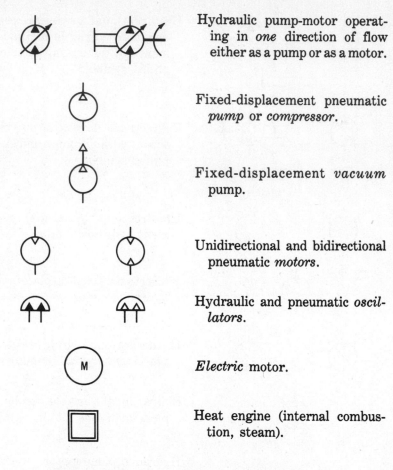

Hydraulic pump-motor operating in *one* direction of flow either as a pump or as a motor.

Fixed-displacement pneumatic *pump* or *compressor.*

Fixed-displacement *vacuum* pump.

Unidirectional and bidirectional pneumatic *motors.*

Hydraulic and pneumatic *oscillators.*

Electric motor.

Heat engine (internal combustion, steam).

INSTRUMENTS AND ACCESSORIES

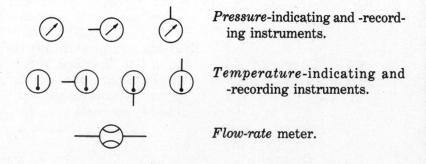

Pressure-indicating and -recording instruments.

Temperature-indicating and -recording instruments.

Flow-rate meter.

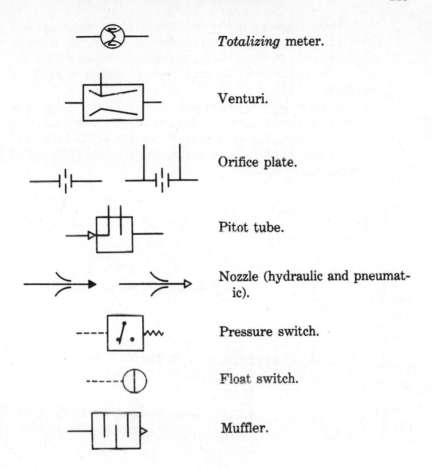

Totalizing meter.

Venturi.

Orifice plate.

Pitot tube.

Nozzle (hydraulic and pneumatic).

Pressure switch.

Float switch.

Muffler.

VALVES

A basic valve symbol is composed of one or more envelopes with lines inside the envelope to represent flow paths and flow conditions between ports. Three symbol systems are used to represent valve types: (1) single envelope, both finite and infinite positions; (2) multiple envelope, finite position; and (3) multiple envelope, infinite position. The symbol systems are:

1. Infinite-position single-envelope valves, the envelope is imagined to move to represent how pressure or flow conditions are controlled as the valve is actuated.

2. Multiple envelopes are used to symbolize valves providing more than one finite flow-path option for the fluid. The multiple envelope moves to illustrate the change of flow paths when the valving element inside the component is shifted to its finite positions.

3. Multiple-envelope valves capable of infinite positioning between certain limits are symbolized as above with the addition of horizontal bars drawn parallel to the envelope. The horizontal bars are clues to the infinite positioning function of the valve represented.

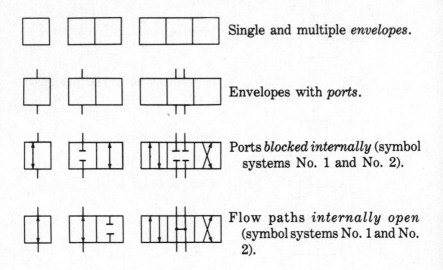

Single and multiple *envelopes*.

Envelopes with *ports*.

Ports *blocked internally* (symbol systems No. 1 and No. 2).

Flow paths *internally open* (symbol systems No. 1 and No. 2).

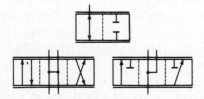

Flow paths internally open (symbol system No. 3).

Two-Way Valves (Two-Ported)

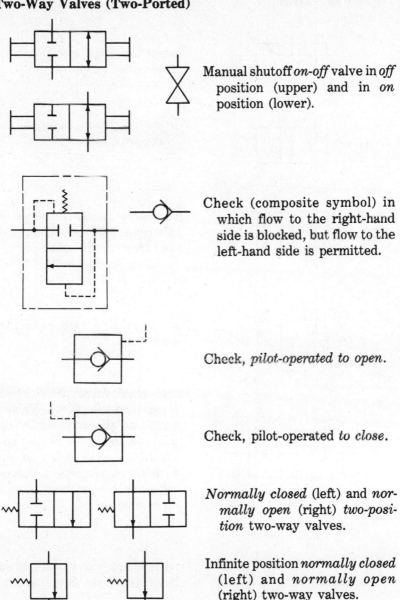

Manual shutoff *on-off* valve in *off* position (upper) and in *on* position (lower).

Check (composite symbol) in which flow to the right-hand side is blocked, but flow to the left-hand side is permitted.

Check, *pilot-operated to open.*

Check, pilot-operated *to close.*

Normally closed (left) and *normally open* (right) *two-position* two-way valves.

Infinite position *normally closed* (left) and *normally open* (right) two-way valves.

Three-Way Valves

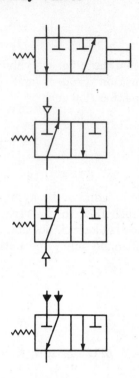

Normally open two-position three-way valve.

Normally closed two-position three-way valve.

Distributor valve. Pressure is distributed first to one port and then to the other port.

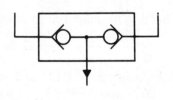

Two-pressure two-position three-way valve.

Double check valve *without cross bleed* (one-way flow). Valves with two poppets usually do not permit pressure to momentarily cross bleed to return during transition. Valves with one poppet may permit cross bleed.

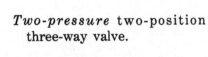

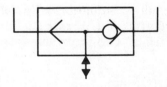

Double check valve *with cross bleed* (reverse flow permitted).

Four-Way Valves

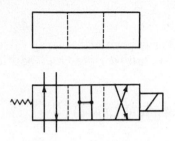

Snap action with transition two-position four-way valve. As valve element shifts positions, it passes through an intermediate position. If this "in-transit" condition is essential to circuit function, it can be shown in the center position, enclosed by broken lines.

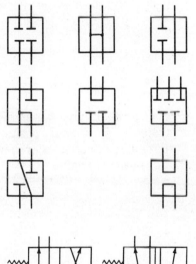

Typical flow paths for *center position* of three-position four-way valves.

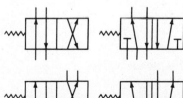

Two-position four-way valve in *normal* position.

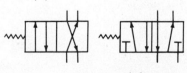

Two-position four-way valve in *actuated* position.

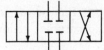

Three-position four-way valve in *neutral* position.

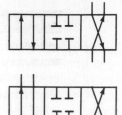

Three-position four-way valve in *actuated left-hand* position.

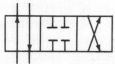

Three-position four-way valve in *actuated right-hand* position.

Infinite Positioning (Between Open and Closed)

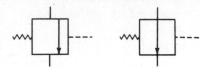

Normally closed valve (left) and *normally open* valve (right).

Pressure Control Valves

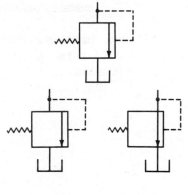

Pressure relief valve (simplified symbol) in *normal* position (left) and in *actuated* (relieving) position (right).

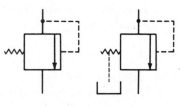

Sequence valve.

Air line pressure regulator (adjustable, relieving).

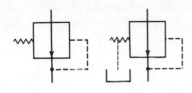

Pressure reducing valve.

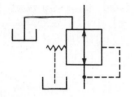

Pressure reducing and relieving valve.

Infinite Positioning Valves

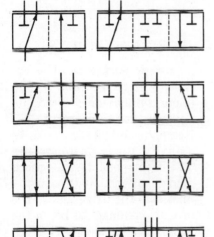

Three-way valves.

Four-way valves.

Flow Control Valves

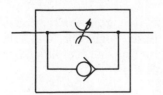

Adjustable *with by-pass* flow control. Flow is controlled in the right-hand direction. Flow in the left-hand direction by-passes the control.

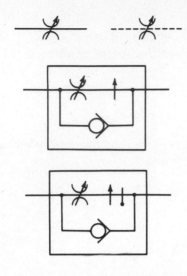

Noncompensated (flow control in each direction) *adjustable* flow control valve.

Adjustable and *pressure-compensated with by-pass* flow control.

Temperature- and *pressure-compensated* adjustable flow control.

REPRESENTATIVE COMPOSITE SYMBOLS

Component enclosure. The enclosure may surround a complete symbol or a group of symbols to represent an assembly, and it is used to convey more information about component connections and functions. The enclosure indicates the extremity of the component or assembly. External ports are assumed to be on the enclosure line, and indicate connections to the component. The flow lines cross the enclosure line without loops or dots.

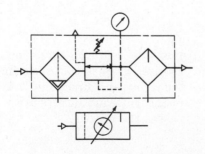

Air line accessories, including filter, regulator, and lubricator (upper). Simplified symbol (lower).

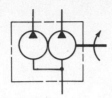

One inlet and two outlets fixed-displacement *double pumps*.

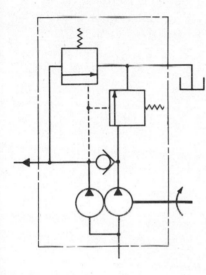

Double pumps with integral *check unloading* and two outlets.

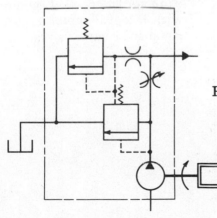

Pump with integral *variable-flow rate control* with overload relief.

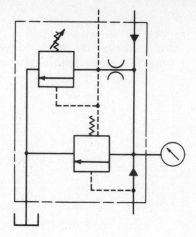

Balanced-type *relief* valve.

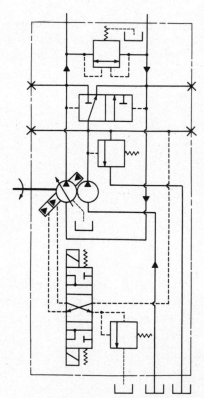

Variable-displacement pump with integral *replenishing* pump and control valves.

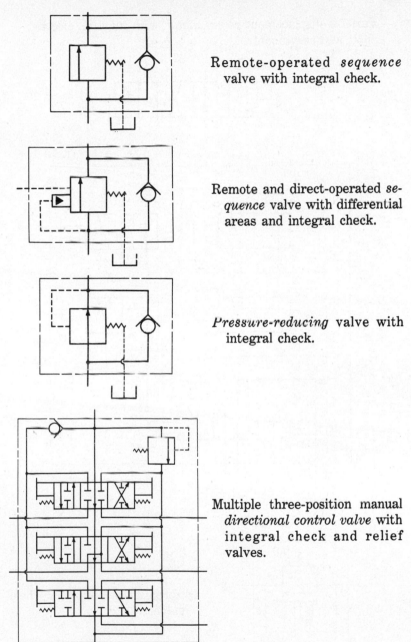

Remote-operated *sequence* valve with integral check.

Remote and direct-operated *sequence* valve with differential areas and integral check.

Pressure-reducing valve with integral check.

Multiple three-position manual *directional control valve* with integral check and relief valves.

Variable-displacement *pump motor* with manual, electric, pilot, and servo control.

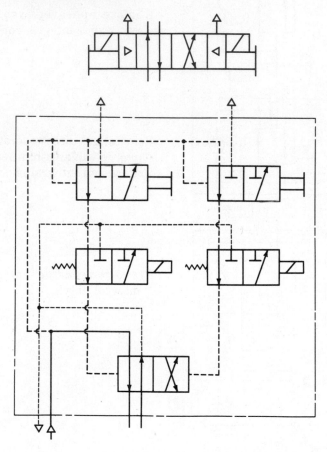

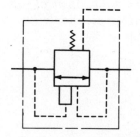

Differential *pilot-opened* check valve.

Four-connection *solenoid-controlled and pilot-operated* two-position valve with manual pilot override. Simplified symbol (upper) and composite symbol (lower) are shown.

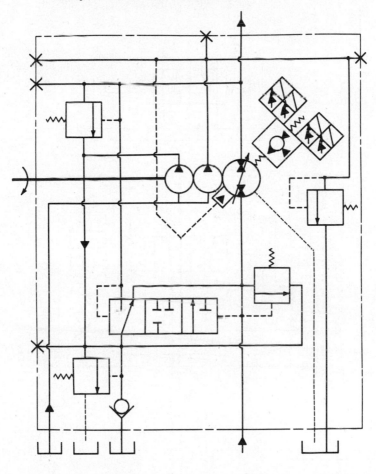

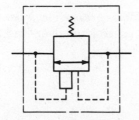

Differential *pilot-opened and -closed* check valve.

Five-connection two-position valve with *solenoid control pilot-operated* with detents and throttle exhaust. Simplified symbol (upper) and composite symbol (lower) are shown.

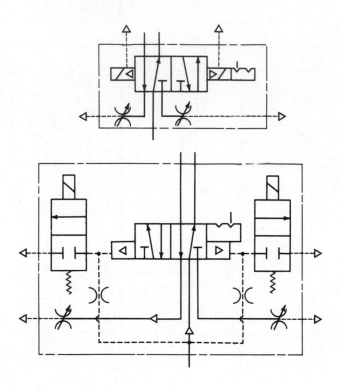

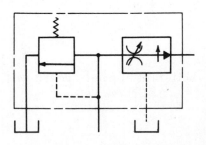

Variable *pressure-compensated* flow control and over-load relief valve.

Five-position *cycle control panel.*

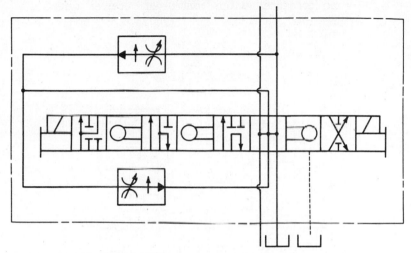

Panel-mounted separate units furnished as a *package* (relief, two four-way, two check, and flow rate valves).

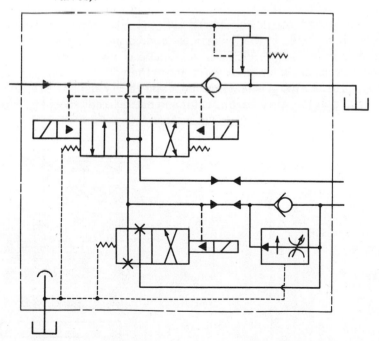

Single-stage compressor with electric motor drive
and pressure switch control of receiver tank
pressure.

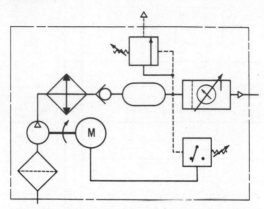

COLOR OR PATTERN CODE

The interconnecting lines in cutaway diagrams are sometimes
colored or patterned to indicate pressure, flow, special functions, and
different fluids during various phases of operation. The code for each
condition used should be identified and illustrated by a note,
preferably located in the lower left-hand corner of the sheet. Only
those lines which perform active functions for the phase shown
should be coded. If a condition fits more than one function, the code
which provides the greatest emphasis and clarity should be selected.
When applicable, the following color and pattern code should be used.

FUNCTION	COLOR	PATTERN
INTENSIFIED PRESSURE	VIOLET	
SUPPLY PRESSURE	RED	
CHARGING PRESSURE REDUCED PRESSURE PILOT PRESSURE	ORANGE	
METERED FLOW	YELLOW	
EXHAUST	BLUE	
INTAKE DRAIN	GREEN	
INACTIVE	BLANK	

Chapter 10

Functional Control Components—Design

In both a pneumatic and a hydraulic system, the functional control components play an important role. The student should study thoroughly and become familiar with the design and function of each of these components if he is to understand how and to what extent they can be employed in a fluid circuit. Functional controls can be divided into two classes—those used for pressure controls and those used for flow controls.

The various types of controls used for pneumatic and hydraulic pressure controls are:

Pneumatic	*Hydraulic*
Pressure regulators	Pressure relief valves
Sequence valves	Pressure reducing valves
Safety valves	Sequence valves
	Unloading valves
	Counterbalance valves

The various types of flow control valves are:

Pneumatic	*Hydraulic*
Speed control valves	Flow control valves
Quick-exhaust valves	Deceleration valves
Check valves	Check valves

Pressure control and flow control components are designed to operate in a specific pressure range. They should always be employed within that range. For example, most pneumatic controls are designed for a 0- to 150-psi (0-1034.2 kPa) range. However, they are used, generally, at only 90 psi (620.5 kPa) or less, since few plants exceed 90 psi (620.5 kPa) in line pressure. In hydraulics a wide pressure range is covered in which a component may be designed for a pressure range of 0 to 500 psi (3447.4 kPa), a second component for 0 to 1500 psi (10,342.2 kPa) and other components for ranges to 5000 psi (34,473.8 kPa) and upward. A wider selection of hydraulic components is available. This also can become more complicated. For example, an application may require an operating pressure of 450 psi (3102.6 kPa) for which a relief valve in the range from 0 to 500 psi (3447.4 kPa) has been selected. Then, after the equipment is installed, it may be found that a part of the operation is performed at 450 psi (3102.6 kPa) and that other applications which require 700 psi (4826.3 kPa) on the relief valve are to be processed on the same machine. This means that a change is necessary in the valving requirements. Sometimes this can be done by changing the springs, etc., in the control. However, in most instances, it becomes quite complicated.

HYDRAULIC PRESSURE CONTROLS

The function of the hydraulic pressure controls is to control the pressure in the hydraulic system. These controls are used to relieve, to reduce, or to begin another function in the hydraulic system.

Pressure Relief Valves

The purpose of the hydraulic relief valve is to control the pressure in the system by rapidly expelling small amounts of fluid. The relief valves protect the hydraulic pump and, in turn, the electric motor from overloading. Also, the other components in the system, such as the piping, cylinders, and directional controls, are protected from overloads which can result from excessive pressure emanating from the pump. How the relief valves are applied is discussed in the next chapter. This chapter is designed to acquaint the student with the design and purpose of these components. Three basic types of relief valves discussed here are: (1) the direct-acting valves; (2) the direct-operated pilot-type valves; and (3) the remote-control valves. A cross-sectional diagram of the direct-acting relief valve is shown in Fig. 10-1. A cross-sectional diagram of the direct-operated pilot-type is shown in Fig. 10-2. A cross-sectional diagram of a remote-control type of relief valve is shown in Fig. 10-3.

Pressure relief controls or valves are built in several different sizes, except for the remote-control pressure relief valve, which is generally available with either ⅛-inch or ¼-inch pipe connections. Sizes to, and larger than, 4-inch port sizes are available for industrial applications. However, most of the demands are for the 1-inch size and smaller. Pressure relief controls use three types of connections, as follows: (1) threaded *NPT* or *SAE*, (see Fig. 10-1); (2) subplate-mounted (Fig. 10-4); and (3) flange-mounted (Fig. 10-5). The threaded and subplate connections are commonly used on the smaller valves. The flange connections are usually used on the larger valves, beginning with 1½-inch port size.

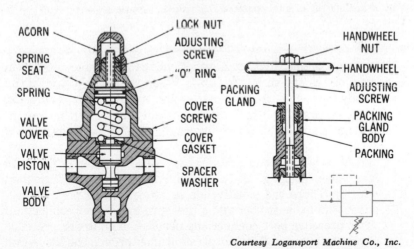

Courtesy Logansport Machine Co., Inc.

Fig. 10-1. Hydraulic relief valve with a screw-type operator. (Courtesy Logansport Machine Co., Inc.)

Fig. 10-2. Cutaway of pilot-operated relief valve. (Courtesy Double A Products Co.)

The direct-acting hydraulic relief valves are quite simple in construction, as shown in Fig. 10-1. When the main piston in the valve body rises, oil is dumped to the exhaust port. This action must take place very rapidly to maintain a constant pressure in the system. The principal parts of this type of hydraulic relief valve are as follows:

1. *Valve body.* The body contains the pipe connections or ports. In the valve (see Fig. 10-1), the pipe connections are threaded (*NPT*). There are usually one or two pressure port connections—an exhaust port and a pressure-gauge port connection. Two pressure port connections permit the oil under pressure from the pump to flow in a straight line through the valve into the system. Relief valves with one pressure port are tee-connected into the line between the pump and the system.

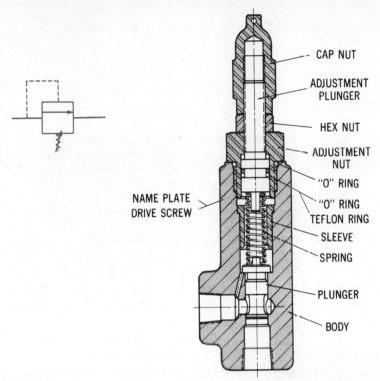

Fig. 10-3. Parts diagram of a remote-control relief valve. (Courtesy Double A Products Co.)

The valve body is usually a casting—high-tensile iron or cast-steel—or it may be made from steel bar stock when it is used for hydraulic oil service. When water is the hydraulic fluid, the valve body may be either bronze or corrosion-resistant cast iron. The mass of the valve body depends on the operating pressure and flow requirements. Flow restrictions through the valve body should be at a minimum. The valve body is bored to receive the valve piston. The tolerance between the body and piston is very close.

2. *Valve piston.* Often referred to as the spool, the valve piston may be lapped into the valve body to reduce leakage. The valve piston is made of alloy steel. It is hardened and ground to a low microinch finish. When water is the hydraulic fluid, the piston may be chrome plated. Or it may be made of stainless steel and

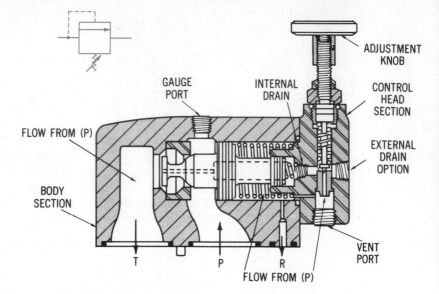

Fig. 10-4. A subplate-mounted relief valve. (Courtesy Double A Products Co.)

then chrome plated. The oil grooves are usually machined in the piston to provide better lubrication.

3. *Valve cover.* The cover contains the spring and the spring-adjustment mechanism. Normally, it is made of the same material as the valve body. In the valve shown in Fig. 10-1, the valve cover is piloted and bolted onto the valve body with socket-head cap screws. In some designs, the valve cover screws onto a threaded valve body. Some valve covers are designed with a protruding section that has a hole near the top of the cover. The hole receives a lock wire which is connected to the cover of the adjusting-screw mechanism. After the pressure has been set and the cover of the adjusting-screw mechanism is locked, a seal is placed on the lock wire. Then the pressure adjustment cannot be changed until the seal is broken.

4. *Valve spring.* The valve spring is designed for high rates of flexing, to provide a longer service life. The movement of the spring is not great, since the piston travels only a short distance to open the exhaust passage. The longer springs tend to provide a longer service life. The ends of the springs should be flat to eliminate "cocking action."

Fig. 10-5. Flange-mounted relief valve. (Courtesy Sperry Vickers Division of Sperry Rand Corp.)

5. *Valve-adjustment screw mechanism.* This mechanism depresses the valve spring and increases the tension on the valve spool. This, in turn, increases the valve operating pressure. The adjusting screw contacts the valve spring seat.

Some of the advantages of the direct-acting hydraulic relief valves are: (1) simple in design with few parts; (2) comparatively easy to service and maintain; (3) no small orifices to clog; (4) no small parts to be damaged; (5) fast response; and (6) inexpensive. A disadvantage that is sometimes encountered is spring breakage, especially where high-shock conditions exist.

The direct-acting pilot-operated hydraulic pressure control (see Fig. 10-2) uses a device that is sometimes referred to as a "control head." The control head consists of the valve cover, which contains a small orifice plug, an orifice bushing, a small adjustment spring, an adjusting screw, and a vent or remote-control connection. A poppet design is used in both the control head and the valve body to shut off the flow.

The materials used in this type of valve are similar to those employed in the valve in Fig. 10-1. Note that the springs in the valve in Fig. 10-2 are much smaller than in the valve in Fig. 10-1. Typical performance graphs for two different relief valves—with six ¾-inch pipe ports and ten 1¼-inch pipe ports—are shown in Fig. 10-6.

The direct-acting pilot-type pressure control is compact and is built with small springs to provide long service. The relatively small orifices in the control present a problem when the oil becomes extremely dirty.

The remote-control type of relief valve is similar to a control head, except for the port connections. This type of control is often used to actuate a control (see Fig. 10-3) from a remote location.

Pressure Reducing Valves

The purpose of the pressure reducing valve is to reduce the pressure in a portion of the circuit in which less operating pressure is required. For example, in a hydraulic press circuit, the full pressure that the pump develops may be required at the blind end of the press cylinder. However, a much lower pressure may be needed to operate the cylinders for either feeding or ejecting the workpiece. Pressure reducing valves are designed to effect as much as a 10-to-1 differential between the upstream and downstream sides of the valve. These controls range in weight from approximately 10 pounds (4.54 kg) for the ⅜-inch size to more than 100 pounds (45.4 kg) for the 2½-inch size.

Pressure reducing valves are built in the same general pressure ranges (for the port types), but in the smaller volume sizes than the pressure relief valves. The pressure reducing valves can be either the direct-acting type (Fig. 10-7) or the direct-acting pilot type (Fig. 10-8). These controls may have built-in check valves for free-flow return. The control shown in Fig. 10-7 is provided with the free-flow return feature. A constant reduced pressure can be maintained on the downstream side of the valve, regardless of the inlet pressure, if the inlet pressure remains higher than the reduced pressure setting. A pressure reducing valve is, in actual practice, a two-way "normally open" valve which must sense the outlet pressure to close. The constant downstream pressure is maintained by permitting a small volume of fluid (reduced pressure) to escape through the valve piston

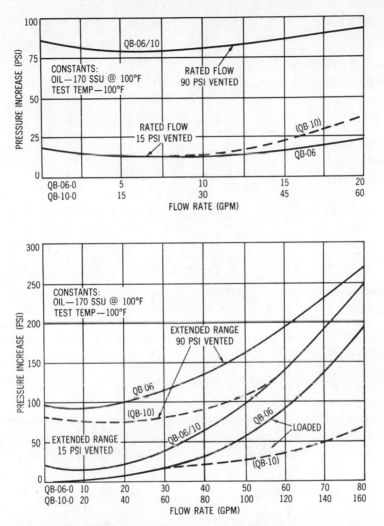

Fig. 10-6. Typical performance graphs for two different sizes of pressure relief valves. (Courtesy Double A Products Co.)

to the reservoir. In calculating the total flow in a system, the volume of fluid permitted to escape in this type of control should be considered. This is especially critical when small-volume pumps are employed in the system. An internal cross-sectional view is shown in Fig. 10-9.

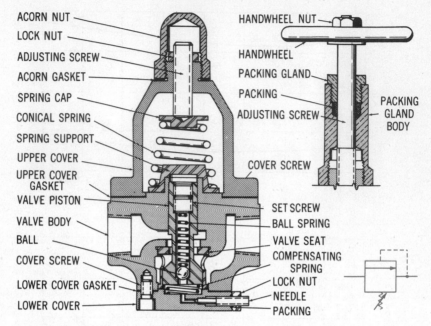

Fig. 10-7. Direct-acting type of pressure reducing valve. (Courtesy Logansport Machine Co., Inc.)

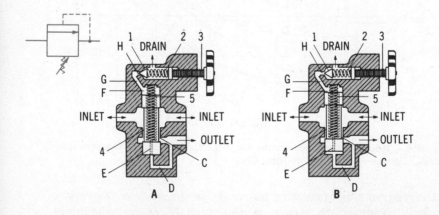

Fig. 10-8. Sectional diagrams showing open (B-1) and closed (A-1) positions of a direct-acting pilot-type pressure reducing valve. (Courtesy Sperry Vickers Division Sperry Rand Corp.)

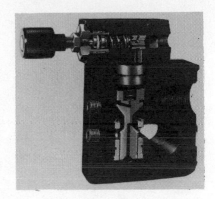

Fig. 10-9. Cutaway view of a pilot-operated pressure reducing valve. The downstream (reduced) pressure is sensed by the pilot ball through the main poppet orifice. (Courtesy Double A Products Co.)

Hydraulic Sequence Valves

The purpose of the hydraulic sequence valves is to direct the fluid from the primary circuit to a secondary circuit when a pressure is built up in the primary circuit. This control can be called a "normally closed" valve. Full line pressure is available at both the primary inlet and the secondary outlet, if the inlet pressure is maintained at a pressure higher than the valve setting. Sequence valves can often be used to eliminate a four-way control valve. This is explained in the next chapter.

Sequence valves may be direct-operated valves (Fig. 10-10), or they may be pilot-type valves. They may be built with or without an integral check. The integral check permits free-flow return of the fluid.

Sequence valves are manufactured in the same general pressure ranges, porting types, and volume sizes as the reducing valves. Other variations in designs, such as the remote-operated sequence valves, are available.

Hydraulic Unloading Valves

The purpose of the unloading valve is to return the hydraulic fluid to the reservoir at very low pressure when the pilot pressure from some other source is greater than the spring force of the unloading valve and remains higher than this setting. These controls are employed in high-low pumping circuits to control the volume of the low-pressure large-volume pump, returning the fluid to the reservoir when the valve setting is exceeded. For example, in a hydraulic press circuit, a high-pressure small-volume pump and a low-pressure

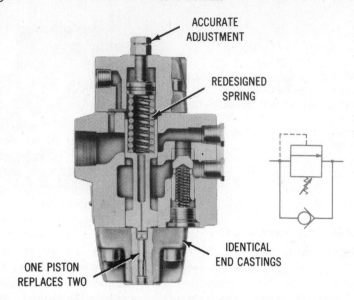

ACCURATE
ADJUSTMENT

REDESIGNED
SPRING

IDENTICAL
END CASTINGS

ONE PISTON
REPLACES TWO

Fig. 10-10. Cutaway diagram of a direct-operated sequence valve. (Courtesy Sperry Vickers Division Sperry Rand Corp.)

large-volume pump may be used to actuate the press ram on the downward stroke. Both pumps begin to pump at low pressure. The ram advances at high speed until the work is encountered. The pressure builds up, the unloading valve opens to dump the large-volume pump, and the high-pressure small-volume pump completes the work cycle. In this arrangement, horsepower, as well as heat, is reduced. This is discussed further in regard to a press circuit in Chapter 41. The loading valves are remote-controlled. They may have threaded, subplate, or flange connections. The unloading valves can be drained internally.

Counterbalance Valves

The counterbalance valve is used to maintain back pressure for counteracting a hydraulically sustained load. This valve is "normally closed." The pressure inlet is blocked until the pressure which is directed through the pilot passage works against the lower end of the pilot piston and overcomes the adjustable spring force. The oil then flows through the discharge outlet at atmospheric pressure, as long as pressure higher than the valve setting is maintained. These valves are also available for remote operation with an internal drain. They may be ported for threaded, subplate, or flange connections.

HYDRAULIC FLOW CONTROLS

A wide variety of designs is available for hydraulic flow controls, ranging from extremely simple to extremely complex designs. It is impossible to cover all the design variations in a single chapter, or in several chapters. Some of the basic designs are dealt with here. In hydraulics, three principal methods are used to control flow with flow control valves, when a relatively constant source of fluid is available. They are: (1) *meter-in;* (2) *meter-out;* and (3) *bleed-off,* as shown in Fig. 10-11. In the "meter-in" method, the fluid is throttled before it reaches the cylinder, fluid motor or actuator, or other device to be controlled. In the "meter-out" method, the fluid is controlled after it leaves the device that is controlled. In this instance, the exhaust fluid is throttled. In the "bleed-off" method, a portion of the fluid is bled off or exhausted before it reaches the device being controlled.

Hydraulic flow controls are either the noncompensating or the compensating type. The noncompensating types are available in port sizes up to 2 inches. The compensating type rarely exceeds 1¼ inches in port size, and they are usually 1 inch and smaller in size.

A *noncompensating flow control* (Fig. 10-12) is equipped with a check valve for providing free-flow return. Note that the check valve slides on the feed needle, and seats on the poppet seat in the valve body. Noncompensating flow controls are commonly used, because they are rugged, low in cost, easily maintained, and readily available. When extreme accuracy in feed is desired, as on machine tools, the noncompensating flow control is not generally used.

A *compensating flow control* (Fig. 10-13) is usually used on machine-tool hydraulic feeds. The automatic temperature compensator provides a constant feed rate for any temperature setting, although temperature changes may occur in the operating fluid. The pressure-compensating device is a built-in pressure hydrostat that compensates automatically for any change in load conditions. The pressure-compensated flow controls are also available without the temperature compensator. They are also available with an overload relief valve. The only load imposed on the pump is the load needed to overcome the work resistance. This reduces the input power and the heat losses in applications where the loads may vary considerably.

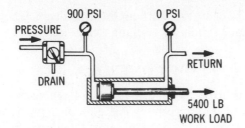

METER-IN CONTROL

Recommended for feeding grinder tables, welding machines, milling machines, and rotary hydraulic motor drives.

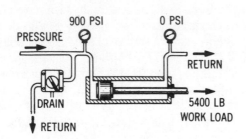

BLEED-OFF CONTROL

Recommended for reciprocating grinder tables, broaching machines, honing machines, rotary hydraulic motor drives.

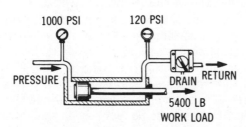

METER-OUT CONTROL

Recommended for drilling, reaming, boring, turning, threading, tapping, cut-off, and cold sawing machines.

Fig. 10-11. Three methods of controlling hydraulic cylinder speeds are: meter-in (top); bleed-off (center); and meter-out (bottom). (Courtesy Sperry Vickers Division Sperry Rand Corp.)

Temperature- and pressure-compensated flow controls are available either with or without the internal check valve. Hydraulic flow controls are available with locking devices, such as keys and closures with lock wires and seals, and with flow designations.

In Fig. 10-14, a cutaway view of a solenoid-controlled pilot-operated relief valve is shown. This valve permits convenient electrical selection of system pressures. This type of valve can be used for unloading system pressure and for creating pressure upon receiving an electrical signal.

A "sandwich"-type double flow control valve can be seen in Fig.

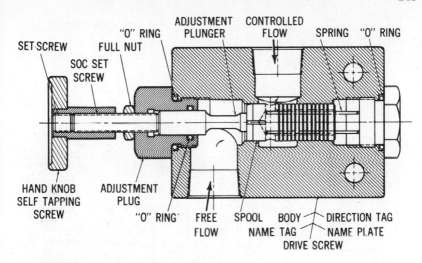

Fig. 10-12. Noncompensating type of hydraulic flow control valve. (Courtesy Double A Products Co.)

10-15. It is designed to be mounted between a four-way directional control valve and its subplate, thereby eliminating the need for additional external plumbing.

Deceleration valves, which are a type of flow control, are available as either "normally open" or "normally closed" valves. The deceleration valves are cam roller operated. They are used to convert

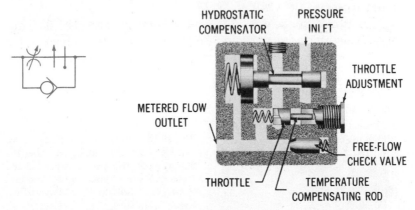

Fig. 10-13. Temperature- and pressure-compensated flow control valve with check valve. (Courtesy Sperry Vickers Division Sperry Rand Corp.)

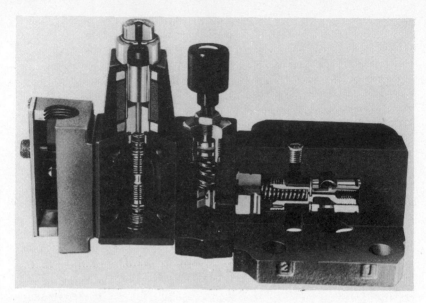

Fig. 10-14. Cutaway view of a solenoid controlled pilot-operated relief valve. (Courtesy Double A Products Co.)

a rapid traverse movement into a slow, controlled feed. These controls are available either with or without integral check valves, which are usually equipped with adjustable-flow orifices. These valves are provided with either threaded pipe ports or subplate connections.

Check valves can be classified as flow controls, although they are not provided with an adjustable orifice. Generally, a check valve provides no flow in one direction and full flow in the opposite direction. All types of check valves, such as in-line checks,

Fig. 10-15. Sandwich-type double-flow control valve. By simply inverting the mounting position, it can serve either a "meter-in" or a "meter-out" function. (Courtesy Continental Hydraulics, Inc.)

right-angle checks, pilot-operated checks, checks with pressure breakers, etc., are available. The checking mechanism may be a ball, poppet, flapper, etc. Some check valves do not open until a given pressure is attained. These valves are usually referred to as back-pressure checks, which keep enough pressure available to operate the pilot-type controls. Check valves are available with threaded, subplate, or flange connections.

Numerous special-function types of functional valves, such as prefill valves, traverse valves, and feed valves, etc., are used in the more complicated hydraulic systems. The various manufacturers' catalogs can be studied to become familiar with these controls.

PNEUMATIC PRESSURE CONTROLS

The purpose of the air or pneumatic *pressure regulator* is to safeguard the pneumatic system. Although the inlet pressure to the pressure regulator may be 150 psi (1034.2 kPa), the outlet pressure can be regulated to as low as zero pressure. If the air pressure between the source and the inlet of the regulator fluctuates considerably, there is little effect on the outlet pressure. The variation is often less than 0.1 psi (0.7 kPa). The mechanism which controls the operation of the pressure regulator may be a diaphragm, a piston, or a bellows. A regulator that uses a diaphragm is shown in Fig. 10-16. It is called a *nonrelieving diaphragm-type* pressure regulator. With a nonrelieving type of regulator, the downstream side of the control must be bled when changing to a lower pressure setting. When using the *relieving-type* pressure regulator (Fig. 10-17), the downstream side of the control need not be bled when changing to a lower pressure setting. A buildup in pressure in the reduced-pressure portion of the closed system is eliminated when the relieving-type regulator is used.

Regulators are built in various port sizes up to 2-inch pipe size. However, most of the regulators that are used in industry range from ¼- to ¾-inch pipe size. The bodies and the covers of the regulators are made from various materials, such as zinc, aluminum, brass, etc. The pressure-adjustment control is usually a screw-type mechanism that works against a spring.

Pressure regulators are often provided with a pressure gauge that is screwed into the reduced-pressure section. This is often available as a part of a triple unit which includes a filter, a regulator with a

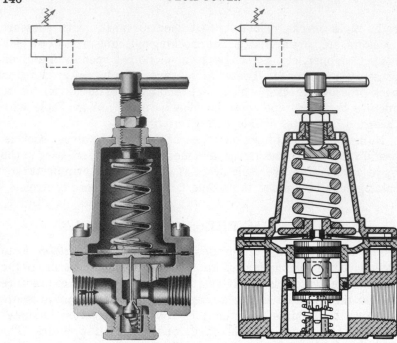

Fig. 10-16. Cross-sectional diagram of nonrelieving diaphragm-type pneumatic pressure regulator. (Courtesy Watts Regulator Company)

Fig. 10-17. Cross-sectional diagram of a relieving type of pneumatic pressure regulator. (This illustration is the courtesy of Parker Hannifin Corporation.)

pressure gauge, and a lubricator. This is a box-type unit (Fig. 10-18) which is often mounted into a panel on machine tools.

The *sequence valves* that are used for air service serve the same purpose as those used for hydraulic service. The sequence valves are *direct-acting* (Fig. 10-19). They are built in sizes that range upward to 1-inch pipe size. The poppet seat is made from a medium-hard synthetic material to effect a good seal. Yet, it provides a long service life when it is subjected to the impurities that are often found in air lines. The opening pressure is controlled by the spring which is compressed by a screw-type adjusting mechanism. In this type of valve, a check provides a free-flow return.

Sequence valves can be troublesome if a wide fluctuation in line

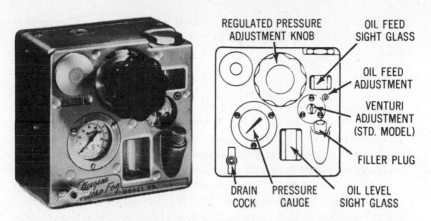

REGULATED PRESSURE
ADJUSTMENT KNOB

OIL FEED
SIGHT GLASS

OIL FEED
ADJUSTMENT

VENTURI
ADJUSTMENT
(STD. MODEL)

FILLER PLUG

DRAIN
COCK

PRESSURE
GAUGE

OIL LEVEL
SIGHT GLASS

Fig. 10-18. Combination regulator, filter, and lubricating unit often employed on a machine tool air circuit. A 50-micron filter element is used on this unit. (Photo courtesy of C. A. Norgren Co.)

pressure occurs. Sequence valve bodies may be made from either brass or aluminum. The working parts should be corrosion resistant.

Safety valves or "pop-off" valves are usually used in pneumatic systems in which excessive pressure buildups are likely to occur. A sequence valve can be used in this type of system. However, it is

Fig. 10-19. An air sequence valve. (Courtesy Logansport Machine Co., Inc.)

usually more costly than a safety valve which bleeds off excessive pressure and closes quickly. A safety valve is a "normally closed" valve.

PNEUMATIC FLOW CONTROLS

In pneumatics, the term *speed control* corresponds to the term "flow control" in hydraulics. As in hydraulics, numerous valve designs are available in pneumatics for controlling speed. Valve designs range from a ⅛-inch port size made of bar stock to a 1-inch port size made of either cast aluminum or cast bronze. A speed control in which the valve body is made from a casting is shown in Fig. 10-20. Note that in this valve an integral check provides free flow in one direction and controlled flow in the opposite direction. The needle setting regulates the volume of air that can pass through the orifice.

Air speed-control valves are employed to meter the air as it is exhausted from a device, such as a cylinder or an actuator. *Cam-operated* speed controls are available in which a "normally open" valve, a speed control, and a check are all built into a single housing. The valve is actuated from a roller mechanism, as shown in Fig. 10-21.

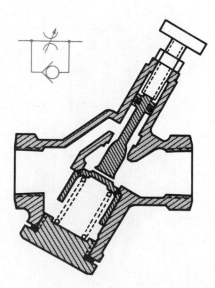

Fig. 10-20. Air speed control valve.

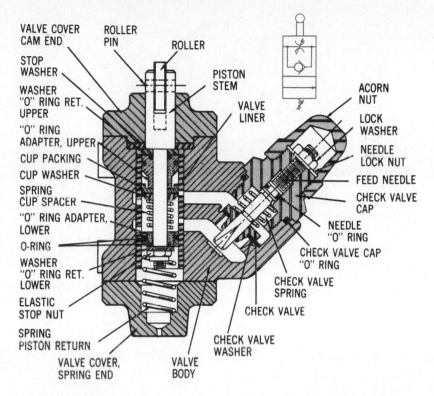

Fig. 10-21. Parts diagram of a cam-operated air speed control valve. (Courtesy Logansport Machine Co., Inc.)

The purpose of *quick-exhaust valves* is to provide a quick exit for a large volume of air without directing it through long lengths of piping. These valves are equipped with a diaphragm, and they act quickly. A cross-sectional diagram of this type of control valve is shown in Fig. 10-22.

Check valves are in limited use by themselves in a pneumatic system. They are usually an integral part of another control valve. As in hydraulics, these valves are used to check the flow of fluid in one direction and to provide free flow in the other direction. Check valves are made of a noncorrosive material. Thus, they cannot be affected by water or other impurities that may be found in the air lines.

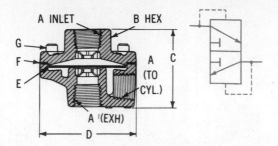

Fig. 10-22. Diagram of a quick-exhaust air valve. (Courtesy Schrader Fluid Power Division)

Chapter 11

Functional Control Components—Application

The application of functional controls can be discussed and understood more easily by using several different circuits for illustration purposes. These circuits are extremely basic. They are only suggested practical circuits, because there are many different ways in which some of the functional controls can be applied.

PNEUMATIC CONTROLS

The *pressure regulator* in the pneumatic system is one component of a combination unit A consisting of the filter, the pressure regulator, and the lubricator (Fig.11-1). The pressure regulator protects the components in the downstream side of the regulator. In this circuit, the regulator provides protection to the four-way valve B, the double-acting cylinder C, and the piping in the system. The pressure regulator is used to regulate the pressure in the downstream side of the system from near-zero pressure to full line pressure.

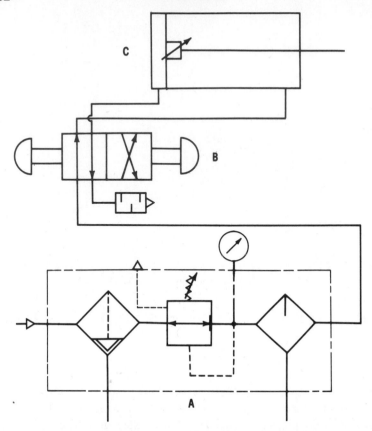

Fig. 11-1. Diagram of a pneumatic circuit with a pressure regulator.

The pressure regulator A in Fig.11-2 sets the primary pressure. The pressure regulator B sets the secondary pressure. The pressure set by the regulator A controls the pressure to the four-way valve C, to the double-acting cylinder D, and to the pressure regulator B. The pressure set by the regulator B controls the pressure to the four-way valve E and to the double-acting cylinder F.

The pneumatic circuit diagrammed in Fig.11-3 consists of two pneumatic *sequence valves*. It functions as described in the following steps: The handle of the two-position four-way valve A is shifted. The air pressure is directed to the blind end of cylinder B. The piston of cylinder B advances to the end of its stroke. Pressure builds up. The orifice in the sequence valve C is opened by the pressure. The air

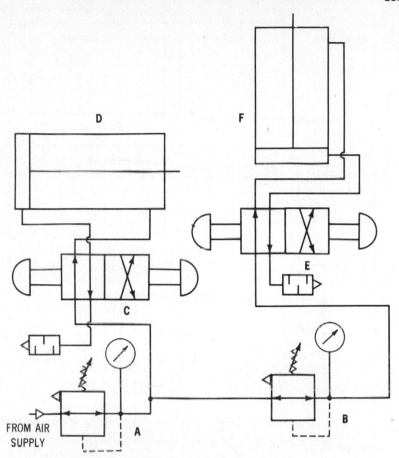

Fig. 11-2. Schematic diagram of a circuit with two pressure regulators.

pressure flows to the blind end of the cylinder D, advancing the piston to the end of its stroke. The handle of the control valve A is shifted to its original position, directing air pressure to the rod end of cylinder D. The piston retracts. Air pressure builds up to open the orifice through the sequence valve E. The air flows to the rod end of cylinder B, which retracts its piston to complete the cycle.

Most pneumatic sequence valves are provided with built-in check valves which permit a free flow of the exhausting air from the cylinder. Sequence valves that are not equipped with built-in check valves are usually provided with check valves to bypass the sequence

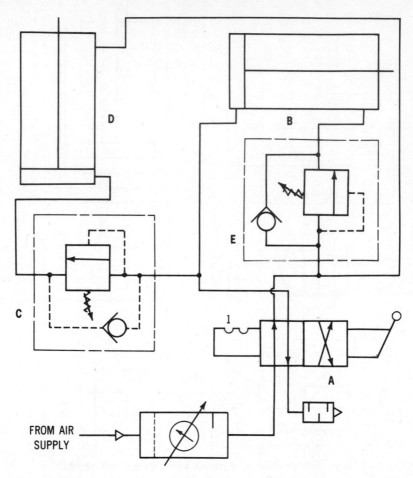

Fig. 11-3. Diagram of a circuit in which two air sequence valves are used.

valve in the exhaust side of the circuit. A sequence valve is used as a pneumatic pressure switch in the diagram in Fig.11-4. The circuit functions as follows: The push button on the three-way valve A is depressed momentarily, directing air pressure to the pilot connection X of valve B. The flow director of valve B is shifted. Air pressure is directed to the blind end of cylinder C. The piston of the cylinder advances to the end of its stroke. The pressure builds up, and the orifice through the sequence valve D is opened. Air pressure is directed to the pilot connection Y of valve B. The flow director in

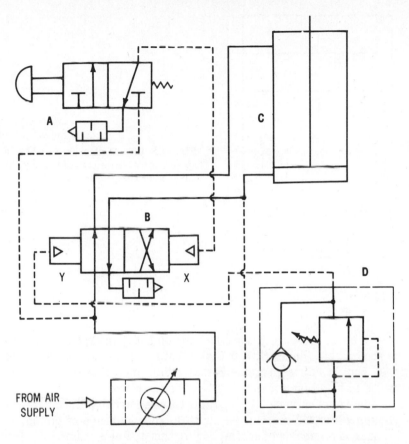

Fig. 11-4. Circuit diagram in which a sequence valve is used as a pressure switch.

valve B is shifted to its original position. Air pressure is directed to the rod end of cylinder C. The piston retracts to complete the cycle. In pneumatic systems with considerable fluctuations in pressure, sequence valves can be a source of trouble, because the spring tension which holds the orifice closed until sufficient pressure is attained may be too high to permit the orifice to open. Or it may permit the orifice to open prematurely.

The application of a pneumatic *speed control* to slow the speed of a cylinder piston as it advances is diagrammed in Fig. 11-5. The speed control A with an adjustable orifice and a built-in check valve meters the air as it exhausts from cylinder B. The four-way control valve C

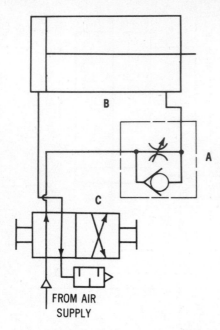

Fig. 11-5. Circuit diagram in which a speed control valve is used to slow the advance of the piston in a cylinder.

FROM AIR
SUPPLY

directs the air pressure first to the blind end of cylinder B and then to the rod end of cylinder B. A mistake that is often made in the use of speed controls for air, service is that intake metering is used. This results in a jumping action of the cylinder piston.

In some applications, it is desirable to quick-exhaust the air from a cylinder without directing the air through a long line of piping. A circuit that is capable of accomplishing this action is diagrammed in Fig.11-6. The circuit functions as follows: The workpiece is placed in the fixture. The handle of the two-position four-way valve A is shifted to direct the air pressure to the blind end of the clamp cylinder B. The piston advances, locking the workpiece in the fixture securely. The flow director in the two-position three-way valve C is shifted. The air pressure that is held on the lower side of the piston of the single-acting cylinder D is released. The air pressure exhausts to the atmosphere from the rod end of the cylinder D as the heavy weight on the end of the piston rod pulls the piston rod downward through the quick-exhaust valve E. This permits the piston to advance at a speed that is nearly equal to a free-falling body. After the cylinder D performs its work, the flow director in valve C is returned to its normal position. The air pressure is directed to the rod

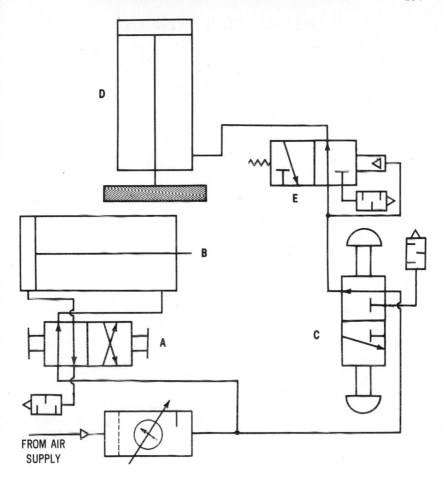

Fig. 11-6. Diagram of a circuit which utilizes a quick-exhaust valve.

end of cylinder D through the nonexhaust orifice of valve E. Then the piston of cylinder D is retracted. The flow director in the control valve A is returned to its normal position. Air pressure is directed to the rod end of cylinder B. The piston of cylinder B retracts, releasing the workpiece in the fixture to complete the cycle. A quick-exhaust valve also may be used with a double-acting cylinder when excessive speeds are desired.

HYDRAULIC CONTROLS

Hydraulic pressure *relief valves* are placed at strategic points in the system to protect the hydraulic system. A hydraulic relief valve A can be installed in a power unit circuit and used in conjunction with a constant-delivery pump (Fig. 11-7). It is important that the relief valve have sufficient capacity to handle the volume of fluid that it can be required to release on demand.

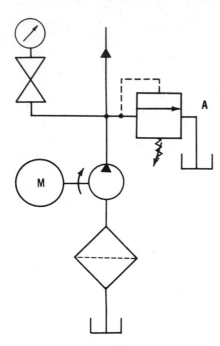

Fig. 11-7. Hydraulic circuit diagram in which a pressure relief valve is used to protect the system.

In Fig. 11-8, the relief valve A is a secondary relief valve that is set at a lower pressure value than the upstream primary relief valve B, which is a part of the power unit. The secondary relief valve A is often used when a lengthy dwell period occurs while the piston of the cylinder C is in the retracted position. The relief valve A is set at a pressure that is barely adequate to return the piston of the cylinder C. This relief valve A is usually set at 50 psi (344.7 kPa).

In many applications, it is desirable to use less pressure in a portion of the system. A simple circuit diagram for two different pressures in the system is shown in Fig. 11-9. In this circuit, the oil pressure

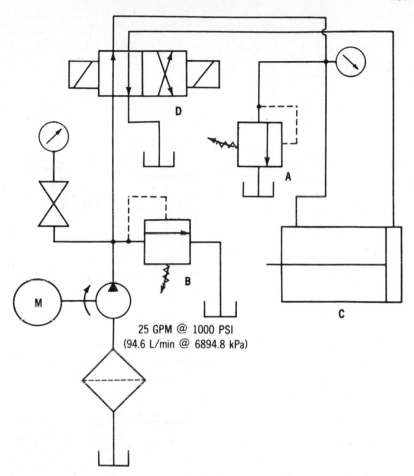

Fig. 11-8. A hydraulic circuit in which two pressure relief valves are used.

through the reducing valve A is reduced, perhaps as much as 10 to 1, depending on the setting of the valve operator of valve A and on the requirements of the system. The flow director in the two-position three-way valve B is shifted. The oil pressure is directed to the blind end of the single-acting spring-return-lock cylinder C. The piston of cylinder C advances at low pressure. The locking device is actuated. The flow director in the two-position four-way valve D is shifted. High-pressure oil is directed to the blind end of the work cylinder E. The piston of cylinder E advances to perform the work. The flow

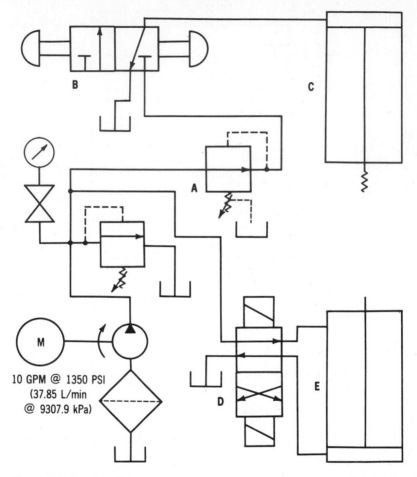

Fig. 11-9. Schematic diagram of a hydraulic circuit in which a pressure reducing valve is used.

director of valve D is returned to its original position. The high-pressure oil is directed to the rod end of cylinder E. The piston of cylinder E retracts. The flow director of valve B is returned to its original position. Spring tension causes the piston in cylinder C to retract, opening the locking device. If it is necessary for the oil to return through the reducing valve, these valves are available with built-in check valves.

The *sequence valves* used in hydraulic systems are similar in function to those used in pneumatic systems. Here, too, it should be remembered that sequence valves are not recommended if a premature pressure surge is likely to occur in the system. This does not indicate that sequence valves are never recommended, because there are many applications in which they perform extremely well (Fig.11-10). The flow director in the two-position four-way valve A is shifted to direct oil pressure to the blind end of cylinder B. The piston of cylinder B advances to the end of its stroke. Pressure builds up.

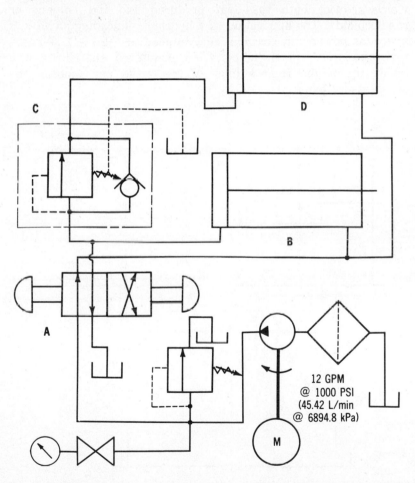

Fig. 11-10. Schematic diagram of a hydraulic circuit illustrating the application of a sequence valve.

The orifice through the sequence valve C opens. The oil pressure is directed to the blind end of cylinder D. The piston of cylinder D advances to the end of its stroke. The flow director of valve A is shifted to its original position. The oil pressure is directed simultaneously to the rod ends of cylinders B and D. Their pistons retract to complete the cycle. If the pistons of cylinders B and D are to retract in sequence, another sequence valve is required in the circuit.

An *unloading valve* can be used to unload the hydraulic pump, which reduces both the heat produced and the horsepower requirements. The application of a hydraulic unloading valve in a hydraulic power unit circuit is diagrammed in Fig.11-11. When a resistance with a given intensity is encountered, the oil pressure builds up at the remote inlet of the valve and unloads the low-pressure pump.

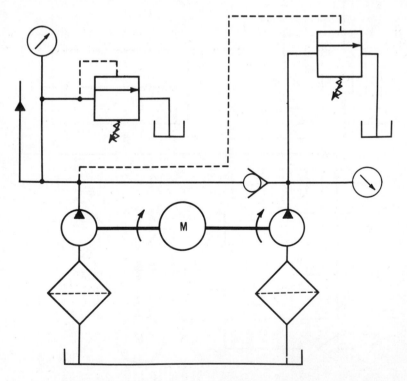

Fig. 11-11. Circuit diagram in which an unloading valve is used to unload the hydraulic pump.

The circuit diagram in Fig.11-12 illustrates the use of one of the many different types of hydraulic *flow controls*. This is the noncompensating flow control. It is recommended for exhaust metering. The flow director in the two-position four-way valve *A* is shifted to direct the oil pressure to the blind end of cylinder *B*. The piston of cylinder *B* advances at a speed set by the orifice in the flow control *C*. The orifice can be changed to suit the application. When the piston of the cylinder *B* reaches the end of its stroke, the flow director in valve *A* is returned to its original position. The oil pressure is directed through the ball check section of the flow control *C* and to the rod end of cylinder *B*. The piston of cylinder *B* returns at a rapid rate to complete the cycle.

Deceleration valves are applied in circuits in which a gradual decrease in speed is required or in which a heavy load is to be stopped without undue shock. A hydraulic circuit in which a worktable is brought to a gradual stop is diagrammed in Fig.11-13. The flow director in the open-center three-position four-way valve *A* is shifted. Oil pressure is directed to the blind end of the transfer cylinder *B* after the load has been moved onto the table. The piston of cylinder *B* begins to advance at a speed set by the pump, until the cam on the slide contacts the roller of the deceleration valve *C*. The valve

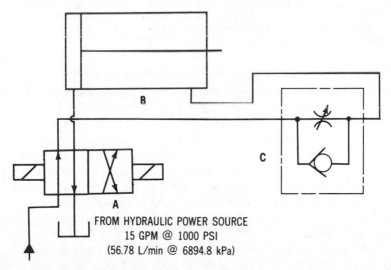

B

C

A

FROM HYDRAULIC POWER SOURCE
15 GPM @ 1000 PSI
(56.78 L/min @ 6894.8 kPa)

Fig. 11-12. Circuit diagram illustrating the application of a hydraulic flow control for exhaust metering.

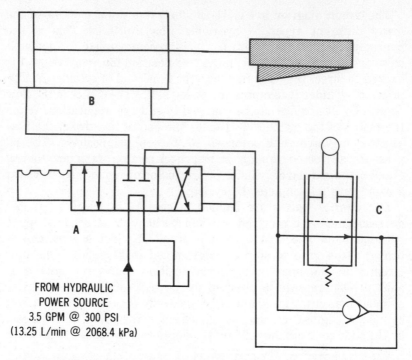

Fig. 11-13. Diagram of a hydraulic circuit in which a worktable is brought to a gradual stop by application of a deceleration valve.

C controls the volume of oil that passes through the valve. As the roller is depressed further, less oil passes through the valve. When the roller is depressed completely, the valve is shut off and the piston of the cylinder B is stopped. After the load has been transferred, the flow director in the valve A is shifted to the extreme opposite end of the valve. The oil pressure is directed through the check valve in valve C and to the rod end of the cylinder B. The piston of cylinder B returns at a rapid rate. When the piston of cylinder B has retracted, the flow director in valve A is shifted to the neutral position. The oil is returned to the tank at a low pressure during the stand-by and loading periods.

The hydraulic *check valves* are used for various purposes in the hydraulic systems. Check valves can be used in conjunction with needle valves to provide a flow control with a free-flow return. Pilot-operated check valves can be used to hold the fluid in a portion of the system which, in turn, may control the position of a heavy load that is held for a long period of time by a cylinder, as shown in Fig. 11-14. When the external pressure is directed to the pilot connection, the check is unseated and the fluid flows through the valve.

In Fig. 11-15, a check valve is employed to create enough back pressure for pilot pressure to be available for the pilot controls. In this instance, the check valve A is connected to the exhaust outlet of valve B. The check-valve passage does not open until 65 psi (448.2 kPa) is attained. This type of check valve is sometimes called a "back-pressure" check.

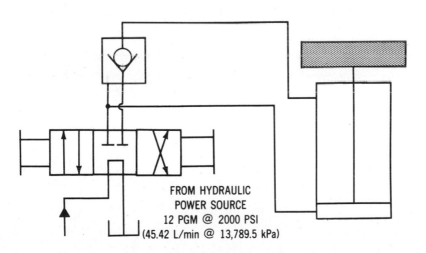

FROM HYDRAULIC
POWER SOURCE
12 PGM @ 2000 PSI
(45.42 L/min @ 13,789.5 kPa)

Fig. 11-14. Circuit in which a pilot-operated check valve controls the position of a heavy load which is held for a long period by a cylinder.

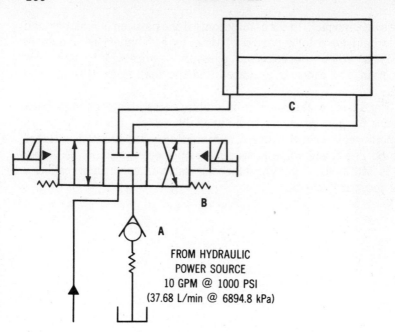

Fig. 11-15. Schematic of a circuit with a spring-loaded check valve.

Chapter 12

Directional Control
Components—Design

Directional controls are used to direct the fluid (air, oil, or water) to
the various places in the fluid power system. These controls range
from simple manually operated two-way shutoff valves to complicated six- or eight-way valves that control the movement of two
cylinders at the same time.

Directional controls of the industrial type are designed for air
service pressures up to 150 psi (1034.2 kPa), sometimes to 250 psi
(1723.7 kPa), even though the air pressure in most plants rarely
exceeds 90 psi (620.5 kPa). On hydraulic applications, directional
controls are designed for specific operating pressures, and their cost
is usually related to this pressure; however, other factors are
involved. Typical hydraulic pressure ranges are 0-1000 (0-6894.8
kPa), 0-1500 (0-10,342.1 kPa), 0-2500 (0-17,236.9 kPa), 0-3000
(0-20,684.3 kPa), 0-5000 (34,473.8 kPa), and 0-10,000 (0-68,947.6
kPa), psi. Few industrial applications exceed 3000 psi (20,684.3 kPa),
and most applications are in the 1000- to 2000-psi (6894.8 to 13,789.5
kPa) range.

Directional controls are built with three different types of port connections: (1) threaded pipe connections with *NPTF* or *SAE* threads; most industrial controls have *NPTF* threaded ports; (2) gasket or subplate mounting in which "0"-rings or a flat gasket provide the seal between the valve and subplate; (3) flange connection in which the mounting flanges are bolted onto the control. Port connections are generally designated in pipe sizes, such as ⅜-inch, ¾-inch, and 1-inch.

Threaded ports are provided on valves ranging in size up to and including 2 inches. However, they are most commonly used in valve sizes up to 1 inch. Subplate-type valves are available to 2 inches in size. The most common size is the ¾-inch size. Flange-connected valves are available up to 4 inches, and sometimes larger. They are not commonly available in the smaller sizes. The flange-connected valves are usually found on the high-pressure oil or water systems. They are seldom used for air service.

It is interesting to note that various materials are used in the production of the bodies of directional control valves. For example, the bodies of pneumatic valves are made from aluminum die castings, sand castings, and permanent-mold castings. They are also made from aluminum and brass bar stock, brass or bronze forgings and castings, and cast-iron castings. The bodies of hydraulic valves are made from high-tensile alloy iron, steel bar stock, cast steel, or high-tensile castings of bronze or stainless steel, or from bar stock of these materials.

Most directional controls are foot-mounted. This means that they are mounted on a subplate or flat surface, depending on the design of the control.

TWO-WAY CONTROLS

A two-way directional control (Fig. 12-1) is provided with an inlet and an outlet. If, in the normal position of the valve actuator, the inlet is connected to the outlet, the valve is labeled a *normally open* valve. If the inlet is closed to the outlet when the valve actuator is in the normal position, the valve is called a *normally closed* valve. The various actuating means for two-way controls are manual, mechanical, electrical, and pilot.

Two-way air controls are available with several different types of flow directors. Among them are a poppet (Fig. 12-2), a sliding spool or

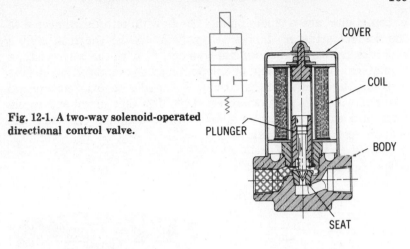

Fig. 12-1. A two-way solenoid-operated directional control valve.

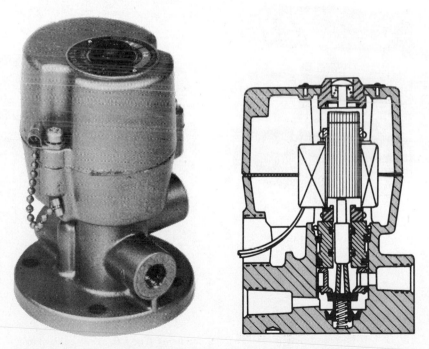

Fig. 12-2. Solenoid-operated, poppet-type, two-way air control (left) with cutaway (right). (Photo courtesy of C. A. Norgren Co.)

piston, a disc, and a tapered plug. The flow director is connected to
the control actuator. In the two-way hydraulic controls, sliding
spools or pistons having a metal-to-metal fit with the valve body, a
rotary seal, a tapered plug, or a poppet may be employed as the flow
director. A needle valve that can be classified a two-way control and
also a flow control is shown in Fig.12-3. The flow director is in the
form of a threaded spool assembly which contains the seal. The
control actuator is a small handle.

Fig. 12-3. High-pressure needle valve (left) can be used as a two-way control or a
flow control. In the exploded view (right), note the seal on the stem. (Courtesy
Gould Inc., Valve and Fittings Division)

THREE-WAY CONTROLS

The three-way directional controls are built with three ports— inlet, outlet or cylinder, and exhaust. The two-position three-way directional controls can be built either "normally open" or "normally closed." The term "two-position" indicates that the valve actuator and the flow director are built with two positions, or two positive stops. The term "three-position" indicates that the valve actuator and the flow director are built with three positions, or three positive stops.

In a "normally closed" two-position three-way valve (Fig. 12-4), the inlet is blocked and the outlet or cylinder port is connected to the exhaust port in the normal position of the actuator. When the flow director is shifted to the second position, the inlet is connected to the outlet and the exhaust port is blocked.

In a " normally open" two-position three-way valve, the inlet is open to the outlet. The exhaust port is blocked in the normal position of the actuator. When the flow director is shifted to the second position, the inlet is blocked. The outlet port is connected to the exhaust port.

In the three-position three-way valve, all the ports are usually blocked when the actuator and flow director are in the middle position. Therefore, no fluid is passing through the directional control. The actuating means for the three-position directional controls are similar to those used in the two-position controls.

The flow directors are somewhat similar to those found in the two-position, three-way valves. A group of actuators for three-way air directional controls is shown in Fig. 12-5A to 12-5G. The actuators on hydraulic controls are generally of more rugged construction.

FOUR-WAY CONTROLS

Four-way directional controls are built with four ports (five ports where two exhausts are used), including an inlet port, two outlet or cylinder ports, and an exhaust port. In the air controls that employ two exhaust ports, speed controls with different orifice settings may be connected to each port. This permits different speeds on the outward and return strokes of a double-acting cylinder or on any other component being actuated.

Four-way directional controls, air or hydraulic, are manufactured

Normally Closed Valve

UNACTUATED – With absence of pilot signal, Inlet to Outlet is blocked, Outlet to Exhaust is open.

ACTUATED – Pilot signal is applied, Poppet moves downward opening Inlet to Outlet and blocking Outlet to Exhaust.

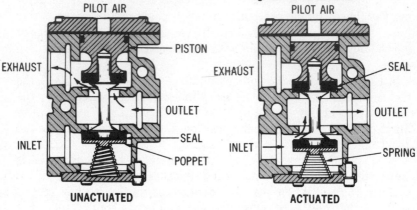

Normally Open Valve

UNACTUATED – With absence of pilot signal, Inlet to Outlet is open, Outlet to Exhaust is blocked.

ACTUATED – Pilot signal is applied, Poppet moves downward blocking Inlet to Outlet and opening Outlet to Exhaust.

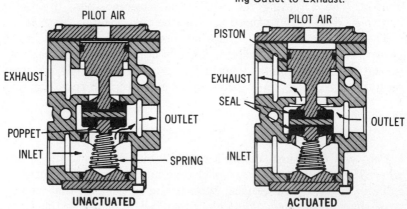

(A) Drawings of normally closed, and normally open three-way air control valves. These are poppet-type controls. (Illustrations courtesy of C. A. Norgren Co.)

Fig. 12-4. Two-position,

in either the two-position or the three-position design. In the two-position design, the valve actuator can be spring-offset. Or it may require an external force to shift the flow director to each position (Fig. 12-6). In the two-position control, in one of the positions of the flow director, the inlet is connected to the No. 1 cylinder port. The No. 2 cylinder port is connected to the exhaust port. In the other position of the flow director, the inlet is connected to the No. 2 cylinder port. The No. 1 cylinder port is connected to the exhaust port.

A single-solenoid pilot-operated four-way air control in which the solenoid operator is in the vertical position is shown in Fig. 12-7. This is a spool-type control in which the spool is returned to the "at rest" position when the solenoid pilot operator is deenergized. Spring force returns the spool. A solenoid pilot-operated air control of the single-solenoid type in which the solenoid is mounted horizontally and in line with the main valve spool is shown in Fig. 12-8. This four-way control uses O-ring type seals on the spool.

Solenoid pilot-operated four-way air controls of the poppet design, along with pressure regulators and gauges mounted on a special valve manifold can be seen in Fig. 12-9. This type of installation conserves space and reduces piping.

In the three-position design, the control actuator can be spring-centered or detent-centered (Fig. 12-10). External pressure

(B) Solenoid-operated, three-way poppet-type control (left) and pilot-operated, three-way poppet-type control (right).

three-way air control valves.

(force) may be required to shift the flow director to each of the three
positions. In Fig.12-10, a manually-operated actuator shifts the flow
director to each of the three positions.

Hydraulic directional controls offer a number of designs for flow
directors. Among them are spool, poppet, and plug. The spool design
is the most common. A number of spool configurations that are most
popular are outlined in Fig.12-11. There are a number of open-and
closed-center types in the three-position design. A cutaway of a
six-section hydraulic control in which the flow directors are
spring-centered is shown in Fig.12-12.

(A) Palm button-operated, with spring   *(B) Bleeder-operated.*
return.

(C) Hand lever-operated, pilot-return
(left), and hand lever-operated,
spring-return (right).

Fig. 12-5. Three-way air controls.

Numerous types of actuators are employed on hydraulic directional controls. Manually-operated controls use hand-lever operators (Fig.12-13), foot-pedal operators, palm-button actuators, etc. The mechanically operated controls employ cam-roller operators (Fig.12-14), toggle operators, offset cam-roller operators, stem operators, etc. Electrically-operated controls use direct-acting solenoid operators (Fig.12-15), solenoid pilot operators (Fig.12-16), motor operators, etc. Pilot-operated controls (Fig.12-17) use the various fluids as an operating means.

Six-way and eight-way directional controls are also available for specific applications. For example, two four-way controls can be assembled to provide an eight-way control which can be operated by means of a "joy-stick" control for moving the pistons of two double-acting cylinders at the same time.

A group of hydraulic control valves mounted to a manifold is shown in Fig.12-18.

(D) Roller cam-operated, spring-return.

(E) Pilot-operated, spring-return.

(F) Cam stem-operated, spring-return.

(G) Solenoid-operated, spring-return.

(Courtesy The ARO Corporation)

Fig. 12-6. Many types of actuators are available for four-way air controls. (Courtesy Schrader Fluid Power Division)

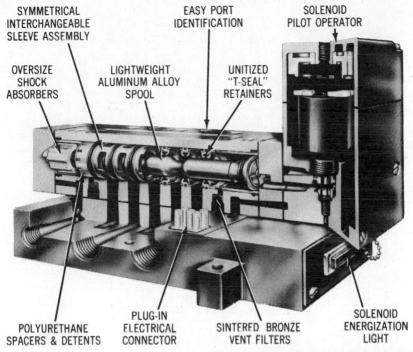

SYMMETRICAL
INTERCHANGEABLE
SLEEVE ASSEMBLY

EASY PORT
IDENTIFICATION

SOLENOID
PILOT OPERATOR

OVERSIZE
SHOCK
ABSORBERS

LIGHTWEIGHT
ALUMINUM ALLOY
SPOOL

UNITIZED
"T-SEAL"
RETAINERS

POLYURETHANE
SPACERS & DETENTS

PLUG-IN
ELECTRICAL
CONNECTOR

SINTERED BRONZE
VENT FILTERS

SOLENOID
ENERGIZATION
LIGHT

Fig. 12-7. Single-solenoid, pilot-operated, two-position, four-way air control with sliding-spool design. (Photo courtesy of C. A. Norgren Co.)

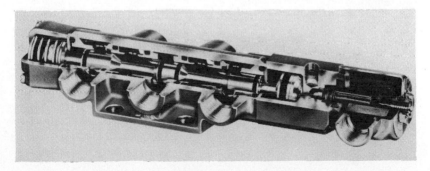

Fig. 12-8. Single-solenoid, pilot-operated, two-position, four-way air control with solenoid operator mounted horizontally and in line with the spool. Spool design has O-ring seals. (Photo courtesy Versa Products Company, Inc.)

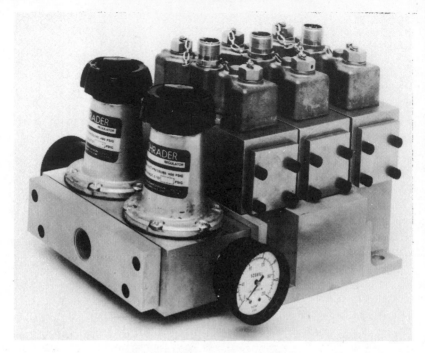

Fig. 12-9. Solenoid pilot-operated four-way air controls mounted on valve manifold. (Courtesy Schrader Fluid Power Division)

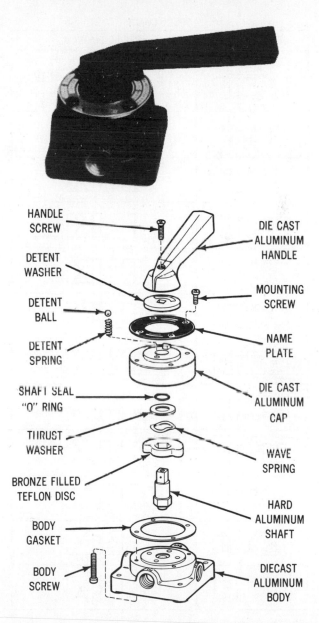

HANDLE SCREW

DIE CAST ALUMINUM HANDLE

DETENT WASHER

MOUNTING SCREW

DETENT BALL

NAME PLATE

DETENT SPRING

SHAFT SEAL "O" RING

DIE CAST ALUMINUM CAP

THRUST WASHER

WAVE SPRING

BRONZE FILLED TEFLON DISC

HARD ALUMINUM SHAFT

BODY GASKET

BODY SCREW

DIECAST ALUMINUM BODY

Fig. 12-10. Four-way, three-position air control with detent in center position (top), with exploded view (bottom). (Courtesy Logansport Machine Co., Inc.)

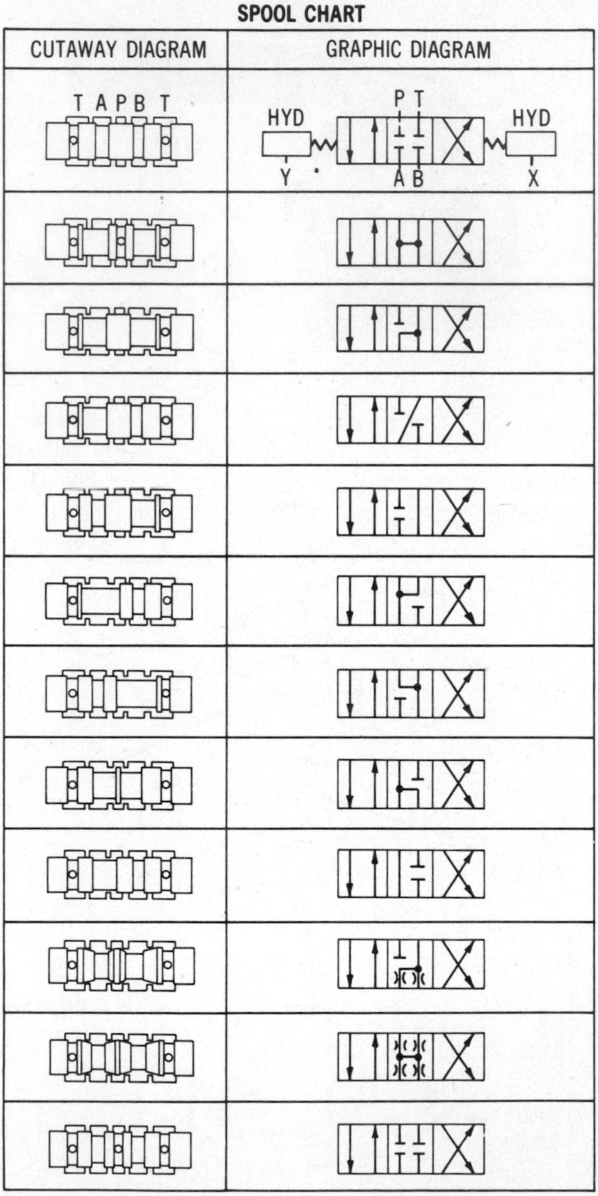

SPOOL CHART

Fig. 12-11. Four-way hydraulic control spool configuration. (Courtesy Double A Products Co.)

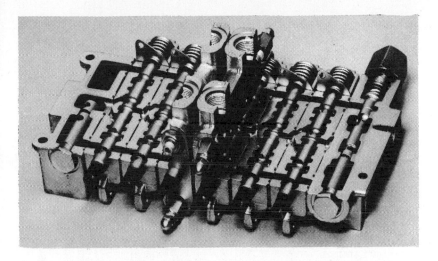

Fig. 12-12. Cutaway of six-section hydraulic control. Note the cored passages in the control bodies. (Courtesy Commercial Shearing, Inc.)

Fig. 12-13. Hydraulic directional controls with hand lever actuators. This type of control configuration is often referred to as a "bank" assembly. With the versatile grouping of different circuit sections, optimum performance is provided for infinite applications. (Courtesy Commercial Shearing, Inc.)

Fig. 12-14. Cam roller-operated, four-way hydraulic control valve with spring return. (Courtesy Continental Hydraulics, Inc.)

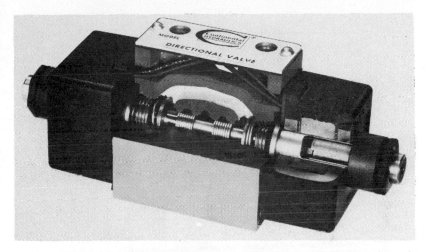

Fig. 12-15. Cutaway of a direct-acting solenoid-operated, four-way hydraulic control valve. (Courtesy Continental Hydraulics, Inc.)

Fig. 12-16. Double-solenoid, pilot-operated, hydraulic four-way control valve which has the pilot section mounted on top of the large directional control. (Courtesy Continental Hydraulics, Inc.)

Fig. 12-17. Air pilot-operated, four-way hydraulic control valve designed for high-cycle operation. This valve has manual overrides, and is spring-centered with all ports blocked in the neutral position. (Courtesy Continental Hydraulics, Inc.)

Fig. 12-18. Hydraulic directionals mounted on manifold eliminates much piping. (Courtesy Double A Products Co.)

Chapter 13

Directional Control Components—Application

Directional control components have many applications. It is wise to study the application thoroughly before making a final selection. Although the cost of a component is always important, the chief consideration is whether the component fulfills the requirements of the application. An inexpensive control is sometimes satisfactory if the actuation of the control is infrequent. However, the same control is entirely unsatisfactory if its failure causes downtime on the production line, resulting in a fouled-up production schedule and considerable loss in man-hours. Then the best control available may not be entirely satisfactory, but it is less likely to fail.

TWO-WAY CONTROLS

Two-way directional controls can be compared to the faucet in a kitchen sink, the damper on a heating pipe, or the accelerator on an automobile. Since the two-way directional control is either a "normally open" or a "normally closed" control, it has many applications in fluid power systems. For example, three two-way controls are used to divert portions of the flow in the diagram in Fig. 13-1. In the circuit diagrammed in Fig. 13-2, two two-way valves are used to actuate a single-acting pneumatic cylinder. The control valve *A* directs the flow of air to the cylinder. The control valve *B* directs the trapped air away from the cylinder after the work has been completed. In the circuit diagram in Fig. 13-3, the air is directed to a second portion of the system when the control is actuated.

Two-way directional controls can be used to actuate fluid motors whose output shafts rotate only in one direction, as shown in Fig. 13-4. They can be used to stop the flow in an entire system when a sudden drop in pressure occurs, as shown in Fig. 13-5.

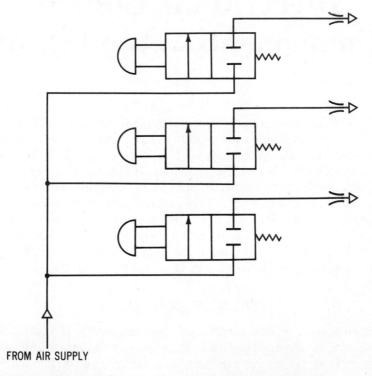

FROM AIR SUPPLY

Fig. 13-1. Schematic diagram of a pneumatic circuit in which three two-way directional controls are used to divert portions of the air flow.

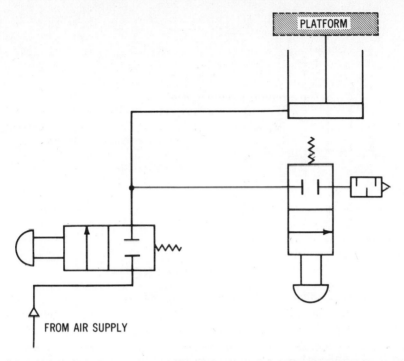

Fig. 13-2. Diagram of a pneumatic circuit in which two two-way control valves are used to actuate a single-acting air cylinder.

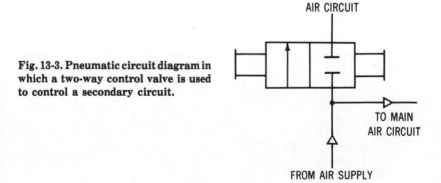

Fig. 13-3. Pneumatic circuit diagram in which a two-way control valve is used to control a secondary circuit.

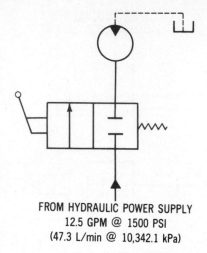

Fig. 13-4. Schematic diagram of a circuit in which a two-way valve is used to control a hydraulic motor.

FROM HYDRAULIC POWER SUPPLY
12.5 GPM @ 1500 PSI
(47.3 L/min @ 10,342.1 kPa)

THREE-WAY CONTROLS

Three-way directional controls can be used in even more applications than the two-way controls. The two two-way controls used in Fig. 13-2 can be replaced by a three-way control. The three-way controls can be used to activate a single-acting cylinder with a spring return or gravity return. A three-position three-way

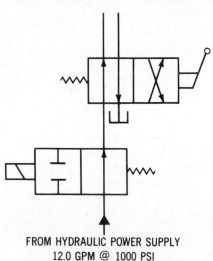

Fig. 13-5. Schematic diagram of a fluid circuit in which a two-way control valve is used as an emergency shutoff to stop the flow when a sudden drop in pressure occurs.

FROM HYDRAULIC POWER SUPPLY
12.0 GPM @ 1000 PSI
(45.4 L/min @ 6894.8 kPa)

directional control can be used to actuate a gravity-return cylinder in which the control can stop piston movement in the cylinder at any desired point in its range of travel (Fig. 13-6).

The application of a three-way directional control valve for actuating a fluid motor which, in turn, actuates a drum arrangement is diagrammed in Fig. 13-7. When the three-way valve shuts off the incoming fluid, the weight on the end of the cable reverses the direction of rotation of the fluid motor, since the inlet port of the motor is then open to the exhaust port of the valve. Two three-way valves are used to actuate a pressure-operated four-way valve in the circuit diagram in Fig. 13-8. The three-way valves are the smaller pilot valves which are used to actuate the larger four-way control valve. This combination is a basic arrangement for many semiautomatic and automatic cycling applications.

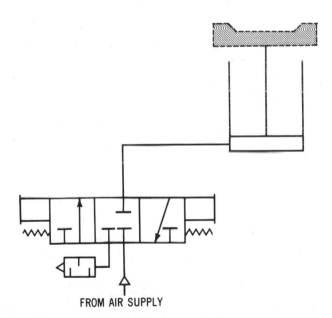

FROM AIR SUPPLY

Fig. 13-6. Schematic diagram of an application in which a three-position three-way directional control valve is used to actuate a gravity-return cylinder in which the control valve can stop the piston movement at any point in its travel.

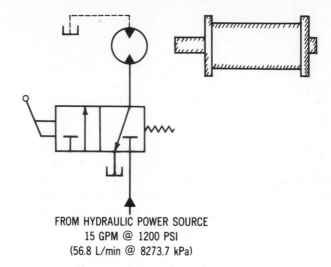

FROM HYDRAULIC POWER SOURCE
15 GPM @ 1200 PSI
(56.8 L/min @ 8273.7 kPa)

Fig. 13-7. Circuit diagram for an application in which a hydraulic three-way directional control valve actuates a fluid motor which, in turn, actuates a drum arrangement.

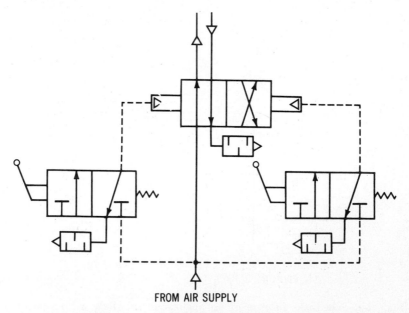

FROM AIR SUPPLY

Fig. 13-8. Circuit diagram for an application in which two three-way directional control valves are used to operate one four-way directional control valve.

FOUR-WAY CONTROLS

Of all the available directional controls, the four-way control undoubtedly has the most applications by far. The four-way controls are designed to direct the pressure either to the rod end or to the blind end of a double-acting cylinder. Also, they receive the exhaust fluid at the same time from the cylinder port opposite the end at which the pressure is applied (Fig. 13-9). Four-way controls are also used to actuate rotary-type actuators, fluid motors, and larger controls. They can be used satisfactorily as a three-way control by plugging one of the cylinder ports.

The two-position four-way control is used in an application where intermediate stops are not required, such as on a double-acting cylinder in which the piston is moved to the end of its stroke when the valve actuator is in one of the positions. When the valve actuator is in the other position, the piston of the cylinder moves in the opposite direction.

A three-position four-way control is used when it is necessary for the piston to make one or more stops before completing either its forward stroke or its return stroke. Various conditions can be

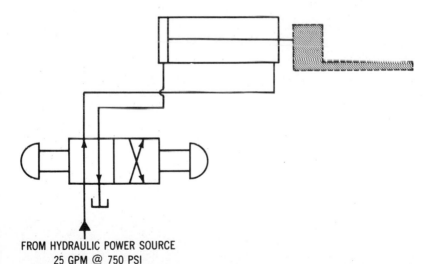

FROM HYDRAULIC POWER SOURCE
25 GPM @ 750 PSI
(94.6 L/min @ 5171.1 kPa)

Fig. 13-9. Diagram of a circuit in which a four-way directional control valve is used to control a double-acting cylinder.

provided in the cylinder while the piston is stopped, such as: (1) pressure can be released from both sides of the piston, so that the piston "floats"; (2) fluid can be blocked on both sides of the piston, so that the piston stays "put"; and (3) one or both cylinder ports can be open. The motions of the fluid motors and actuators can be halted when the controls are placed in the neutral position. Six-way and eight-way directional control valves for either hydraulic or pneumatic service are employed on special-purpose applications which require more than one device to be actuated at the same time.

In high-speed applications for hydraulic service, air pilot-operated valves are often used. The pilots can be actuated by inexpensive air controls which provide a rapid response in the hydraulic control. Examples of these applications are high-speed coining operations, compacting devices, punching operations, etc.

Hydraulic pilot-operated directional control valves are also in demand on machine tools and machine applications. A definite trend has been noted toward the use of more solenoid-operated and pilot-operated directional controls in industry.

Chapter 14

Control Components—
Installation
and Maintenance

Since many students are aiming toward the engineering or design phase of fluid power, it is important that they should receive basic instructions in the installation and maintenance of fluid power components. This knowledge can aid in solving some of the design problems they are likely to encounter.

PRESSURE CONTROL VALVES

Pressure controls or hydraulic relief valves serve purposes other than to protect the pumps in a system. They should be installed in locations where they can be adjusted for the pressure requirements and for servicing. Although the pressure settings do not require changing in most installations, a pressure setting may need to be changed after each operation, such as on a testing device in some instances. In those installations which require the pressure to be

193

changed only rarely, some type of locking mechanism should be installed on the pressure-adjustment device. The locking device may range from a locking wire, in which a seal is broken to change the pressure setting, to a padlock and key. A responsible person should be in charge of the keys for those components. If the pressure setting requires frequent changing, a handwheel attached to the pressure-adjusting mechanism is desired by most operators.

Pressure relief valves should be connected to suitable exhaust lines, without restrictions, to keep the back pressure at a minimum. If more than one relief valve is connected to a common exhaust line, the line should be a size that is more than ample for meeting the peak requirements.

When subplate-mounted relief valves are employed in a system, the mounting plate should be flat to eliminate undue stress on the valve body. This type of stress can cause a binding action between the valve body and the piston. Also, care must be exercised to prevent shearing the mounting gaskets or "O" rings between the body and the mounting plate.

Relief valves should be selected that are capable of meeting the pressure and flow requirements of the application. Too often, problems arise from improper selection of valves in regard to their capacities.

When pipe line mounted valves are used, the pipes or pipe connections should be turned tightly into the valve body, to eliminate external leakage. However, if the connections are turned in too tightly, the valve body may be distorted, causing a binding condition.

The presence of dirt and heat is of primary concern in fluid power systems, resulting in conditions that cause serious maintenance problems. The fluid power control components, especially those designed for hydraulic service, are precision devices with parts fitted to extremely close tolerances.

Small particles of dirt in a system can clog the small orifices in a valve, especially in the direct-acting pilot-operated valves. This results in erratic pressure readings. Dirt particles may cause the piston to stick and may also score the smooth finish of the bore in the valve body, causing excessive leakage. Leakage may increase until the valve cannot hold the desired pressure setting.

The presence of excessive heat in a system can cause the seals to become brittle, resulting in either internal or external leakage. Excessive heat also changes the tolerances between the valve body and the piston. This may cause either a binding action or an excessive internal leakage problem.

In addition to the presence of dirt and heat in a system, relief-valve failure can be caused by various conditions. The presence of air in the hydraulic system can cause erratic action in the relief valve. Air is often entrapped in the hydraulic fluid. It is a common remark that the presence of air can be heard, because the sound is similar to someone shaking a "bucket of bolts."

Broken valve springs present another problem that may be encountered. This problem is more prevalent in the direct-acting relief valves using the larger springs. However, the problem can occur in direct-acting pilot-operated relief valves using the smaller valve springs. Broken springs usually can be recognized, since a change in the setting on the control head does not change the operating pressure.

A problem with the relief valve is sometimes suspected, although another valve is involved, when an open-center valve is used in the system. Until this valve is closed, pressure cannot be built up in the relief valve.

Since relief valves are rather simple in their design, they are usually quite easy to maintain. If there is uniform wear in the bore of the valve body, these valves can be repaired either by installing oversize pistons or by chrome plating the original pistons. Even when the packing appears to be in good condition, it is wise to replace it when the valve is repaired or overhauled.

After the valve has been repaired, it should be tested thoroughly before being placed in operation. A valve can be tested by connecting the inlet to a hydraulic pump, plugging the outlet, and connecting the exhaust port to a reservoir. Place a pressure gauge between the pump and the valve. Then move the pressure-adjustment screw through its entire pressure range to determine whether the valve fulfills the requirements of the system. The capacity of the test pump must be adequate for the job.

Hydraulic pressure reducing valves require care similar to that of a hydraulic relief valve. Pressure reducing valves with external drains should be installed with the drain connections piped either to the reservoir or to an adequate return line. It is important that the drain section be kept free from back pressure at all times.

Since many reducing valves are constructed with several moving parts and small orifices, it is imperative that they be kept free from dirt particles. A reducing valve that fails during operation places an extra load on the components in the portion of the system in which reduced pressure is required. Leaky internal parts, broken springs, damaged seats, and wear are the chief factors that cause these valves to malfunction.

In servicing reducing valves, all parts should be cleaned thoroughly with a good solvent and then dried. Varnish deposits and other types of oil deposits sometimes cause malfunctions.

In testing a reducing valve after it has been repaired, place one pressure gauge on the inlet side and another on the outlet side. Then direct pressure to the valve. Move the pressure-adjustment screw to determine whether the entire pressure range for which it was designed can be obtained on the outlet port.

Hydraulic sequence valves require nearly the same installation and maintenance procedures as the other hydraulic pressure control valves. The pneumatic sequence valves are designed with greater tolerances than the hydraulic controls. In the hydraulic controls, most of the sealing occurs between metal-to-metal fits. However, in the pneumatic sequence valves, nonmetallic seals are generally used. If these valves are used in high ambient temperatures, the common seals often deteriorate. Therefore, it may be necessary to use high-temperature types of seals. Pneumatic sequence valves are similar to their hydraulic counterparts in installation procedures. Most pneumatic sequence valves are pipe-line mounted. Failures in these valves usually result from a broken spring, a leaky seal, a damaged seat, or a misapplication. Either lack of lubrication or improper lubrication can also cause a failure in a valve.

When servicing the sequence valves, the seat in the valve body and the bore should be inspected thoroughly. It is always good practice to change all packings when servicing the valve. If the valve contains a built-in check, be sure the check seals properly.

The procedure in installing and maintaining the other types of pressure controls is similar to that already discussed. Pressure controls with fixed orifices should be checked for the presence of dirt or other foreign particles. Pneumatic pressure-regulating valves should be installed downstream from the filter. Excessive dirt in the lines without filters can foul the action of the pressure regulator. In maintaining a regulator, the springs, seals, diaphragms, etc., should be inspected carefully. After the regulator has been serviced, it should be checked before being placed in the circuit. Place a pressure gauge on each side of the unit. Then move the pressure-adjustment screw.

FLOW OR SPEED CONTROLS

Proper installation of flow or speed control valves is an important step for the system to function properly. Valves designed to control the exhaust flow should be located where only the exhaust flow is controlled. In pneumatic systems, nearly all the speed controls are designed to control the exhaust fluid. If these controls are installed to control the incoming pressure, an erratic feed rate is often encountered. In too many instances, these controls are installed improperly.

Most flow controls and speed controls are relatively simple in design and can be serviced easily. The seats of speed controls should be inspected, since dirt particles sometimes cut them.

When these controls are overhauled, it is a good practice to install new packings, seals, gaskets, and even springs, since all these parts are relatively inexpensive and are costly to install later. Other types of valves that control flow are relatively easy to service. Some of the more expensive parts that become worn can often be built up to original size by chrome plating, without great expense.

Before attempting to overhaul the various pressure, flow, or directional controls, it is wise to obtain a parts diagram from the manufacturer. The manufacturer also should be consulted in regard to a repair kit for the component. Typical parts diagrams and parts lists for the various pressure and flow components are shown in Figs. 14-1, 14-2, and 14-3.

Before returning the flow or speed control components to the system, they should be tested thoroughly for flow, pressure, and operational performance characteristics. If the repaired components are to be placed in storage, their ports should be plugged, and the components should be placed in a clean, dry location in which the temperature is kept nearly constant.

DIRECTIONAL CONTROL VALVES

Since numerous designs of directional control valves are used in both air and oil service, it is impossible to discuss their installation and maintenance here. Only a few of the more important points in

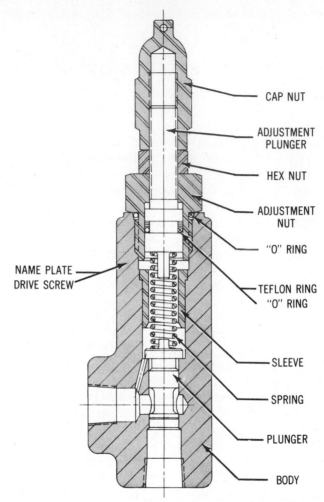

Fig. 14-1. Diagram of a hydraulic relief valve with part names. (Courtesy Double A Products Co.)

regard to installation and maintenance are covered. As with the other fluid power components, it is good practice to have available a parts diagram and parts list for study before dismantling one of these components.

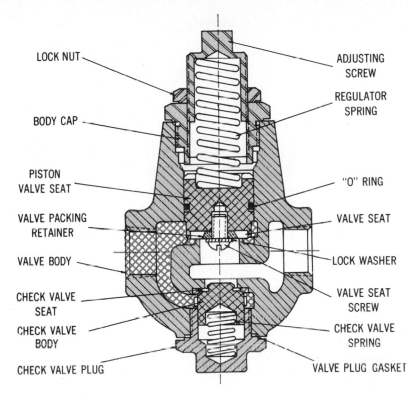

LOCK NUT

ADJUSTING SCREW

REGULATOR SPRING

BODY CAP

PISTON VALVE SEAT

"O" RING

VALVE PACKING RETAINER

VALVE SEAT

VALVE BODY

LOCK WASHER

CHECK VALVE SEAT

VALVE SEAT SCREW

CHECK VALVE BODY

CHECK VALVE SPRING

CHECK VALVE PLUG

VALVE PLUG GASKET

Fig. 14-2. Diagram of an air sequence valve with part names. (Courtesy Logansport Machine Co., Inc.)

Installation

The use of good judgment and a few simple rules aid in the installation of two-way valves.

Manual Operation—Here are some of the important rules for installation of two-way valves:

1. If the valve has mounting feet, mount them securely on a flat surface. Do not throw a strain on the valve body by mounting the valve on an uneven surface.
2. Install the valve so that the actuator is readily accessible to the operator. Do not require him to reach up and over a moving mechanism to get at the valve actuator.

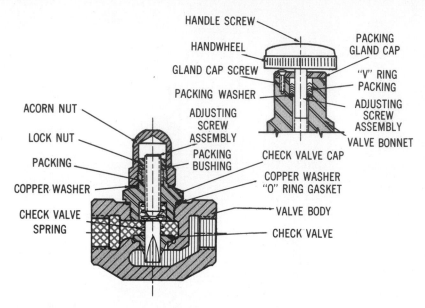

Fig. 14-3. Diagram of a hydraulic speed control valve with part names. (Courtesy Logansport Machine Co., Inc.)

3. Install the valve so that it is not subject to hot blasts. Extreme heat has a detrimental effect on most packings.
4. Install the valve so that it is not covered with dirt. Valves often become covered up and are difficult to locate.
5. Make connections to valve ports so that the connections do not leak, but do not apply so much effort in tightening the connections that the ports are broken.
6. Install valves so that they can be easily accessible to the operator, yet high enough above the floor that they cannot be bumped by lift trucks and carts.
7. When installing valves, do not use excessive pipe compound in the threaded ports. Be sure that no dirt has entered the valves before they are installed.

Mechanical Operation—When mechanically operated valves are installed these rules should be followed:

1. The valve should be mounted on a good flat surface. This valve

must withstand thrust from cams, trip arms, pins, and other mechanical operating means. The mounting screws must be pulled down evenly and tightly.

2. If the valve is of the manifold type, be sure the seal is in place.
3. Be certain that the valve is clean before it is installed.
4. Install the valve so that it works best into the piping layout. In other words, eliminate bends in the piping as much as possible.
5. Make certain that all pipe connections are tight. Oil and air leaks are costly.
6. Check mechanical trips to see that they are in proper alignment. Also check for overtravel.
7. Lubricate the actuating mechanisms, and be sure that they operate freely.

Solenoid-Operated Valves—When installing solenoid-operated valves, here are some of the points to follow:

1. Mount the solenoid valve per manufacturer's recommendations. Some designs are recommended for horizontal movement of the plunger, and other designs are recommended for vertical movement of the plunger.
2. If the valve is a type that is designed for pipe-line mounting, make certain that the valve is piped so that it does not leak. Since many in-line mounted valves have cast iron or brass bodies, make certain that the pipes are not screwed into the valve body to the extent that they crack the valve body.
3. Check carefully the current specification on the nameplate of the valve or on the solenoid coil, before connecting the wires. Never use a different electrical current from that specified on the valve. The manufacturer's recommendations should be followed.
4. Keep solenoid valves away from regions where the temperature is high. Solenoid coils should be kept at normal temperatures if possible.
5. In explosive atmospheres, install solenoid valves which have explosion-proof housings that meet with Underwriter's approval for the conditions involved.
6. Protect solenoid valves, especially the type that do not have weatherproof housings, from water spray and excessive residue in the air.

7. The valve interior should be blown with compressed air to insure the removal of dirt and particles.
8. Mount the solenoid valve so that it is protected from shop trucks, tote pans, and other items which may damage the valve.
9. Be sure that the solenoid covers are in place before the valve is placed in service; this keeps out dirt and foreign particles.

Pilot-Operated Valves—Some important points in installations of these valves are:

1. If the valve has mounting feet, mount it securely to a smooth flat surface. If the valve is pipe-line mounted, the pipes should be securely tightened into the valve body.
2. Secure the pilot connections so that there is no leakage.
3. Make certain that the pilot connections are not restricted.
4. Keep the valve away from hot blasts.
5. See that lubrication is available to the valve.
6. Mount the valve, whenever possible, in a horizontal position. Although this is not necessary for some valves, it is necessary for others.
7. Do not hammer on the valve either when installing it or after it is installed.
8. If, when mounting the valve, the piping is not ready to install, do not remove the pipe plugs.
9. Be sure to select the correct pilot valve for actuating the pilot-operated valve.
10. Do not connect pilot pressure into the spring end of a spring-offset valve.

Repair and Maintenance

Most two-way valves can be repaired. Some important points to observe are:

1. The repairs should be made on a clean workbench that is neat and orderly.
2. Since some exertion is required to disassemble most valves, the valve should be placed in a vise with soft jaws which should be closed against the valve body with only enough pressure to prevent the valve body from slipping while parts are being removed.

3. Place the smaller metal parts of the valve in a container and wash them thoroughly with a solvent until all foreign matter is removed. Lubricate the metal parts and, if the valve is not to be reassembled immediately, place these parts in a clean, closed container. Nonmetallic parts, such as rings, seals, gaskets, etc., should be replaced. If the component contains springs, check them thoroughly; if in doubt about their condition, they should be replaced.

4. Replace all worn linkages.

5. If a spool-type valve is being repaired, check the valve bore to determine whether there are score marks which may have been caused by dirt, lack of lubrication, wire drawing, etc. If the score marks are not too deep, they can be honed out and an oversized spool can be used. The spool in the valve can be made oversize by chrome plating. If a valve body is badly scored, it is usually less expensive to replace the valve body.

6. If a poppet valve is being repaired, check the poppet seats for imperfections. It is often possible to rework the poppet seat or to bore out and braze in a new seat. When the poppet is made from a synthetic material, such as those used for air service, the valve may require new poppets. Hardened poppets, such as those used in hydraulic service, can withstand considerable abuse, but it is sometimes necessary to replace them.

7. If a plug-type valve is involved, the valve body may be scored, and these score marks can sometimes be removed if they are not too deep. When the bore of a valve with a tapered-type plug is reworked, the plug generally seals, if it is not damaged. When a valve with a straight plug is reworked, it is necessary either to install a new oversized plug or to make the old plug oversized by chrome plating it.

8. When reassembling the valve, a sufficient amount of lubricant should be used to prevent having to force the parts together.

9. If a foot-mounted control is being reassembled, perform the final assembly of the covers on a flat surface.

10. If it is necessary to replace the solenoids of solenoid-operated valves, be sure that the new solenoids are of correct voltage, stroke, etc.

11. Never substitute the parts. This applies to springs, gaskets, etc. Do not attempt to stretch a spring.

12. Thoroughly test the directional control after it has been repaired and before it is returned to the circuit.

13. If the rebuilt control does not pass a bench test, it should be discarded. Then, a new control should be put in the circuit.

Chapter 15

Nonrotating Cylinders—Design

The cylinder is a fluid power component which depends on the operating pressure, the force requirements, the service life, and the application for its design. A cylinder is a force component that converts fluid energy to mechanical energy, and its motion is linear.

There are numerous cylinder designs. Their operating medium can be hydraulic fluid, water, air, or other gases. The cylinder is designed accordingly. Cylinders may be either single-acting or double-acting. They may be equipped with either single-end or double-end rods and with oversize rods. The porting in the cylinders may be *NPT*, *SAE*, or flange-type connections.

OPERATING PRESSURES

Hydraulic (oil) cylinders are designed to operate in the various pressure ranges, such as 0-500 psi (0-3447.4 kPa), 0-1000 psi (0-6894.8 kPa), 0-1500 psi (0-10,342.2 kPa), 0-2500 psi (0-17,236.9 kPa), 0-5000 psi (0-34,473.8 kPa), etc. In designing a cylinder for the various operating pressure ranges, the shock load must be considered. Many hydraulic cylinders are designed with a 5:1 or 4:1 safety factor. If the safety factor is lower, the application should be analyzed carefully to be sure that the peak loads do not exceed the safety factor. Some specifications for cylinders include both an operating pressure reading and a test pressure reading. Cylinders that operate in a water medium are designed for pressure ranges that are similar to those for the oil in hydraulic cylinders.

The pneumatic cylinders are usually designed for pressures ranging from 0 to 150 psi (0 to 1034.2 kPa). However, they are seldom used for pressures higher than 90 psi (620.5 kPa). Some air cylinders (industrial) are designed for pressures upward to 250 psi (1723.7 kPa). Air or gas cylinders designed for higher pressures are seldom used in industrial applications.

Most of the air cylinders are designed with a rather high safety factor, such as 12:1 and 15:1, etc. The safety factor of an air cylinder should be checked against the manufacturer's specifications.

CONSTRUCTION

A pneumatic cylinder (Fig. 15-1) consists of several different parts, including a cylinder tube, two end covers, a piston, a piston rod,

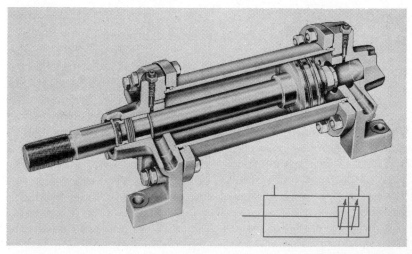

Fig. 15-1. Cutaway view of a double-acting mill-type pneumatic cylinder. (Courtesy The Sheffer Corp.)

seals, nuts, bolts, screws, and washers. Of course, there are many different types of construction, depending on the design. The cylinder in Fig. 15-1 is a double-acting cylinder in which fluid pressure is applied in either direction to create alternate "push" and "pull" forces.

In some types of cylinder construction, a cylinder cover may be a part of the cylinder tube. The piston and piston rod may be of one-piece construction.

The piston seals can vary widely in design (see Table 15-1), ranging from the metallic automotive-type piston rings (Fig. 15-2) to the synthetic U-type packings (Fig. 15-3). Other types of piston seals often found are "O" rings, quad rings, block vees, chevron-type packings, vee-cups, etc. Three advantages of the automotive-type rings on hydraulic pistons are that they: (1) provide longer service life; (2) withstand higher ambient temperatures; and (3) provide less starting friction.

The pistons in hydraulic cylinders are made from a material that is compatible with the cylinder tube. Millions of cycles can occur without scouring the cylinder tube or causing undue wear. Steel or chrome-plated steel may be used in the cylinder tube. However, the piston may be made from a high-tensile iron or steel that is bronze-faced on its outer surface. The piston serves as a bearing for supporting the piston rod.

In some instances, the metal in pistons used for air service bears against the cylinder walls. Therefore, the different metals, such as the brass in the cylinder tube and the cast iron in the piston, must be compatible. In other piston designs, the metal in the piston does not bear against the cylinder walls. But some type of synthetic material does bear on the surface (Fig. 15-4). When some types of gases are used to operate single-acting and double-acting cylinders at extremely low temperatures (-75 to -100 degrees F, or $-59°$ to $-73°$ C), special types of piston seals, such as degreased leather, etc., are required. The pistons are sometimes made from a synthetic material. Their periphery is coated with a lubricating material.

Pistons used in water service are often made from a high-tensile bronze material that resists "wire drawing," which is caused by high-pressure, high-velocity water. The cylinder tubes are usually made from a hard chrome-plated material. The piston seals may be bronze automotive-type rings, cup-type leathers, chevron-type packings, leather or impregnated materials, etc.

Table 15-1. Materials Commonly Used in Cylinder Parts

Cylinder Part	Service Medium		
	Hydraulic	Air	Water
Covers	Steel bar High-tensile 　cast iron Cast steel Forged steel	Steel bar High-tensile 　cast iron Aluminum Brass	Bronze Plated steel
Tube	Steel Chrome-plated 　steel	Brass Chrome-plated 　steel Aluminum- 　coated 　reinforced 　plastic	Bronze Chrome-plated 　steel
Piston rod	Steel Chrome-plated 　steel Hardened and 　chrome- 　plated steel	Steel Chrome-plated 　steel	Stainless steel Chrome-plated 　stainless steel
Pistons	Cast iron Bronze-faced 　steel	Cast iron Synthetic	Bronze
Packings and 　seals	Cast iron 　(automotive- 　type piston 　rings) "O" rings Quad rings Chevron rings Vee-type 　packing Hat-type 　packing Cup-type 　packing	Block vee Cup-type 　packing "O" rings Quad rings Hat-type 　packing "V"-cup	Bronze 　(automotive- 　type piston 　rings) Cup-type 　packings Vee-type rings

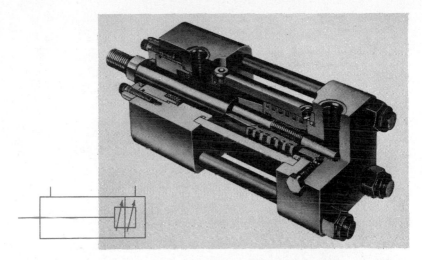

Fig. 15-2. Cutaway view of a hydraulic cylinder with metal automotive-type piston rings. Note the tapered cushions. (Courtesy The Sheffer Corp.)

Many different construction designs are used to attach the cylinder cover to the cylinder tube. Some of these designs are:

1. *Tie-rod construction (see Fig. 15-4).* This is one of the simpler methods, but it is not recommended on the long-stroke cylinder, because of tie-rod stretch. The tie rods are made from a good

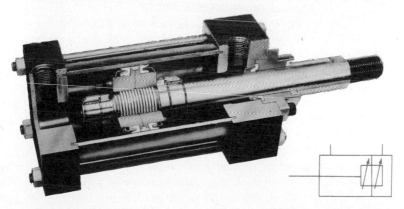

Fig. 15-3. Cutaway view of a hydraulic cylinder having a piston with a wear ring and synthetic piston seals. Seal on piston rod is a "filled" *Teflon* seal. (Courtesy Miller Fluid Power Corp.)

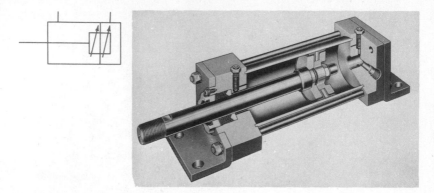

Fig. 15-4. Cutaway view of a pneumatic cylinder with vee-type rod packing and cup-type piston packing. (Courtesy The Sheffer Corp.)

grade of steel with fairly high tensile strength. When tightening the nuts of the tie rods, a torque wrench should be used to maintain equal torque on the tie rods.

2. *Ring construction on the tube (see Fig. 15-1).* Here the cylinder tube is grooved on the outside diameter (OD) to receive a retainer ring. Then the tube ring is retained by the retainer ring. The cover screws extend through the cover and the tube ring. They are retained with lock washers and nuts. In this type of construction, a tube pilot on the cover is advisable. Then if the screws are drawn too tightly, there is no tendency for the end of the tube to "pull in," causing the piston to bind when it reaches the end of the stroke. Another type of ring construction in which the ring groove is cut on the inside diameter (ID) of the cylinder tube and on the outside diameter (OD) of the cylinder cover pilot is shown in Fig. 15-5. The ring retains the cover to the tube and allows for a compact design. With this design, cylinders can be spaced close to each other.

3. *Tapped-tube construction (Fig. 15-6).* The tube is provided with a number of tapped holes on a bolt circle. The cylinder cover is provided with the same number of holes on the same bolt circle. The holes in the cover are usually 1/64 to 1/32 inch (0.3969 to 0.7938 mm) larger in diameter than the cover screws. For example, the hole for a 1/2 inch (12.7 mm) screw may be either 33/64 inch (13.0969 mm) or 17/32 inch (13.4938 mm) in

Fig. 15-5. Cylinder with ring construction which attaches cylinder tube to cylinder covers. (Courtesy Carter Controls, Inc.)

diameter. High tensile strength screws are usually used as the cover screws. Cap screws are often used.

4. *Threaded-tube construction (Fig. 15-7).* Both the cylinder tube and the cylinder covers are threaded. Depending on the design, some designs thread the outside diameter of the cylinder tube and the inside diameter of the recess in the cylinder cover. In other designs the inside diameter of the tube is threaded and the outside diameter of the cover pilots are threaded. There are many small-diameter cylinders that use the threaded construction. Various types of seals are used between the cylinder covers and the cylinder tube.

Cushion designs vary considerably. The chief purpose of the cushion is to eliminate or reduce shock at the end of the piston travel. The cushions can be either adjustable or nonadjustable. The adjustable cushions are more prevalent. The cushions are designed with a close fit between the cushion nose and the metal cushion orifice (Fig. 15-8). In Fig. 15-9, a heavy-duty hydraulic cylinder makes use of a tapered cushion collar and cushion nose. The cushion collar is the floating type which has a seal underneath it. All cushion collars and noses have some relief on the end that enters the cushion recess, some have a short taper, others a long taper, and some have a large radius. In some cushion designs, the cushion collar or nose enters a synthetic seal to provide positive shutoff in the cushion orifice (Fig. 15-10).

Special long-stroke cylinders operating at a high velocity often require a much longer cushion to bring the piston to a halt properly,

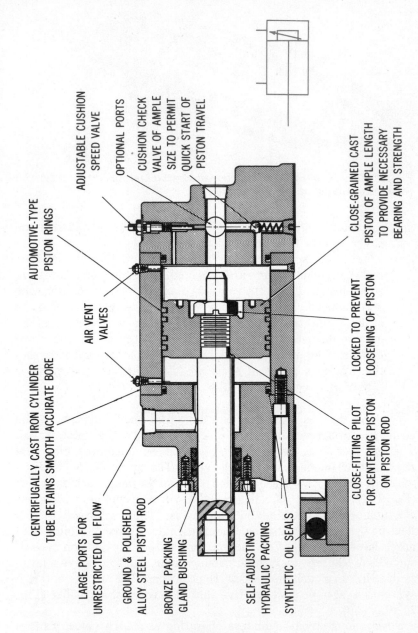

Fig. 15-6. Tapped-tube cylinder construction.

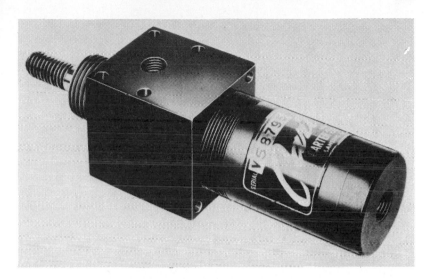

Fig. 15-7. Cylinder with threaded-tube construction. (Courtesy Carter Controls, Inc.)

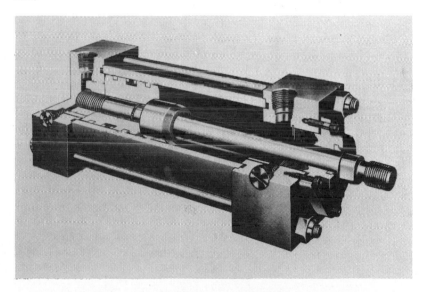

Fig. 15-8. Heavy-duty hydraulic cylinder with adjustable cushions. The fit in the cushioning must be precise to effect proper cushioning. Note that the cylinder has a rod wiper to remove any contaminants that might collect on the piston rod when it is in its extended position. (Courtesy Galland Henning Nopak Inc.)

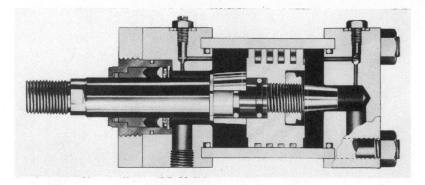

Fig. 15-9. Cutaway view of hydraulic cylinder with tapered cushion collar and nose. (Courtesy Carter Controls, Inc.)

without excessive shock. These cushions can be 6 or 8 inches (15.2 to 20.3 cm) in length. Three cushion designs that are sometimes used are shown in Fig. 15-11. The cushions should not be used as flow controls. They are strictly shock alleviators.

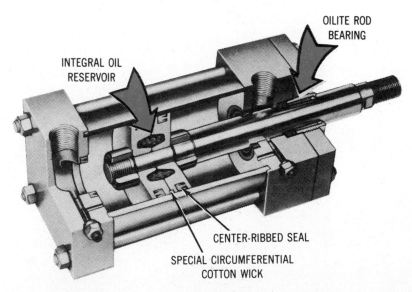

Fig. 15-10. An air cylinder that uses synthetic seals to effect cushioning at either end of the cylinder. Note the oil reservoir in the piston of the cylinder. This keeps the interior of the cylinder lubricated, yet there is no oil mist expelled with the exhaust. (Courtesy Lehigh Fluid Power, Inc.)

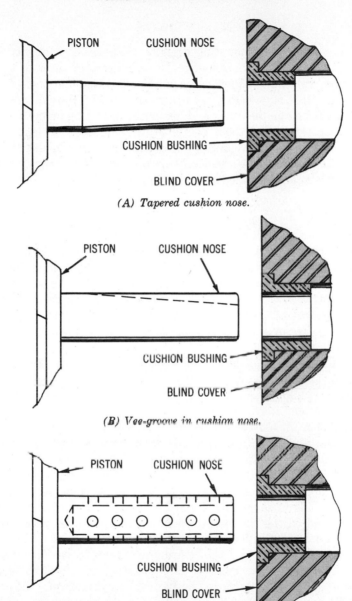

(A) Tapered cushion nose.

(B) Vee-groove in cushion nose.

(C) Holes in the cushion nose.

Fig. 15-11. Three different designs for the cushion in a cylinder.

MOUNTINGS

Numerous cylinder-mounting designs are necessary for meeting the wide variety of requirements for cylinders. Several of these mounting styles are shown in Fig. 15-12. Combinations of the mounting styles, such as foot mounting on one end of the cylinder and flange mounting on the other end, or flange mounting on both ends, are available. Some of these combination mountings are shown in Fig. 15-13. The various mounting styles that are now available should be studied. Note that the mountings are usually a part of the cylinder cover, except for intermediate trunnion mounting.

SIZE OF CYLINDERS

Cylinders now available range in size from the miniature cylinders which are 1/2 inch and 3/4 inch (1.3 and 1.9 cm) in diameter to cylinders which are 20 inches (50.8 cm) in diameter and even larger. In addition to the miniature cylinder, cylinders ranging from 1 ½ to 8 inches (3.8 to 20.3 cm) are in great demand. However, cylinders ranging upward in size to a 14-inch (35.6 cm) bore, inclusive, are in fair demand.

The finish in the cylinder bore should be smooth. Hydraulic cylinders are usually finished to a 15- to 20-microinch finish. Pneumatic cylinders usually require a smooth bore, since synthetic seals are usually used on the piston.

The length of stroke in a cylinder depends, of course, on its application. Cylinders are available with strokes ranging from a fraction of an inch to several feet in length. A long-stroke cylinder is shown in Fig. 15-14. It is usually necessary to provide longer rod bearings, longer pistons, and larger piston rods on long-stroke cylinders. This is somewhat dependent on how the cylinders are mounted and on the external support that is given the piston rod.

One type of application uses both air and hydraulics in a single cylinder, as shown in Fig. 15-15. The air cylinder is located in the rear portion. The hydraulic cylinder is located at the front. The air cylinder provides the power. The hydraulic cylinder provides the feed. The hydraulic cylinder is in a closed hydraulic circuit in which the oil flows from the face of the hydraulic piston to its other side through a bypass line, as the pneumatic piston advances the hydraulic piston. The purpose of the hydraulic cylinder is to provide

MOUNTING STYLE

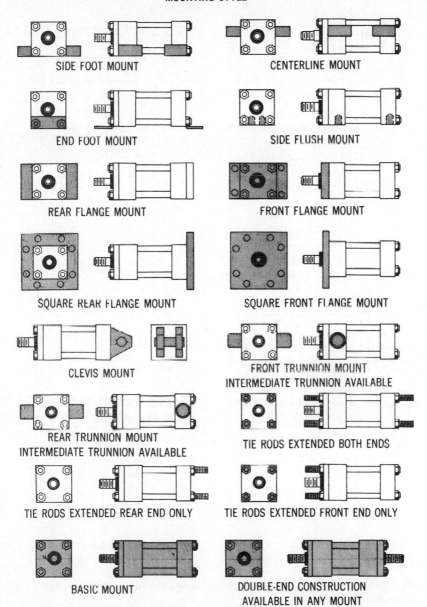

Fig. 15-12. Optional mounting details for cylinders. (Courtesy Carter Controls, Inc.)

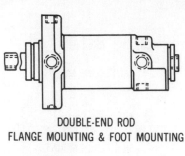

DOUBLE-END ROD
FLANGE MOUNTING & FOOT MOUNTING

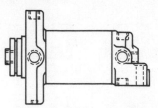

FLANGE MOUNTING AT ROD END,
FOOT MOUNTING AT BLIND END

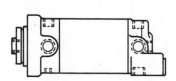

DOUBLE-END ROD
FLANGE MOUNTING & CENTERLINE MOUNTING

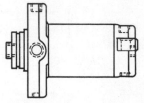

FLANGE MOUNTING AT ROD END,
CENTERLINE MOUNTING AT BLIND END

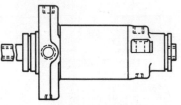

RABBETED MOUNTING AT ROD END,
FOOT MOUNTING AT BLIND END

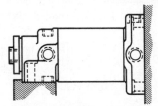

FOOT MOUNTING AT ROD END,
FLANGE MOUNTING AT BLIND END

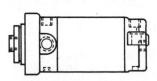

RABBETED MOUNTING AT ROD END,
CENTERLINE MOUNTING AT BLIND END

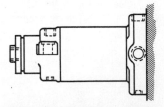

CENTERLINE MOUNTING AT ROD END,
FLANGE MOUNTING AT BLIND END

Fig. 15-13. Some of the many combinations of mounting styles available for hydraulic cylinders.

Fig. 15-14. A heavy-duty hydraulic cylinder, 20-inch (508 mm) bore by 200-inch (5,080 mm) stroke. (Courtesy Miller Fluid Power Corp.)

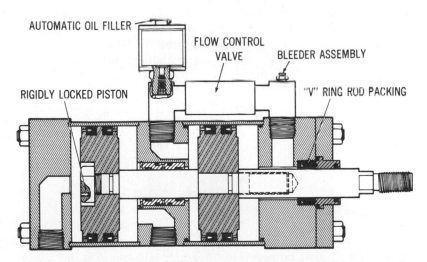

Fig. 15-15. A compressed-air system can be used for power in conjunction with a hydraulic system for the feed control in a machine application.

smooth feed movements without the expense of a complete hydraulic system. The force required or exerted by the hydraulic cylinder is no greater than the area of the pneumatic cylinder multiplied by the air line pressure. Some friction loss results from the seals required in both the air and oil portions.

Depending on the construction of the air-hydraulic cylinder, feed movements of 1/4 inch (6.35 mm) per minute and less can be controlled accurately over extended periods of time. Heat buildup in the hydraulic section rarely occurs. The change in oil viscosity is extremely small.

Components also are available in which the oil section is a completely separate component, as shown in Fig. 15-16. This component can be either parallel-mounted or tandem-mounted on an air cylinder, and provides accurate feeds. Skip feeds also are available.

PISTON-ROD PROTECTION

The piston rod probably is the most vulnerable part of an air or hydraulic cylinder, when exposed to the elements in which the cylinder is performing. Since dirt, cement, weld spatter, and other foreign matter can collect on the piston rod, when it is in the extended position, to cause score marks or to cause the rod packing to become damaged to the point where it will leak, some protection must be provided. One method of piston-rod protection is an accordion-type boot (Fig. 15-17). These boots are manufactured from various

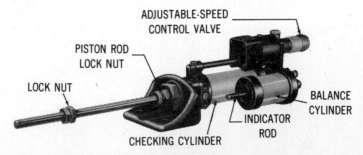

Fig. 15-16. A checking cylinder containing hydraulic fluid. The unit can be connected to an air cylinder to provide accurate feed. (Courtesy Bellows International)

Fig. 15-17. Cylinder equipped with a rod boot to provide protection for the highly polished finish of the piston rod.

materials. In hot spots, the boots often are of asbestos material. Other means of protecting the piston rods are wipers, scrapers, and metal protectors.

CYLINDERS AND ACCESSORIES

A number of attachments and accessory devices are designed for both pneumatic and hydraulic cylinders to meet the application requirements. By taking a standard cylinder and applying certain attachments, new concepts can be created. For example, adding a valve package to a standard square-end cylinder creates a power package that affords only a single air line installation (Fig. 15-18). The speed of the piston is controlled from the valve mounting block. The four-way valve may be single-solenoid, double-solenoid, single pilot-charge type, double pilot-charge type, or bleeder type.

In Fig. 15-19, mechanically operated limit switches are located in the cylinder covers. Here again, these mechanical devices can be attached to standard air or hydraulic cylinders. This eliminates the need for cams and trip dogs on a machine.

In Fig. 15-20, a heavy-duty cylinder is equipped with a complete hydraulic power package and automatic valving. Therefore, piping is not required in the field. It is necessary only to make the electrical connections to the unit. This package is a completely closed, prefilled system.

In Fig. 15-21, the magnetic limit switches provide an inexpensive stroke adjustment feature and eliminates the need for external cams, cogs, switch dogs, linkages, etc. The cylinders must be designed for

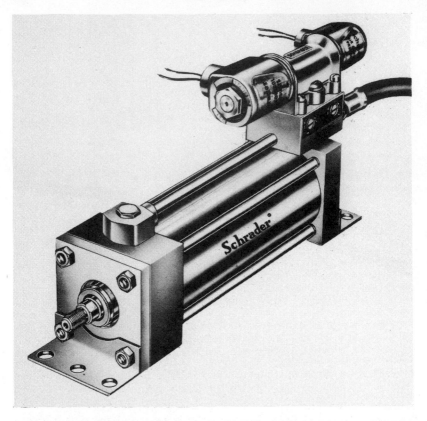

Fig. 15-18. Air cylinder with four-way control valve mounted on it requires only a single low-cost air line installation. Built-in speed controls for controlling the speed of the piston in either direction are located in the valve mounting block. (Courtesy Schrader Fluid Power Division)

magnetic operation. Standard cylinders without magnetic construction will not work.

In order to allow for some misalignment between the piston rod of a cylinder and the mating connecting part, self-aligning cylinder rod couplers are often employed. This type of coupler is shown in Fig. 15-22.

Other quite common accessories are rod clevises, female rod eyes, eye brackets, clevis brackets, trunnion mounting brackets, and other items to assist the user of cylinders.

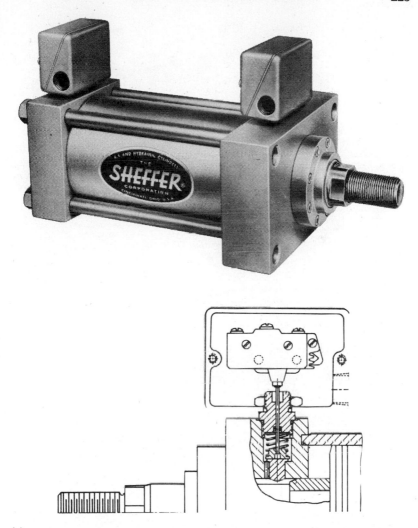

Fig. 15-19. Hydraulic cylinder with mechanically operated limit switches. (Courtesy The Sheffer Corp.)

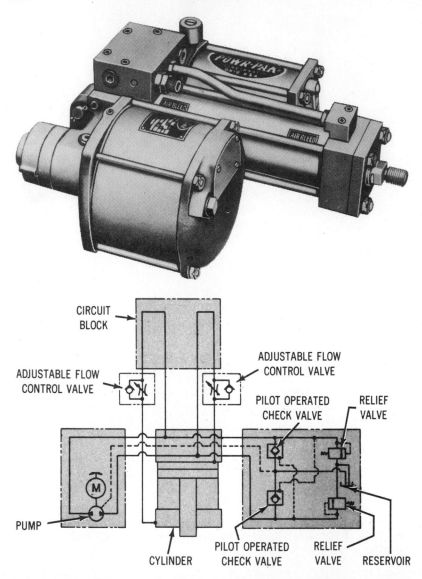

Fig. 15-20. A compact self-contained hydraulic package that supplies a force where a minimum size and maximum power are required. (Courtesy The Sheffer Corp.)

Fig. 15-21. Air cylinder with magnetic limit switches which provide simplified stroke adjustment. (Courtesy Schrader Fluid Power Division)

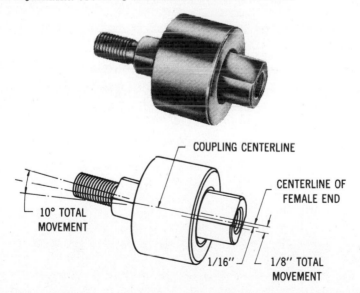

Fig. 15-22. Self-aligning cylinder rod coupler. (Courtesy Schrader Fluid Power Division)

Chapter 16

Nonrotating Cylinders—Application

The applications for nonrotating cylinders, both pneumatic and hydraulic, are so numerous that it is impossible to list all the possibilities here. If enough cylinders were strategically placed, the largest buildings in the world could be lifted. If an application requires lifting, pushing, pulling, tilting, clamping, etc., it is likely that a cylinder can do the job satisfactorily. Some of the basic uses for the nonrotating type of cylinder are shown in Fig.16-1.

FORCE CALCULATIONS

In the application of cylinders, it should be remembered that some cylinders can deliver extremely high forces. For example, an 8-inch (20.3 cm) hydraulic cylinder operated at a pressure of 3000 psi (20,684.3kPa) is capable of developing approximately 150,795 pounds of force (670,769.3N).

The force developed by nonrotating cylinders of the common bore size and operating pressures can be found in Table 16-1.

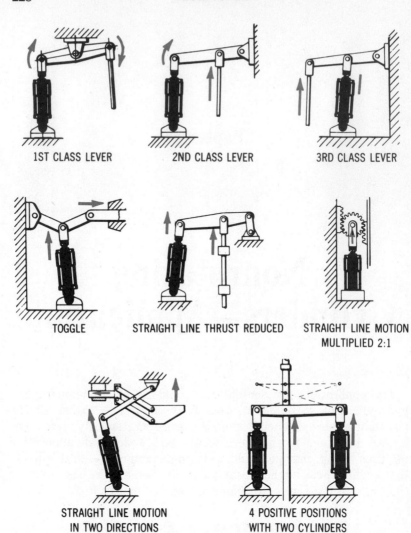

1ST CLASS LEVER   2ND CLASS LEVER   3RD CLASS LEVER

TOGGLE   STRAIGHT LINE THRUST REDUCED   STRAIGHT LINE MOTION MULTIPLIED 2:1

STRAIGHT LINE MOTION IN TWO DIRECTIONS   4 POSITIVE POSITIONS WITH TWO CYLINDERS

Fig. 16-1. Applications for the nonrotating type of

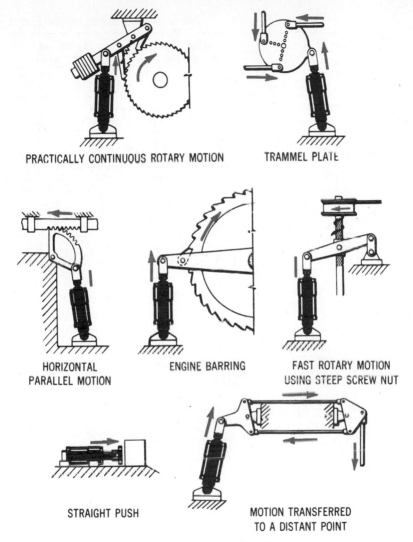

PRACTICALLY CONTINUOUS ROTARY MOTION

TRAMMEL PLATE

HORIZONTAL
PARALLEL MOTION

ENGINE BARRING

FAST ROTARY MOTION
USING STEEP SCREW NUT

STRAIGHT PUSH

MOTION TRANSFERRED
TO A DISTANT POINT

cylinder. (Courtesy Rexnord, Inc., Cyl. Div.—HANNA)

Table 16-1. Approximate Force

[2]Cyl. Bore— in. (mm)	[3]Area of Cyl. Bore— in.[2] (cm.[2])	Dia. Rod— in. (mm)	At 80 psi (551.6 kPa)		At 100 psi (689.5 kPa)	
			[5]Push—lb (N)	Pull—lb (N)	Push—lb (N)	Pull—lb (N)
2 (50.8)	3.142 (20.27)	1 (25.4)	251 (1117)	188 (837)	314 (1397)	235 (1046)
2½ (63.5)	4.909 (31.66)	1⅜ (34.93)	392 (1744)	273 (1215)	490 (2181)	342 (1522)
3 (76.2)	7.069 (45.6)	1⅜ (34.93)	565 (2514)	443 (1971)	706 (3142)	558 (2483)
3½ (88.9)	9.621 (62.06)	1⅜ (34.93)	769 (3422)	650 (3892)	962 (4281)	813 (3618)
4 (101.6)	12.566 (81.05)	1¾ (44.45)	1005 (4474)	812 (3613)	1256 (5589)	1016 (4521)
5 (127)	19.635 (126.65)	2 (50.8)	1570 (6986)	1319 (5870)	1963 (8735)	1649 (7338)
6 (152.4)	28.274 (182.41)	2 (50.8	2261 (10,061)	2010 (8945)	2827 (12,580)	2513 (11,183)
7 (177.8)	38.485 (248.23)	2½ (63.5)	3078 (13,697)	2682 (11,953)	3848 (17,124)	3357 (14,939)
8 (203.2)	50.265 (324.21)	2½ (63.5)	4021 (17,893)	3628 (16,145)	5026 (22,366)	4535 (20,181)

[1]Does not include reduction for friction.   [2]1 in. = 25.4 mm   [3]1 in.[2] = 6.4516 cm[2]

[4]1 psi = 6.894757 kPa   [5]1 lb (force) = 4.448221 newtons (N)

The calculations in the table are based on the formula:

$$F = PA; \text{ and } F = P (A - A_1)$$

in which;
   F is force developed (not including friction),
   P is line pressure (psi),
   A is area of cylinder bore (sq in.),
   A_1 is area of cross section of piston rod (sq in.).

Developed by Nonrotating Cylinders

At 500 psi (3447.4 kPa)		At 1000 psi (6894.8 kPa)		At 1500 psi (10,342.1 kPa)	
Push—lb (N)	Pull—lb (N)	Push—lb (N)	Pull—lb (N)	Push—lb (N)	Pull—lb (N)
(1571) (6991)	(1178) (5242)	3142 (13,982)	2357 (10,489)	4713 (20,973)	3535 (15,731)
2454 (10,920)	1712 (7618)	4909 (21,845)	3424 (15,237)	7363 (32,765)	5136 (22,855)
3534 (15,726)	2792 (12,424)	7069 (31,457)	5584 (24,849)	10,603 (47,183)	8376 (37,273)
4810 (21,405)	4068 (18,103)	9621 (42,813)	8136 (36,205)	14,431 (64,218)	12,204 (54,308)
6283 (27,959)	5080 (22,606)	12,566 (55,919)	10,164 (45,230)	18,849 (83,878)	15,241 (67,822)
9817 (43,686)	8246 (36,695)	19,635 (87,376)	16,493 (73,394)	29,457 (131,084)	24,739 (110,089)
14,137 (62,910)	12,566 (55,919)	28,274 (125,819)	25,132 (111,837)	42,411 (188,729)	37,698 (167,756)
19,242 (85,627)	16,788 (74,707)	38,485 171,258)	33,576 (149,413)	57,727 (256,885)	50,364 (224,120)
25,132 (111,837)	22,678 (100,917)	50,265 (220,079)	45,356 (201,834)	75,397 (335,368)	68,034 (302,751)

The formula $F = PA$ can be used to determine the force developed when the fluid pressure is applied at the blind end of the cylinder. The formula $F = P(A - A_i)$ is used to determine the force created when fluid pressure is applied to the rod end of the cylinder. When equal operating pressures are applied to each end of the cylinder, the higher force is exerted by the blind end, because its area is greater. In a double-end cylinder with piston rods having equal diameters, equal forces are exerted at each end. This principle should be remembered in solving fluid power problems, in regard to both the force and volume factors.

MOUNTING STYLES

In selecting the mounting style for an application, it should be remembered that the cylinder movement should not cause excessive side thrust on either the piston rod or the cylinder tube. In press applications, it is good practice to use a cylinder with reverse-flange mounting, as shown in Fig. 16-2. This mounting style eliminates thrust on the mounting screws of the cylinder. Also, the mounting plate on the press should be strong enough to resist the forces created by the cylinder. If the mounting plate flexes, undue stress on the mounting bolts results.

The blind-flange mounting style is used to mount a cylinder on the base of a machine. All the force is resisted by a solid mounting, and there is no stress on the mounting screws.

The foot-mounted cylinders use one of the most common mounting styles. There are various foot-mounting designs. Some of these designs are side-foot, front-foot, and the front-foot style that is connected to the tie rods. To anchor these cylinders properly and to remove undue stresses on the cylinder mounting bolts, especially on the larger bore sizes, it is advisable to provide back-up blocks similar to those shown in Fig. 16-3. This practice also provides extra

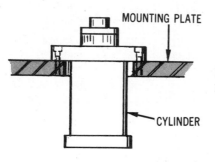

Fig. 16-2. Reverse-flange mounting for press applications.

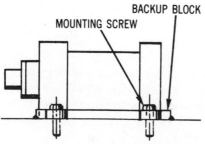

Fig. 16-3. A commonly used foot-mounting style.

protection if the mounting bolts should become loosened. Since the center line of the piston rod and the center line of the cylinder can be inches apart, depending on the bore size of the cylinder, some torque is created.

The center-line mounting in which the center lines of the thrust and the mounting surface are in the same plane is one of the most ideal mountings for a cylinder (Fig. 16-4). Here again, back-up blocks similar to those shown in Fig. 16-3 should be used.

The clevis- or trunnion-mounted cylinder is available for moving an object through an arc. The use of an intermediate trunnion-mounted cylinder to control a dumping device is diagrammed in Fig. 16-5. It is imperative that the trunnion bearing be rugged enough to handle the load. It is advantageous to place the two bearings that support the trunnion fairly wide apart. Trunnion lugs that are built into the front

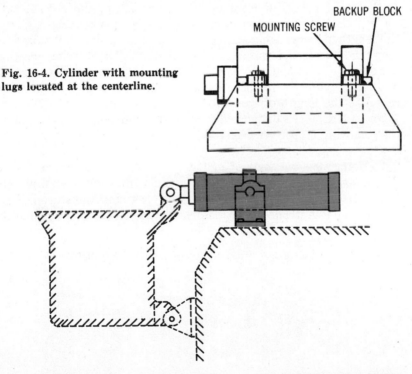

Fig. 16-4. Cylinder with mounting lugs located at the centerline.

Fig. 16-5. Diagram of an intermediate trunnion-mounted cylinder used to control a dumping device.

or blind covers are also available. If clevis-mounted cylinders are used, the clevis bracket should be rugged. It should be fastened to a solid foundation for satisfactory results.

On long-stroke cylinders, adequate support should be provided for both the cylinder tube and the piston rod. A center support for the cylinder tube is shown in Fig. 16-6. When this type of support is used, extreme care should be taken in fastening the support to the mounting surface to place no stress in the center portion of the tube.

The end of the piston rod should be supported adequately to keep the end of the rod from sagging and causing undue wear on the rod bearing and the packings. Although the end of an extra-long piston rod may be supported adequately, the middle of the rod may sag from its own weight. Therefore, the diameter of the piston rod should be given careful consideration in all applications.

Stop tubes are valuable in providing less bearing wear. The length of the stop tube is determined by the length of the stroke, the application, etc. A cylinder with a stop tube is usually designated by the length of its working stroke or net stroke. It does not take the stop tube into consideration. To determine the total stroke, add the net stroke to the length of the stop tubes. Many manufacturers of hydraulic cylinders publish a chart and instructions on how to calculate the length of stop tubes, mounting, etc. (Fig. 16-7).

The selection of a piston rod for thrust (push) conditions requires the following steps (see Fig. 16-7):

1. Determine the type of cylinder mounting style and rod-end connection to be used. Then consult the chart and find the "stroke factor" that corresponds to the conditions used.
2. Using this stroke factor, determine the "basic length" from the equation:

$$basic\ length = actual\ stroke \times stroke\ factor$$

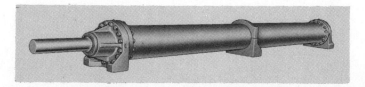

Fig. 16-6. Long-stroke cylinders require a center support for the cylinder tube. (Courtesy Logansport Machine Co., Inc.)

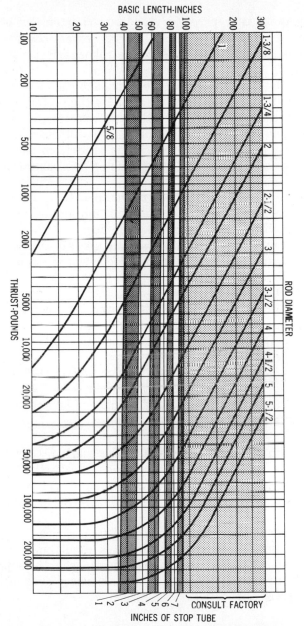

Fig. 16-7. Piston-rod stroke selection chart for a hydraulic cylinder application. (This illustration is the courtesy of Parker Hannifin Corporation)

The graph is prepared for standard rod extensions beyond the face of the gland retainers. For rod extensions greater than standard, add the increase to the stroke in arriving at the "basic length."

3. Find the load imposed for the thrust application by multiplying the full-bore area of the cylinder by the system pressure.

4. Enter the graph along the values of "basic length" and "thrust" as found above, and note the point of intersection:

   a. The correct piston-rod size is read from the diagonally curved line labeled "Rod Diameter" next above the point of intersection.

   b. The required length of the stop tube is read from the right-hand side of the graph by following the shaded band in which the point of intersection lies.

   c. If the required length of the stop tube is in the region labeled "consult factory," submit the following information for an individual analysis:

      (1) Cylinder mounting style.

      (2) Rod-end connection and method of guiding load.

      (3) Bore, required stroke, length of rod extension (Dim."LA") if greater than standard, and series of cylinder used.

      (4) Mounting position of cylinder. (Note: If at an angle or vertical, specify direction of piston rod.)

      (5) Operating pressure of cylinder if limited to less than standard pressure for cylinder selected.

Double-end cylinders also provide less bearing wear within the cylinder, but the extra rod also decreases the effective area on the piston. In the double-end cylinder, the second piston rod can be used to perform other work, such as tripping limit switches, cam rollers on valves, etc.

When the cylinders are subjected to eccentric loading, provision should be made to compensate for this condition. A suggested method is shown in Fig. 16-8. Longer bearings on the supports are advantageous in this type of application. Eccentric loading is often a factor in press applications.

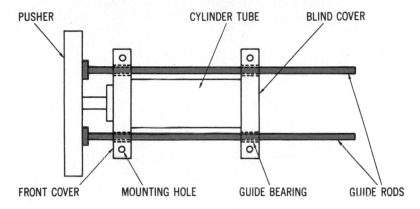

Fig. 16-8. Diagram of a cylinder with guide rods to compensate for eccentric loading.

STROKE ADJUSTMENTS

Cylinders with stroke adjustments have many industrial applications. This type of mechanism is shown in Fig. 16-9 in which the adjustment is limited in the mechanical linkage. In the cylinder diagrammed in Fig. 16-10, the forward stroke of the cylinder is limited. It is possible to adjust the stroke to such a fine point that an extremely fragile object, such as an egg, can be contacted and held without breaking it. Of course, it is essential that these objects arc all exactly the same size for a given setting of the cylinder. In a few applications, the cylinders are provided with an adjustment on the return stroke to provide: (1) limited movement on press applications in which a short stroke can be used for one type of work and a longer stroke can be used for another type of work to reduce time and volume requirements; (2) accurate positioning of slides and other mechanisms which are controlled accurately by a cylinder on the return stroke; and (3) positioning of the workpiece on the return stroke of the cylinder.

Applications in which the forward stroke of the cylinder may require adjustment are: (1) press applications in which a change in stroke may be necessary, because of the type of material being processed, such as for deep-drawing applications; (2) measuring applications in which the end of the piston rod is connected to a device in which each 1 inch of stroke causes a given volume of fluid to be emitted from the measuring device (examples of this type of

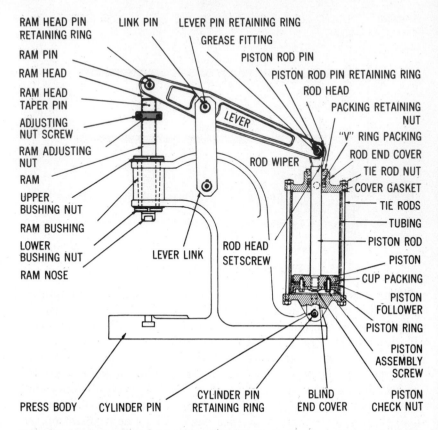

RAM HEAD PIN RETAINING RING
LINK PIN
LEVER PIN RETAINING RING
GREASE FITTING
RAM PIN
PISTON ROD PIN
RAM HEAD
PISTON ROD PIN RETAINING RING
ROD HEAD
RAM HEAD TAPER PIN
PACKING RETAINING NUT
ADJUSTING NUT SCREW
LEVER
"V" RING PACKING
RAM ADJUSTING NUT
ROD WIPER
ROD END COVER
RAM
TIE ROD NUT
UPPER BUSHING NUT
COVER GASKET
TIE RODS
RAM BUSHING
TUBING
LOWER BUSHING NUT
LEVER LINK
ROD HEAD SETSCREW
PISTON ROD
RAM NOSE
PISTON
CUP PACKING
PISTON FOLLOWER
PISTON RING
PISTON ASSEMBLY SCREW
PRESS BODY
CYLINDER PIN
CYLINDER PIN RETAINING RING
BLIND END COVER
PISTON CHECK NUT

Fig. 16-9. Stroke adjustment in this cylinder is limited in the mechanical type of linkage. (Courtesy Logansport Machine Co., Inc.)

application are the automatic measuring of antifreeze, oil, or grease for new cars); and (3) for accurate positioning of workpieces on high-cycling automatic equipment.

Some applications in industry use the *tandem-type* cylinders (Fig. 16-11). This type of cylinder is often used in applications that require more force than can be exerted by a cylinder whose diameter permits installation in a given space. If length is not a factor, tandem cylinders that contain two or three pistons can be used to obtain the additional force. A typical application of this type is in a die cushion in which the space is limited by diameter, and length is not a limiting factor.

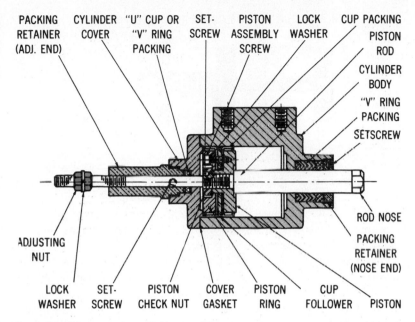

PACKING RETAINER (ADJ. END) | CYLINDER COVER | "U" CUP OR "V" RING PACKING | SET-SCREW | PISTON ASSEMBLY SCREW | LOCK WASHER | CUP PACKING | PISTON ROD | CYLINDER BODY | "V" RING PACKING | SETSCREW

ADJUSTING NUT | LOCK WASHER | SET-SCREW | PISTON CHECK NUT | COVER GASKET | PISTON RING | CUP FOLLOWER | PISTON | ROD NOSE | PACKING RETAINER (NOSE END)

Fig. 16-10. Diagram of a cylinder in which the forward stroke can be adjusted or limited.

Duplex cylinders (Fig. 16-12) are often used where the space available is important. Here, one piston rod can be used to perform the clamping action, and the other piston rod can be used to perform the work operation. Since duplex cylinders possess two pistons, the rear piston exerts more pressure on the forward stroke, because it has the greater area. Duplex cylinders are rather expensive to build and repair. But their compactness may be advantageous.

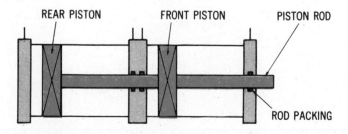

REAR PISTON | FRONT PISTON | PISTON ROD | ROD PACKING

Fig. 16-11. A tandem-type cylinder.

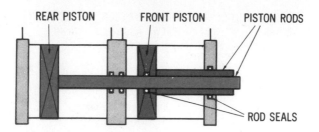

Fig. 16-12. A duplex-type cylinder.

The *telescoping cylinders* (Fig. 16-13) are used on some applications in industry. They are used more often on mobile applications in which space limitations are involved. These cylinders are usually operated hydraulically. Their piston strokes may be 10 to 15 feet, (3.0 to 4.6 meters), or more.

There are many applications for the *air-hydraulic cylinder*. These cylinders are usually mounted in the same way that air cylinders are mounted, but their applications are unique. This type of cylinder permits extremely fine feeds on machine tools, drilling machines, processing equipment, etc. Also, these cylinders lend themselves readily to synchronization, as shown in Fig. 16-14. The piston rods of both cylinders can be synchronized, regardless of loading conditions. It is possible to tie several air-hydraulic cylinders together to synchronize piston movement. This eliminates costly metering

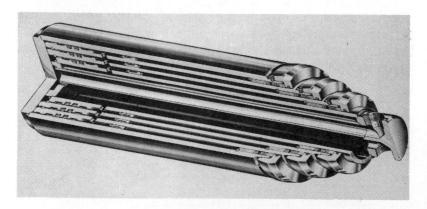

Fig. 16-13. Cutaway view of a telescopic cylinder. (Courtesy Ward Hydraulics Division A-T-O, Inc.)

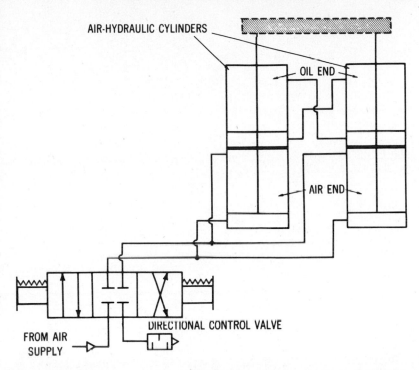

Fig. 16-14. Diagram in which synchronization of movements is achieved in air-hydraulic cylinders.

devices which cannot provide the accuracy that can be obtained with the air-hydraulic cylinder. When the air-hydraulic cylinder is used strictly as a metering cylinder, various types of feeding arrangements are possible with the proper auxiliary valving.

Chapter 17

Rotating Cylinders—Design

Persons who are not engaged actively in the metalworking field and who are not often involved with the uses of rotating cylinders usually do not understand their important function. A knowledge of their possibilities opens new avenues for the application of these cylinders.

Rotating cylinders are designed primarily for application to a rotating hollow shaft or spindle which rotates with the cylinder inside the distributor.

The rotating cylinders are designed to be operated with air, petroleum base fluids, or synthetic base fluids. Each of these mediums usually requires a different cylinder design or at least a modification. Rotating cylinders that operate on air are usually available for pressures from 0 to 150 psi (0 to 1034.2 kPa). The hydraulic cylinders are used for pressures ranging from 0 to 500 psi (0 to 3447.4kPa) and from 0 to 1000 psi (0 to 6894.8 kPa).

PNEUMATIC CYLINDERS

The approximate air consumption and the force delivered at the various operating pressures can be found in Table 17-1. When the

Table 17-1. Approximate Air Consumption and Force Delivered

[1]Bore Diam.—in. (mm)	Piston Stroke—in. (mm)	[2]Compressed Air Consumed in Full Piston Stroke—ft³ (m³)	[3]Thrust at 60 psi (413.7 kPa) [4]Pounds Force (lbf) [Newtons (N)]	Thrust at 80 psi (551.6 kPa) Pounds Force (lbf) [Newtons (N)]	Thrust at 100 psi (689.5 kPa) Pounds Force (lbf) [Newtons (N)]
3 (76.20)	1 (25.4)	0.0040 (0.0001132)	424 (1,882.6)	565 (2508.6)	706 (3134.6)
4½ (114.30)	1 (25.4)	0.0092 (0.00026)	954 (4235.8)	1272 (5647.7)	1509 (6670.0)
6 (152.40)	1½ (38.1)	0.0244 (0.00069)	1696 (7530.2)	2261 (10,038.8)	2827 (12,551.9)
8 (203.2)	2 (50.8)	0.0580 (0.0016)	3015 (13,386.6)	4021 (17,853.2)	5026 (22,315.4)
10 (254.00)	2 (50.8)	0.0908 (0.00256)	4712 (20,921.3)	6283 (27,896.5)	7854 (34,871.8)
12 (304.80)	2 (50.8)	0.1308 (0.0037)	6786 (30,129.8)	9048 (40,173.4)	11,310 (50,216.4)
14 (355.60)	2 (50.8)	0.1708 (0.0048)	9236 (41,007.8)	12,315 (54,678.6)	15,394 (68,349.4)
16 (406.40)	2 (50.8)	0.2326 (0.00066)	12,063 (53,559.7)	16,084 (71,412.96)	20,106 (89,270.6)
18 (457.20)	2 (50.8)	0.2944 (0.0083)	15,268 (67,789.9)	20,357 (90,385.1)	25,447 (112,984.7)
20 (508.00)	2 (50.8)	0.3636 (0.0103)	18,849 (83,689.6)	25,132 (111,586.1)	31,416 (139,487.0)

[1]1 in. = 25.4 mm　　[2]1 ft³ = 0.0283 m³　　[3]1 psi (lb/in²) = 6.894757 kilopascals (kPa)　　[4]1 pound force (lbf) = 4.448221 newtons (N)

cylinder is used for the pull stroke, the cross-sectional area of the piston rod should be subtracted from the effective area of the cylinder and the result multiplied by the operating pressure.

Rotating cylinders operated on air pressure are manufactured in standard bore sizes ranging upward to, and including, 20 inches (50.8 cm) in diameter. However, the rotating cylinders used for hydraulic (oil or synthetic-base fluids) service rarely exceed 14 inches (35.56 cm) in diameter (Table 17-2). The standard stroke lengths of the cylinders range upward to 2 inches (5.08 cm).

Many rotating cylinders are being designed for speeds upward to 3600 rpm. Machine tool builders are demanding high speeds in order to meet their customer's requirements. These high-speed cylinders are usually static and dynamically balanced. The rotating hydraulic cylinders in the higher pressure range usually operate at much lower speeds, since the bulk of their applications are on heavy-duty mill equipment. Port sizes for the rotating cylinders rarely exceed 1/2 inch NPTF.

A rotating air cylinder with a safety air lock device is shown in Fig. 17-1. This will cause the air to be retained in the cylinder if an air failure such as a broken air supply, a compressor failure, etc. should occur outside the cylinder. The air trapped inside the cylinder will keep force on the holding device being operated by the cylinder until the machine can be shut down.

A cross-sectional diagram of a rotating cylinder used for air is shown in Fig. 17-2. The various parts are described as follows:

1. *Cylinder body.* The rotating cylinder body can be either one-piece or two-piece construction. The cylinder body shown in Fig. 17-2 is one-piece construction, and it is a casting. Two-piece construction is usually used for long-stroke rotating cylinders in which the tube is heavy-wall tubing and the front

Fig. 17-1. Rotating air cylinder with safety air lock. (Courtesy Cushman Industries, Inc.)

Table 17-2. Fluid Displacement and Pressure Delivered on Rotating Hydraulic Cylinders

[1]Size Dia. of Bore—in. (mm)	[2]Area of Bore—in.² (cm²)	Stroke of Piston—in. (mm)	Dia. of Piston Rod—in. (mm)	[3]Cyl. Disp.—in.³ (Pull) (cm³)	[4]Approximate Power in Pounds Pressure (Pull)— Pounds Force (lbf) [Newtons (N)]				
					[5]100 psi	150 psi	200 psi	250 psi	500 psi
3 (76.2)	7.068 (45.59)	1 (25.4)	1 (25.4)	6.28 (102.91)	628 (2788.3)	942 (4182.5)	1256 (5576.6)	1570 (6970.8)	3141 (13,946.0)
4½ (114.3)	15.904 (102.58)	1 (25.4)	1 (25.4)	15.11 (247.61)	1511 (6708.8)	2267 (10,065.5)	3023 (13,422.1)	3770 (16,778.8)	7559 (33,662.0)
6 (152.4)	28.274 (182.37)	1½ (38.1)	1½ (38.1)	39.76 (651.55)	2650 (11,766.0)	3976 (17,653.4)	5301 (23,536.4)	6626 (29,419.4)	13,253 (58,843.3)
8 (203.2)	50.265 (324.21)	1½ (38.1)	1 3/4 (44.45)	71.79 (1176.42)	4786 (21,249.8)	7179 (31,874.8)	9572 (42,499.7)	11,965 (53,124.6)	23,930 (106,249.2)
10 (254.0)	78.540 (506.58)	1½ (38.1)	2 (50.8)	113.09 (1853.21)	7539 (33,473.2)	11,308 (50,207.5)	15,079 (66,950.8)	18,849 (83,689.6)	37,699 (167,383.6)
12 (304.8)	113.100 (729.49)	2 (50.8)	2½ (63.5)	216.40 (3546.15)	10,820 (48,040.8)	16,230 (72,061.2)	21,640 (96,081.6)	27,050 (100,102.0)	54,100 (240,204.0)
14 (355.6)	153.940 (992.91)	2 (50.8)	2 3/4 (69.85)	296.00 (4850.55)	14,801 (65,716.4)	22,201 (98,572.4)	29,602 (131,432.9)	37,002 (164,288.9)	74,005 (328,582.2)

[1]1 in. = 25.4 mm [2]1 in.² = 6.4516258 cm² [3]1 in.³ = 16.387162 cm³ [4]1 pound force (lbf) = 4.448221 newtons (N) [5]1 psi (lb/in²) = 6.894757 kilopascals (kPa)

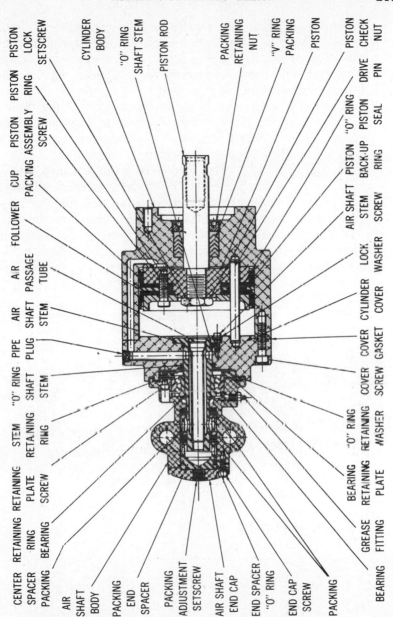

Fig. 17-2. Cross-sectional diagram of an air rotating cylinder. (Courtesy Logansport Machine Co., Inc.)

cover is made either from a casting or from plate material. If the cylinder body is made from a casting, the material may be aluminum or cast iron. Only a few air cylinder bodies are made from cast iron, because the trend is to reduce weight as the speed requirements are increased.

A cylinder body contains air passages for directing the air through the walls of the cylinder body to the front side of the piston. The body also houses the piston-rod packing, which is held in place by a retaining means.

The cylinder body is machined on both the outside diameter (OD) and the inside diameter (ID). The body is machined on the ID to provide a smooth finish for the piston packing. The OD is machined to improve its appearance and to assist in balancing.

The rod end of the cylinder body is designed for mounting onto an adapter. The recess for the adapter is made either to American Standards specifications (Fig.17-3), or to the manufacturer's specifications.

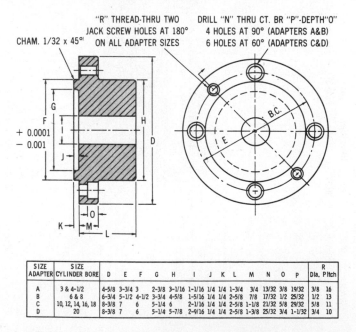

SIZE ADAPTER	SIZE CYLINDER BORE	D	E	F	G	H	I	J	K	L	M	N	O	P	R Dia.	Pitch
A	3 & 4-1/2	4-5/8	3-3/4	3		2-3/8	3-1/16	1-1/16	1/4	1/4	1-3/4	3/4	13/32	3/8	19/32	3/8	16
B	6 & 8	6-3/4	5-1/2	4-1/2	3-3/4	4-5/8	1-5/16	1/4	1/4	2-5/8	7/8	17/32	1/2	25/32	1/2	13
C	10, 12, 14, 16, 18	8-3/8	7	6		5-1/4	6	2-1/16	1/4	1/4	2-5/8	1-1/8	21/32	5/8	29/32	5/8	11
D	20	8-3/8	7		6	5-1/4	5-7/8	2-9/16	1/4	1/4	2-5/8	1-3/8	25/32	3/4	1-1/32	3/4	10

Fig. 17-3. American Standard Specifications for an adapter used for mounting a rotating cylinder. (Dimensions also show metric conversion.)

Provision is usually made inside the front section of the cylinder body for a pin which passes either through the piston or into the piston, and is anchored within the cylinder body. If the pin extends through the piston (see Fig. 17-2), one end of the pin is anchored in the back cover. If the pin extends into the piston, two short pins—one in the cylinder cover and one in the cylinder body—are used.

2. *Cylinder cover.* The cylinder cover provides a closure for the cylinder body and an anchor for the air-shaft stem. The cover is usually made from the same material as the cylinder body. It can be either an alloy casting or a plate. One or more holes are drilled in the cover from its center point to its rim to serve as air passages which connect the cylinder body to the air shaft. Those passages in the cylinder body serve as air passages to the front side of the piston.

3. *Piston rod.* The piston rod is made from a high-tensile steel which is ground, polished, and usually chrome plated. Unlike nonrotating cylinders, only one standard-size rod diameter is available for each standard cylinder bore. The threaded end of the piston rod is normally a female thread. Male threads are available as special equipment.

4. *Piston assembly.* The piston assembly uses several different styles of seals to effect a seal between the piston and the cylinder walls as the piston is moved or contacted by the air pressure. The cylinder shown (see Fig. 17-2) uses synthetic cup-type packings. However, O-rings, quad rings, and block-vee packings are other resilient seals often used. The pistons can be one-piece construction, or they can be constructed from several pieces (Fig. 17-4). Pistons are made from an aluminum alloy or from a high-tensile cast iron. The pistons are designed for maximum strength, but they should be lightweight. In the piston assembly in Fig. 17-2, the follower, which contacts the cylinder walls, is made from a good-grade bearing material that resists wear.

5. *The air-shaft assembly.* This assembly consists of several different parts, as shown in Fig. 17-5. It is an important part of the rotating cylinder. It should be capable of withstanding high speeds and sudden stops.

The air shaft is made in several different designs. One of these designs is shown in Fig. 17-5. A two-piece construction in

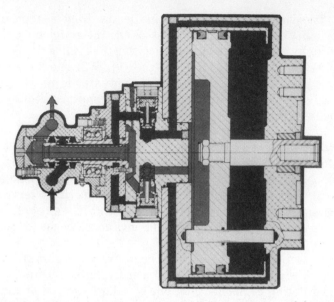

Fig. 17-4. Rotating air cylinder having a piston of one-piece construction and resilient piston seals and wear strip. Cylinder has a lock feature in case of air-pressure failure. (Courtesy Logansport Machine Co., Inc.)

which a hollow, formed tube is brazed inside the air-shaft stem assembly is shown in Fig. 17-5.

The air-shaft stem is fastened securely to the cylinder cover, which provides a substantial bearing surface for supporting the stem. The air-shaft seal (see Fig. 17-5) is required to withstand high temperatures and high speeds. However, it often receives only minimum lubrication. Various types of seals are used in other air-shaft assembly designs, such as the hat-type seal and other formed seals.

The air shaft assembly (Fig. 17-5) is equipped with a double-row bearing for high-speed applications.

The air-shaft body can be made from either cast aluminum or bronze with thin-wall sections for dissipating heat. The shaft body, however, should be sturdy enough to withstand the external forces that are caused by flexible hose and pipe connections. The two ports of entry are located in the shaft body. The shaft body serves as a housing for the shaft packings and as a retainer for the bearing. It also houses the method of

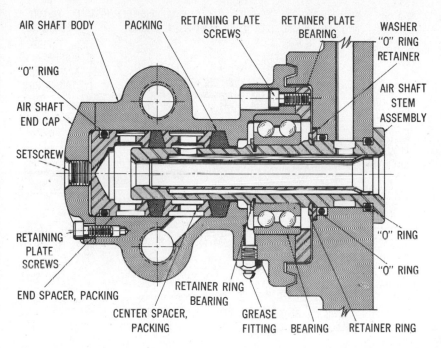

Fig. 17-5. Air-shaft assembly with a hollow tube in the air shaft. (Courtesy Logansport Machine Co., Inc.)

lubrication for the bearing. In special applications that use hollow air shafts, the air-shaft housing may be provided with water passages within to cool the shaft and the packings.

HYDRAULIC CYLINDERS

Rotating cylinders that operate on either hydraulic oil or water are slightly different in design from the air rotating cylinder. However, they work on the same basic principle (Fig. 17-6). Since higher operating pressures are involved, the materials should be selected accordingly. For hydraulic-oil service, the cylinder body is usually made from a high-tensile iron alloy. The cover is usually made from steel plate, and the oil-shaft body is made from a bronze casting. The oil-shaft stem is usually larger in diameter than the air-shaft stem, since larger passages are required to handle the oil. The piston may

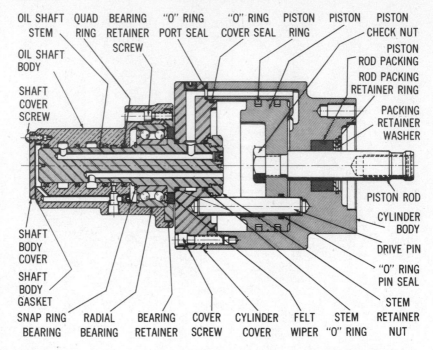

OIL SHAFT | QUAD | BEARING | "O" RING | "O" RING | PISTON | PISTON | PISTON
STEM | RING | RETAINER | PORT SEAL | COVER SEAL | RING | | CHECK NUT
| | SCREW | | | | | PISTON
OIL SHAFT | | | | | | | ROD PACKING
BODY | | | | | | | ROD PACKING
| | | | | | | RETAINER RING
SHAFT | | | | | | | PACKING
COVER | | | | | | | RETAINER
SCREW | | | | | | | WASHER

PISTON ROD

CYLINDER
BODY

DRIVE PIN

"O" RING
PIN SEAL

STEM
RETAINER
NUT

SHAFT
BODY
COVER

SHAFT
BODY
GASKET

SNAP RING | RADIAL | BEARING | COVER | CYLINDER | FELT | STEM
BEARING | BEARING | RETAINER | SCREW | COVER | WIPER | "O" RING

Fig. 17-6. A hydraulic rotating cylinder. (Courtesy Logansport Machine Co., Inc.)

be made from a steel or iron alloy, and automotive-type rings may be used, as shown in Fig. 17-6. Other types of seals, such as O-rings, quad rings, or block-vee packings, may be used. Piston-rod diameters should be adequate for handling the loads. The seals on the oil shaft can be quad rings, as shown in the diagram, or other types of seals with low friction can be used. Provision is made in the oil-shaft body for a drain connection, so that minute quantities of oil that pass by the seals are drained to the oil reservoir. A lengthy period of time may be required before any of the oil seeps to the drain port.

If water is the medium for a rotating cylinder, the parts should be treated accordingly. Corrosion-resistant materials should be used. Since the required material is expensive, these cylinders can become quite expensive. Stainless steel is one of these materials. It should be chrome plated to provide better wearing qualities.

Rotating hydraulic cylinders (Figs. 17-7 and 17-8) are made with a hole through the center for accepting bars of steel, aluminum, etc.

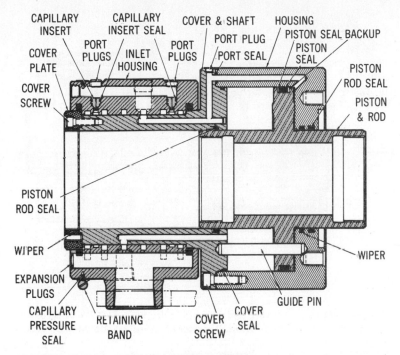

CAPILLARY INSERT · CAPILLARY INSERT SEAL · COVER & SHAFT · HOUSING · PORT PLUG · PISTON SEAL BACKUP · COVER PLATE · PORT PLUGS · INLET HOUSING · PORT PLUGS · PORT PLUG · PORT SEAL · PISTON SEAL · COVER SCREW · PISTON ROD SEAL · PISTON & ROD · PISTON ROD SEAL · WIPER · WIPER · EXPANSION PLUGS · CAPILLARY PRESSURE SEAL · RETAINING BAND · COVER SCREW · COVER SEAL · GUIDE PIN

Fig. 17-7. Hollow rotating hydraulic cylinder.

They are quite different in design from those cylinders with closed centers. The rotating hydraulic cylinder is designed to accommodate 3-inch (7.62cm) bar stock. Other available designs can accommodate bar stock up to 8 inches (20.32 cm) in diameter.

In the design in Fig. 17-7, a double hydrostatic inlet has no conventional shaft seals or roller bearings. In the past, the conventional shaft seals have been a source of trouble, due to the friction created over a large area. This caused extreme heat buildup and deterioration of the shaft seals.

The hollow-center rotating cylinder in Fig. 17-8 is equipped with thermal relief valves. Thus, if a buildup in heat occurs on long-holding applications, there will not be a pressure buildup. Also the cylinder is equipped with leakage compensators which are a type of small accumulator.

Special rotating cylinders, such as those designed with tail rods for tripping switches, those with small holes for feeding fluids, and those

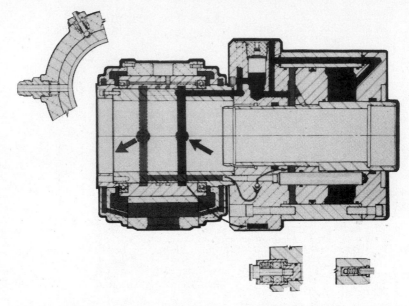

Fig. 17-8. Hollow-center rotating hydraulic cylinder with *SAF-T-LOC*, leakage compensator, and thermal relief valve. (Courtesy Logansport Machine Co., Inc.)

with two pistons either on a common piston rod or on two piston rods are discussed in the following chapter. These special types of rotating cylinders often are used on unusual applications where space, force, or other conditions are a problem.

Chapter 18

Rotating Cylinders— Application

In general, rotating cylinders are found in a much more narrow range of applications than the nonrotating cylinders. Rotating cylinders are used chiefly in two industrial categories—machine tools and heavy metalworking equipment. In the machine-tool industry, rotating cylinders are used on lathes, grinders, gear hobbers, etc. They are attached to these machines by means of adapters. The piston rod on the rotating cylinders is attached to a draw rod which is projected through the hollow machine spindle. The opposite end of the draw rod is attached to a holding device, such as a power chuck, collet, mandrel, or fixture. Unlike a nonrotating cylinder, most of the work is accomplished on the pull or retracting stroke. The work is usually accomplished with a short stroke that is often less than 2 inches (5.1 cm) in length. In the heavy metal-working-equipment industry, rotating hydraulic cylinders are used to actuate mandrels and large collets. They are used on coilers, uncoilers, slitters, etc. Rotating hydraulic cylinders also are used on other equipment in steel, aluminum, or brass mills.

ADAPTATIONS

Although there are exceptions, rotating air cylinders (Fig. 18-1) and low-pressure rotating hydraulic cylinders are chiefly used on machine tools. The high-pressure rotating hydraulic cylinders are found chiefly on heavy-duty mill equipment. Many machine tools use a medium-pressure hydraulic system within the confines of the machine tool for performing certain functions. The same system is often used to operate the rotating hydraulic cylinder. The more solid effect provided by the hydraulics is sometimes preferred to the compressibility of air. Also, hydraulics allows for the use of

Fig. 18-1. Large rotating air cylinder mounted on the spindle of a single-spindle automatic chucker. (Courtesy Warner & Swasey Turning Machine Division)

smaller-diameter cylinders which reduce the cylinder weight on the machine spindle (Figs. 18-2 and 18-3).

If the rotating cylinder is designed with a small hole extending through the piston rod and if a small rotary joint is connected to and extends beyond the air or oil-shaft distributor, the fluid—air, hydraulic oil, or coolant solution—can be passed through the piston rod to an outside point. The fluid can be used to eject a workpiece, to distribute coolant onto the workpiece, to blast chips, etc. The hole extending through the piston rod can be 1/4 to 1/2 inch (6.35 to 12.7mm) in diameter, depending on the volume of fluid needed.

Hollow-center rotating hydraulic cylinders now are used extensively on the newer designs of turning centers, lathes, etc., in conjunction with collets and hollow-center power chucks. With this setup, turning centers and lathes can be used as chuckers or bar machines. Bar stock can be fed through the rear of the rotating cylinder, as shown in Fig. 18-4. In Figs. 18-5, 18-6, and 18-7, rotating

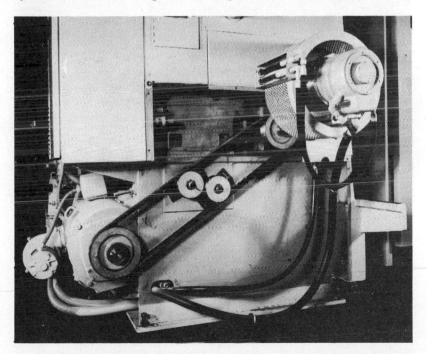

Fig. 18-2. Rotating hydraulic cylinder with solid center mounted on the spindle of a heavy-duty turning machine. (Courtesy The Monarch Machine Tool Company)

Fig. 18-3. A rotating hydraulic cylinder mounted on the spindle of an NC turning machine. (Courtesy Waterbury Farrel Division of Textron Inc.)

hydraulic cylinders can be seen that are capable of accommodating large-diameter bar stock. Note the hydraulic lines to the distributor on the cylinders. The lowest line on each is the drain line, and it must be unrestricted for best results.

Hollow-center rotating air cylinders are available for high-speed applications. However, they do not have the available thrust that can be obtained with a hydraulic cylinder.

The turning machines are often equipped with a bar stock feed mechanism like the one in Fig. 18-4. A bar stock feed mechanism is a type of feed device for moving the bar stock through the hollow center of the rotating cylinder, the machine spindle, and into the holding device, whether it is a collet or a chuck. These mechanisms are often operated by compressed air, nonrotating cylinders, mechanical devices, or other means to properly position the bar.

Fig. 18-4. Bar stock feed mechanism used in conjunction with a hollow-center rotating hydraulic cylinder that is mounted to the spindle of a modern turning machine. (Courtesy The Monarch Machine Tool Company)

OTHER ROTATING CYLINDERS

A *tandem* cylinder (Fig. 18-8) is often used to increase the force of a rotating cylinder without increasing its diameter. The force is approximately doubled with the two pistons in these cylinders. The length of a tandem cylinder is considerably longer than the length of a single-piston cylinder.

Duplex rotating cylinders (cylinders with two pistons and a piston rod for each piston) are used to operate special holding devices. One of the pistons can be used to operate one of the two sets of jaws in a four-jaw chuck. The second piston can be used to operate the remaining set of jaws in the chuck. The duplex rotating cylinders are quite expensive. However, they are designed to perform certain operations more efficiently.

Fig. 18-5. Hollow-center rotating hydraulic cylinder with a large-diameter hole. Note the three pipe connections; the drain is at the bottom. (Courtesy Cushman Industries, Inc.)

In Fig. 18-9, a power chuck is used in conjunction with a rotating hydraulic cylinder on an NC turning machine. Since there is a large hole in the center of the chuck, the workpiece can be partially positioned in the hole. Thus, a long overhang from the face of the chuck can be eliminated. The jaws perform much more efficiently when their gripping areas are near the face of the chuck.

Fig. 18-6. Hollow-center rotating hydraulic cylinder that is directly adapted to the spindle of an NC Turning and Chucking Machine. (Courtesy Cincinnati Milacron)

Fig. 18-7. This actuator consists of a hollow rotating, double-acting hydraulic cylinder, surrounded by a stationary housing. The cylinder is mounted on the machine spindle by a suitable adapter, and it revolves with the spindle. (Courtesy The Monarch Machine Tool Company)

Fig. 18-8. A tandem rotating air cylinder mounted to the machine spindle by a cylinder adapter. This machine is a single-spindle automatic chucker. (Courtesy Warner & Swasey Turning Machine Division)

Fig. 18-9. A three-jaw power chuck with a large hole in center operated by the rotating hydraulic cylinder shown in Fig. 18-3. (Courtesy Waterbury Farrel Division of Textron Inc.)

Chapter 19

Using the Fluid Power Calculator

The fluid power calculator (Figs. 19-1 and 19-2) is a device which can be used to solve quickly the various mathematical calculations involving fluid power components and their applications. Although the handy calculator is slightly more inaccurate than the mathematical slide rule which can produce calculations that are accurate within a few decimal places, it is accurate for all practical purposes. As thousands of users have discovered, the calculator is a timesaver. The calculator is designed primarily for computations in regard to hydraulic oil and water. Also, it can be used for some pneumatic calculations. Calculators are available that can be used only for pneumatics.

THE CALCULATOR

A thorough working knowledge of the calculator can be a valuable asset to fluid circuit designers, plant engineers, and others who are involved in fluid power applications. The data needed to determine the force exerted by a cylinder and the volume displayed during the

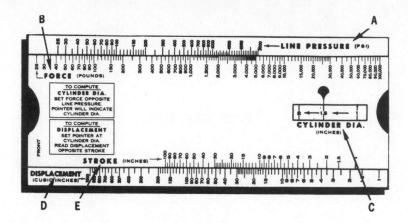

Fig. 19-1. The FRONT of the fluid power calculator.

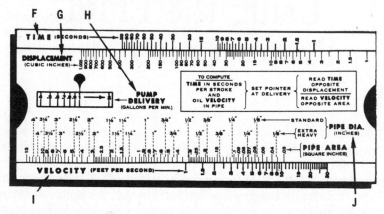

Fig. 19-2. The BACK of the fluid power calculator.

cylinder stroke is found on the *FRONT* of the calculator (Fig. 19-1). On examining this side of the calculator, the following scales can be found:

1. On the slide, scale *A* indicates the *line pressure*, ranging from 25 to 3000 psi.
2. On the calculator case and directly below scale *A*, the *force* ranging from 25 to 100,000 pounds is indicated by scale *B*.
3. Scale *C*, which is located on the slide, provides the *cylinder*

diameter, ranging from 1 to 20 inches. The arrow in the window on the calculator case points to the cylinder diameter.

4. The lower scale *D* on the slide indicates the *displacement*, ranging from 1 to 1000 cubic inches. It is used in conjunction with the cylinder bore and stroke.
5. The *stroke* scale *E* is located on the calculator case directly above the displacement scale *D*. The scale ranges from 1 to 100 inches.

Further computations, such as time, pump delivery, and velocity of flow in the pipe, can be determined on the *BACK* of the calculator (Fig. 19-2). The scales located on the backside are as follows:

1. At the top of the slide, scale *F* gives the *time*, in seconds, ranging from 1 to 100 seconds.
2. On the case and directly below scale *F*, the *displacement* is indicated by scale *G*. This scale is identical to scale *D* on the front side of the slide, and, of course, ranges from 1 to 1000 cubic inches.
3. In the center scale *H* on the slide, the *pump delivery* is indicated, ranging from 0.4 to 200 gallons per minute.
4. The lower scale *I* on the slide provides the *velocity* of the liquid. These values range from 1 to 30 feet per second.
5. On the slide case and directly above scale *I*, the *pipe areas*, in square inches, and the *pipe diameters*, in inches, are found in scale *J*. The pipe diameters are for both *standard* and *extra-heavy* pipe. The pipe areas range from 0.03 to 15 square inches. The pipe diameters range from 1/8 inch to 4 inches.

PROBLEMS

The calculator should be used to solve the following problems, which are typical of the problems encountered in the fluid power industry. Solving these problems with the calculator provides an opportunity to achieve a better understanding of the calculator and its use.

Problem No. 1: Calculate the volume of oil required to advance the piston of a 6-inch bore by 20-inch stroke cylinder to the end of its stroke.

Solution: On the *FRONT* of the calculator:

1. Move the slide until the cylinder diameter *6* on scale *C* coincides with the arrow on the window.
2. Then, opposite the length of stroke *20* on scale *E* read the piston displacement on scale *D*, which is *550* cubic inches.

This means that a volume of 550 cubic inches of oil is required to advance the piston of a 6-inch bore by 20-inch stroke hydraulic cylinder.

Problem No. 2: In a hydraulic cylinder with the same bore and stroke (6 inches × 20 inches) as the cylinder in Problem No. 1, calculate the volume of oil required to return the piston, if the piston rod is 4 inches in diameter.

Solution: The volume of oil required to return the piston is the same as the volume required to advance the piston (550 cubic inches from Problem No. 1) less the volume or displacement of the piston rod. Then, the problem is to find the volume displaced by the piston rod, as follows:

1. On the FRONT of the calculator, set the piston rod diameter *4* on scale *C* at the arrow in the window.
2. Then, opposite the length of stroke *20* on scale *E* read the piston-rod displacement on scale *D*; it is *250* cubic inches.

Therefore, the volume of oil required to return the piston in the hydraulic cylinder is:

volume of oil required to advance piston = 550 cubic inches
volume of oil displaced by 4-inch di-
ameter piston rod = 250 cubic inches
volume of oil required to return piston = 300 cubic inches

Problem No. 3: In regard to Problems No. 1 and No. 2, what volume of oil is required for one complete cycle of the piston in a 6-inch bore by 20-inch stroke cylinder with a 4-inch diameter rod?

Solution: The total volume of oil required for one complete cycle is:

outstroke = 550 cubic inches
return stroke = 300 cubic inches
total volume required = 850 cubic inches

Problem No. 4: Calculate the force exerted by the piston of a 6-inch
   bore cylinder on its forward stroke if the operating pressure of the
   system is 700 psi.
Solution: Set the cylinder diameter *6* of scale *C* in the window.
   Opposite the line pressure *700* psi on scale *A*, read the force *20,000*
   pounds on scale *B*. Therefore, a 6-inch bore cylinder operating at
   700 psi exerts a force of 20,000 pounds on its forward stroke.

To calculate the force exerted by a cylinder on the return stroke,
determine the reduction in force caused by the cross-sectional area of
the piston rod. Thus for Problem No. 4, the force on the return stroke
is 20,000 pounds, less the area of the piston rod times the operating
pressure. This calculation is illustrated in the following problem.

Problem No. 5: Calculate the force exerted by a 6-inch bore cylinder
   on the return stroke, if the piston rod is 4 inches in diameter.
Solution: Using the calculator to determine the reduction in force
   caused by the cross-sectional area of the piston rod, set the rod
   diameter *4* on scale *C* in the window. Opposite the line pressure *700*
   on scale *A*, read the force *8750* pounds on scale *B*. Therefore, the
   total force exerted on the return stroke is (20,000 − 8750), or 11,250
   pounds.

If the pump delivery, in gallons per minute, is known, the speed of
the piston in a hydraulic cylinder can be determined. This can be of
value in checking a circuit to determine whether the pump is
delivering at full capacity or whether the cylinder is producing at the
correct speed.

Problem No. 6: Determine the speed on the forward stroke of the
   piston in a 6-inch bore by 20-inch stroke hydraulic cylinder used in
   conjunction with a 15-gallon-per-minute pump (see rating on the
   nameplate of the pump).
Solution: Using the calculator:

1. Set the cylinder diameter *6* on scale *C* in the window.
2. Opposite the stroke *20* on scale *E*, read the displacement on
   scale *D*, which is *550* cubic inches.
3. Turning the calculator to the opposite side, set the pump
   delivery *15* gallons per minute in the window.

4. Opposite the displacement *550* on scale *G* read the time *9.5* seconds on scale *F*.

Therefore, if the piston rod does not complete its forward stroke in approximately 9.5 seconds, the lines or valves should be checked for restrictions; or the pump should be checked to determine whether it is producing its rated capacity.

The velocity of the oil passing through a portion of the system can be determined on the calculator. To do this, the capacity of the pump and the size of the lines should be known.

Problem No. 7: Assuming that the pump is delivering 30 gallons per minute and that 1¼-inch diameter extra-heavy piping is used, determine the velocity of the oil in the system.

Solution: On the back of the calculator, set the pump delivery *30* gallons per minute on scale *H* in the window. Opposite the extra-heavy pipe diameter 1¼ inches on scale *J* read *7.6* feet per second on scale *I*, which is the velocity of the oil in the system.

In some instances, it may be desirable to determine the required capacity for a pump to do a given job. If the size of the cylinder and the time are known, the calculator can be used to solve the problem.

Problem No. 8: If the cylinder is 5-inch bore by 25-inch stroke and if the piston must advance the entire stroke in 5 seconds, what pump capacity or delivery is required?
Solution: Using the calculator:

1. Set the cylinder diameter *5* on scale *C* in the window.
2. Opposite the stroke *25* on scale *E*, read the displacement, which is *490* cubic inches, on scale *D*.
3. Turning the calculator to the opposite side, set the time *5* seconds on scale *F* opposite the displacement *490* cubic inches on scale *G*.
4. Then, read the required pump delivery *25* gallons per minute on scale *H* in the window.

Therefore, the pump delivery or capacity needed in the installation is 25 gallons per minute.

Chapter 20

Design of Basic Pneumatic Circuits— Including Speed Control

A working knowledge of circuit design is essential for anyone who deals with fluid power components or machines which use fluid power circuits. The designer of a fluid power circuit should be familiar with the capabilities of the components involved and be able to visualize the desired end results of the system. Many circuits are overdesigned, which results in higher costs and, often, in excessive maintenance. All circuits should be as simple as possible. Yet, they should accomplish the desired results. It should be remembered in designing a pneumatic circuit that air is a compressible fluid and that provisions are required to compensate for this compressibility.

MOVING AN OBJECT

Some of the basic principles of pneumatic circuitry are discussed in this chapter. One of these basic principles is involved in moving an object from a position *1* to a position *2*. This movement can be either in a straight line or in an arc-like path. To perform the movement in a satisfactory manner, the compressed air should be filtered, regulated, and lubricated properly. This conditions the compressed air before it is permitted to enter the other components in the circuit. All pneumatic circuits should be provided this protection.

In a pneumatic circuit in which the piston of a cylinder is used to move an object from position *1* to position *2* when air pressure is applied to the back of the piston, either a three-way or a four-way directional control valve can be used. If a three-way directional control valve is used (Fig. 20-1), an external means is required to return the piston of the cylinder, when the work has been completed. If the cylinder is mounted in a vertical position with the piston rod placed upward, the piston in the cylinder can be returned by gravity, if enough weight is available on the end of the piston rod to overcome the friction when the air pressure is released from the back of the

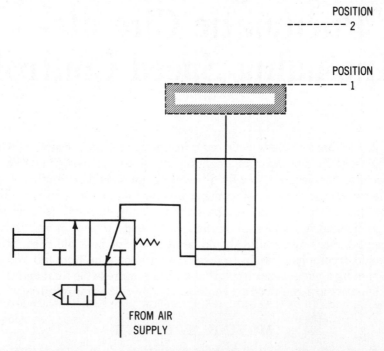

Fig. 20-1. Diagram of a pneumatic circuit in which a three-way directional control valve is used to actuate a single-acting cylinder.

piston. If the piston is to be returned by spring pressure when the air pressure is released from the back of the piston, it should be remembered that a portion of the theoretical force created by the air pressure acting against the piston is lost in compressing the spring.

When an air three-way directional control valve is used in conjunction with a single-acting air cylinder, it can be either a two-position or a three-position valve. In the two-position valve design, there are two positions for the flow director and the control operator. In the three-position valve design there are three positions. Although the two-position three-way directional control valves can be either "normally open" or "normally closed," those valves used with a single-acting cylinder in which the normal position of the cylinder piston is in the retracted position are usually "normally closed." When a three-position directional control valve is used, the center position is blocked. This permits stopping the piston at intermediate positions. It is usually more difficult to control the intermediate stops in a single-acting cylinder than in a double-acting cylinder.

Either one four-way directional control valve or two three-way directional control valves are required to cause the piston of a double-acting air cylinder to move an object from position 1 to position 2 and then to return the object to position 1 by applying pressure first on the blind end of the piston, as the object is moved from position 1 to position 2, and then to the rod-end side of the piston, as the object is moved from position 2 to position 1. The use of the two-position four-way directional control for this purpose is shown in the diagram in Fig. 20-2. The speed at which the piston of the cylinder moves from position 1 to position 2 depends largely on the volume of compressed air that can be directed to the blind side of the piston in the cylinder and on the ability to exhaust the air from the rod-end side of the piston. The size of the ports in the cylinder, the diameter and the nature of the piping, the port and orifice sizes in the directional control valve, and the ability of the exhaust muffler to vent the air quickly are important factors in establishing the speed of the piston movement in a cylinder. If one of these components is inadequate in capacity, the movement of the piston will be hindered.

The cylinder piston can be made to move an object from position 1 to position 2 at an extremely rapid rate for effecting an impact or hammer-like blow near position 2, such as that required in the action of a drop hammer. A quick-exhaust valve is used in this type of circuit

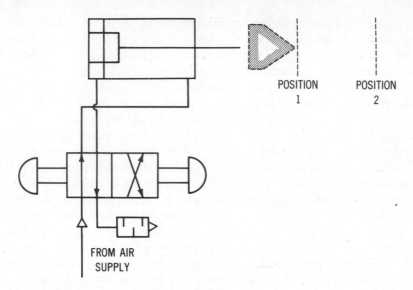

POSITION 1   POSITION 2

FROM AIR SUPPLY

Fig. 20-2. Diagram of a pneumatic circuit in which a four-way directional control valve is used to control an air cylinder.

(Fig. 20-3), so that the exhausting air at the rod end of the cylinder is not required to pass through lengthy piping and through the four-way directional control valve. When the air is exhausted from the cylinder, it enters the quick-exhaust valve and then passes to the atmosphere.

When moving an object from position *1* to position *4*, it is sometimes desirable to halt the object at intermediate stops, such as positions *2*, *3*, etc. Then a three-position four-way directional control valve should be used (Fig. 20-4). The flow director in the directional control valve should block all the ports when the valve operator is in the neutral position. The valve operator can be spring centered. Or it can be equipped with ball detents to "feel" each position of the valve operator.

The intermediate stops at positions *2*, *3*, etc. (see Fig. 20-4) are much more difficult to accomplish with air pressure than with hydraulic oil pressure, as a result of the compressibility of the air. If extremely accurate positioning is required, pneumatics alone should not be considered.

In the circuit diagrammed in Fig. 20-5, the piston of cylinder *A* moves the object from position *1* to position *2*. Then a mechanical

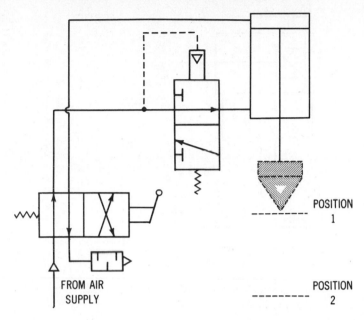

POSITION
1

FROM AIR
SUPPLY

POSITION
2

Fig. 20-3. Circuit in which a quick-exhaust valve is used to increase the speed of the piston in the cylinder.

pickup carries the object from position 2 to position 3. In this circuit, a three-position directional control valve with solenoid operators is used in conjunction with electrical holding relays. In the three-position valve, the exhaust port is connected to both cylinder ports when the flow director is in the neutral position. This permits the piston in the cylinder to float, since the pressure is released from both sides of the piston. In the neutral position, the pressure port of the control valve is blocked.

If an obstruction is likely to be encountered when the piston of an air cylinder A moves an object from position 1 to position 2, the circuit shown in the diagram in Fig. 20-6 can be used. In this circuit, the workpiece is moved when the push button PB-1 is depressed momentarily. This momentarily energizes solenoid S-1 of the control valve B, shifting the flow director to direct the air pressure to the blind side of the piston in cylinder A. As the piston advances, resistance is encountered, and the "normally open" pressure switch C closes. This momentarily energizes the solenoid S-2 of the control valve B. The flow director in the valve shifts. Air pressure is directed

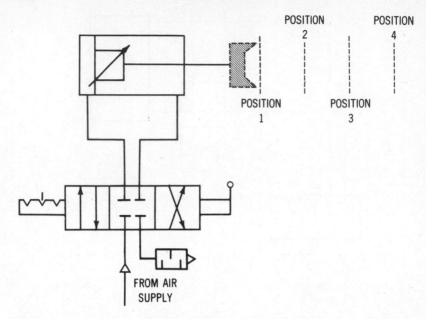

Fig. 20-4. Diagram of a circuit in which a three-position four-way directional control valve is used to halt the piston movement of the cylinder at intermediate stops.

to the rod end of the cylinder. The piston retracts to the starting position. If an obstruction is not encountered, the piston moves the object to position 2. Pressure then builds up, the pressure switch closes, and solenoid S-2 is energized. The piston in the cylinder then retracts automatically.

In some applications, electrical controls are not practical, either from the standpoint of the materials being handled or from the standpoint of the atmospheric conditions. A circuit that can be used and which does not require electrically operated controls is diagrammed in Fig. 20-7. In this circuit, pilot-operated controls are used in conjunction with an air sequence valve. To advance the piston in the cylinder A, the pilot-operated valve B is actuated momentarily. The air pressure is directed to the pilot chamber R-1 of the directional control valve C. The flow director of valve C is shifted. Air pressure is directed to the blind end of cylinder A. If an obstruction is encountered during the travel from position 1 to position 2, or when position 2 is reached, pressure builds up in the line between the

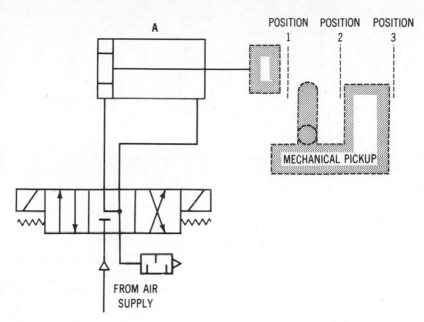

Fig. 20-5. Diagram of a circuit in which a solenoid-operated three-position directional control valve is used in conjunction with a mechanical pickup to move an object.

four-way valve C and the blind end of cylinder A. Then the sequence valve D opens, directing air pressure to the pilot chamber R-2 of valve C. The flow director shifts immediately. Air pressure is directed to the rod end of the cylinder. The piston retracts to its original position. By installing an overload protection in the circuit, much damage can be prevented if an unexpected obstruction is encountered during the operation of the circuit.

CONTROLLING SPEED OF MOVEMENT

If it is desirable for the piston in the cylinder to move an object at a given speed from position 1 to position 2, and then move at a slower speed from position 2 to position 3, the circuit shown in the diagram in Fig. 20-8 can be used. A cam-operated speed control is engaged at position 2. This restricts the exhaust air leaving the cylinder, thereby causing a slower speed. The cam-operated speed control can be

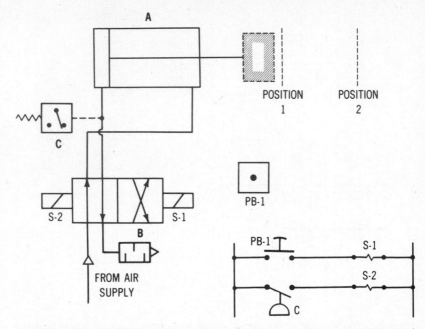

Fig. 20-6. Diagram of a circuit in which a solenoid-operated four-way directional control valve is used to reverse the piston movement when an obstruction is encountered.

triggered by a cam bar located on a machine table or by some other member that is moved by the cylinder. A built-in check valve prevents the flow being restricted when the piston returns the object from position *3* to position *1*.

A control similar to that effected in Fig. 20-8 can be accomplished with electrical controls, as shown in the diagram in Fig. 20-9. A limit switch is used in conjunction with a cam bar. When the cam bar contacts the limit switch, the "normally open" two-way valve *A* closes, trapping the exhaust air which is then required to pass through the orifice set by the speed control valve *B*. On the return stroke, the ball check in valve *B* is opened to permit a free-flow return.

If a cylinder is required to move an object from position *1* to position *2* in an arc-like path, the success of the movement is dependent on the mounting style of the cylinder. To accomplish this movement, a flexible mounting, such as a pivot or a trunnion

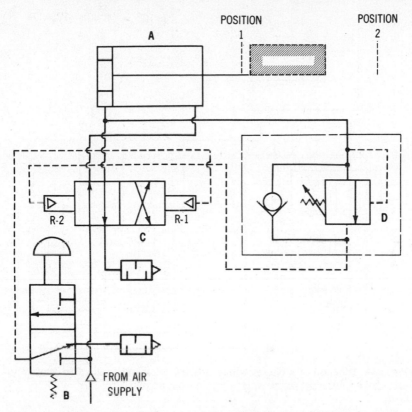

Fig. 20-7. Diagram of a circuit in which a pilot-operated four-way directional control valve is used in conjunction with a sequence valve to reverse the piston movement when an obstruction is encountered.

mounting, is required. These mountings are used in loading, dumping, transferring, and similar applications.

Other devices used for moving objects in an arc-like path from one position to a second position are actuators and air motors. The basic circuits discussed in this chapter are often used with these devices.

The selection of the operators for the directional control valves is governed by the application on which the operator is to be used. If it is necessary to "feel" the action that is taking place, a manual operator may be required. If more than one person is involved and if safety may become a problem, operators with no-tiedown devices may be advantageous. A thorough study of the application should be

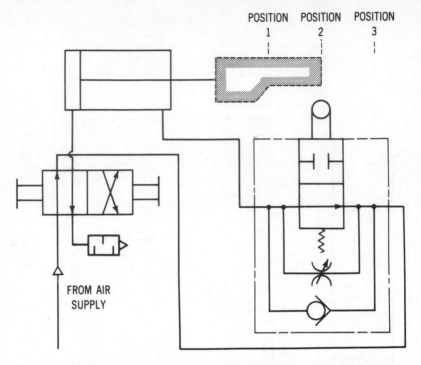

Fig. 20-8. Diagram of a circuit which utilizes a cam-operated speed control to provide two different piston speeds for moving an object.

accomplished before selecting the valve operators. The most expensive operator may be the most practical.

In selecting the piping for the basic pneumatic system, a wide range of selections is available. Some of the newer, more flexible piping permits the saving of considerable time and labor in the installation of a system. It is important that the capacities of the piping and the components are adequate. This prevents restrictions in the lines. Fittings should be installed that can withstand the vibrations and shock loads often encountered in pneumatic systems as a result of high-speed movements.

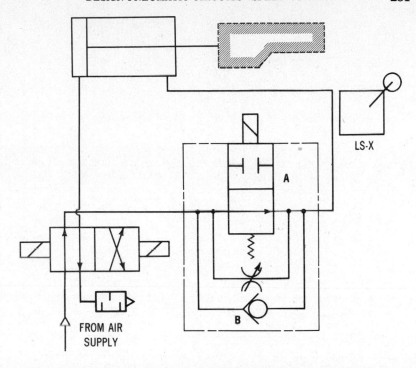

Fig. 20-9. Diagram of a circuit which utilizes a solenoid-operated speed control to provide two different piston speeds for moving an object.

Chapter 21

Design of Basic Hydraulic Circuits— Including Flow Control

The basic hydraulic circuits are used widely in industry. A good circuit designer should endeavor to keep each circuit as simple as possible, thereby using a minimum number of components to accomplish the desired results.

CIRCUIT DESIGN

In this chapter, only the basic circuits and components needed to satisfy the requirements are discussed. For example, if it is required to move an object from position *1* to position *2* in a straight line by hydraulic means, one solution includes the following:

1. A power unit to supply oil under pressure.
2. A directional control valve to direct the oil flow from the power unit.

3. A cylinder large enough to move the object from position *1* to position *2*.
4. Piping to transfer the hydraulic fluid within the circuit.

The required capacity of the power unit is determined by the volume of fluid and the force required to move the object at a given speed. The size of the other components is usually determined by the size of the cylinder that is to move the object. For example, it can be assumed that a force of 12,000 pounds (53,378.64 N) is required to move the object, and the object is to be moved a distance of 24 inches (60.96 cm). If the cylinder selected is to operate at a pressure of 1000 psi (6894.8 kPa), the bore size of the cylinder should be large enough to move 12,000 pounds (5443.1 kg). This requires a cylinder with a 4-inch (10.16 cm) bore (check this size on the fluid-power calculator). In stroking the 4-inch (10.16 cm) bore cylinder through a distance of 24 inches (60.96 cm), approximately 295 cubic inches (4834.19 cm^3) of oil is required on the forward stroke. If the forward stroke is to be accomplished in 5 seconds, the requirements of the pump on the power unit can be determined. Using the calculator, the pump is required to produce an oil volume of 15 gallons (56.8 liters) per minute at 1000 psi (6894.8 kPa). A 15-hp motor is required to actuate the pump.

The volume of oil required to pass through the control valve determines the port size of the control valve. If a maximum velocity of 15 feet per second (4.572 m/sec) is desired, a 3/4-inch (19.05 mm) port size is needed in the control. This also means that the lines and porting throughout the system should be 3/4-inch (19.05 mm) minimum size.

The above reasoning is typical, but this can result in a faulty design. If the cylinder is equipped with a 2:1 ratio piston rod, the pump is producing 15 gallons (56.8 liters) per minute for the return stroke. Since the cross-sectional area of the piston rod is one-half the area of the cylinder bore, the blind side of the cylinder is required to exhaust 30 gallons (113.6 liters) per minute. Usually, this requires a larger control valve and larger piping. Even the experienced designers often forget this principle, especially on press circuits, and restriction problems are the result.

In selecting the directional control valve, the designer is required to decide which type of valve operator best fits the requirements of the circuit. The designer also decides whether to use a two-position four-way control valve or a three-position valve. If a three-position valve is selected, either an open-center or a closed-center valve must be chosen. A two-position four-way control valve is used to control the cylinder in the circuit diagram in Fig. 21-1, and a three-position four-way valve is used in the circuit in Fig. 21-2. The advantage of the three-position valve is that the piston in the cylinder can be stopped at any point in its travel. This is especially advantageous on "inching" applications.

If a short cylinder stroke is required, a three-way directional control valve can be used in conjunction with a single-acting spring-return cylinder, as shown in the diagram in Fig. 21-3. If the piston movement is vertical (Fig. 21-4), a three-way directional control valve can be used in conjunction with a gravity-return

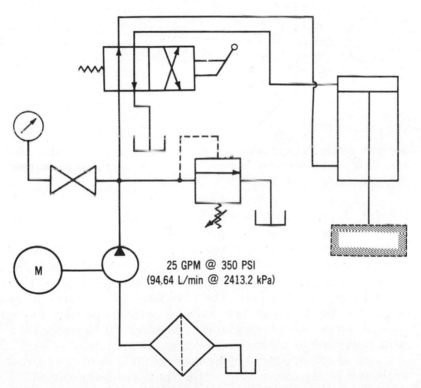

25 GPM @ 350 PSI
(94.64 L/min @ 2413.2 kPa)

M

Fig. 21-1. Diagram of a circuit in which a two-position four-way directional control valve is used to control the piston movement in the cylinder.

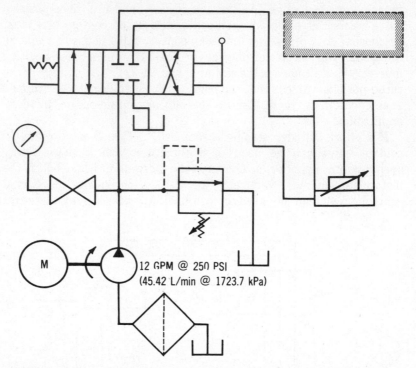

Fig. 21-2. Diagram of a circuit in which a three-position four-way directional control valve is used to control the piston movement in the cylinder for "inching" applications.

cylinder. The three-way control valve can be either a two-position or a three-position directional control valve.

If the movement of an object through an arc-like path is desired, a rotary actuator can be used, rather than a cylinder (Fig. 21-5). As discussed in a previous chapter, the rotary actuator can move an object through a limited arc. The directional control valve in the circuit can be a two-position valve or a three-position valve, depending on whether the actuator in its travel is to be stopped at a point other than at the extreme ends of the arc.

A basic circuit in which a hydraulic motor is used to drive a reel mechanism is diagrammed in Fig. 21-6. The motor can be rotated in a clockwise or a counterclockwise direction, depending on which motor port the hydraulic pressure is applied. In the diagram, a

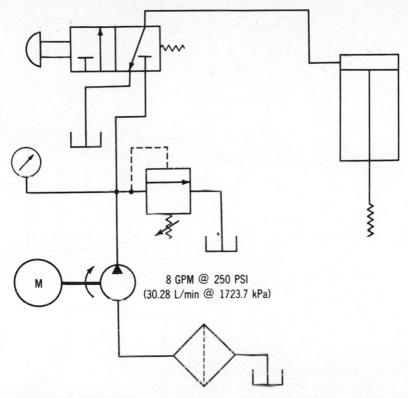

Fig. 21-3. Circuit diagram in which a three-way directional control valve is used in conjunction with a single-acting spring-return cylinder.

8 GPM @ 250 PSI
(30.28 L/min @ 1723.7 kPa)

closed-center three-position four-way directional control valve is used either to start or to stop the motor at any time, and the speed of the motor can be controlled for "inching" applications.

CONTROLLING FLOW

In Figs. 21-7, 21-8, and 21-9, an object is moved from position *1* to position *2* at a controlled speed and by hydraulic means. Then the piston in the cylinder is retracted at high speed, also by hydraulic means. This type of movement can be accomplished by three different methods when a constant-displacement pump is used. The controlled-exhaust method which uses an inexpensive flow control

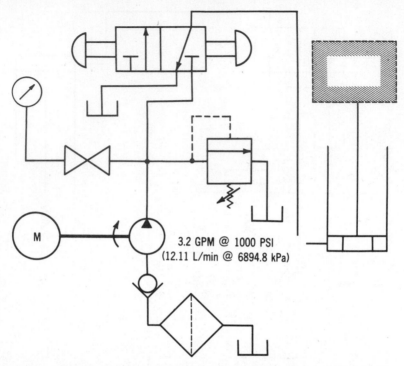

Fig. 21-4. Circuit diagram in which a three-way directional control valve in conjunction with a gravity-return cylinder is used to control vertical piston movement in a cylinder.

can be employed (see Fig. 21-7). The oil that is exhausted from the cylinder A on its forward stroke is controlled by the adjustable orifice in the flow control B. In Fig. 21-8, the inlet flow to the cylinder A is controlled by the orifice in the flow control B which is placed between the four-way valve C and the blind end of the cylinder. In Fig. 21-9, a portion of the fluid flowing from the directional control valve A to the blind end of the cylinder B is bled off through the flow control C. This is termed the "bleed-off" method for controlling piston movement. It can be found on honing machines, broaching machines, grinders, etc.

In the pneumatic systems, in nearly all instances the exhausting air from the cylinder is controlled. Controlling the inlet air to the cylinder usually results in an erratic piston movement.

A generally accepted method for controlling the speed of the piston travel in a cylinder A is shown in the circuit diagram in Fig. 21-10.

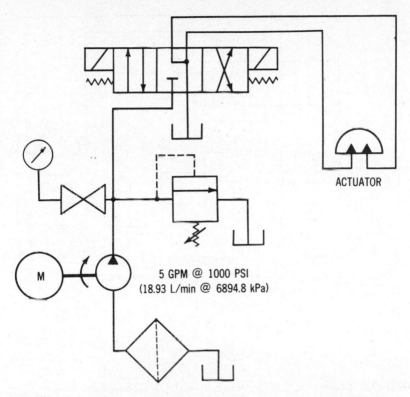

Fig. 21-5. Circuit diagram in which a three-position four-way directional control valve is used to control a hydraulic actuator for moving an object in an arc-like path.

The piston moves at full speed from position *1* to position *2*, and then moves at controlled feed from position *2* to position *3*. From position *3* to position *1*, the piston travels at full speed. A cam-operated flow control valve, which is in reality a combination shutoff valve and flow-control valve, is used in the circuit. This circuit is typical of circuits used on machine tools where skip feeds are desirable; on long-stroke cylinders where deceleration is essential, but cannot be accomplished by the usual cushioning; and on workpiece feeding arrangements, such as on the larger saws.

A solenoid-operated valve, rather than a cam-operated valve, can be used to actuate the shutoff section of the valve. Limit switches can be positioned along the travel of the piston rod. This arrangement

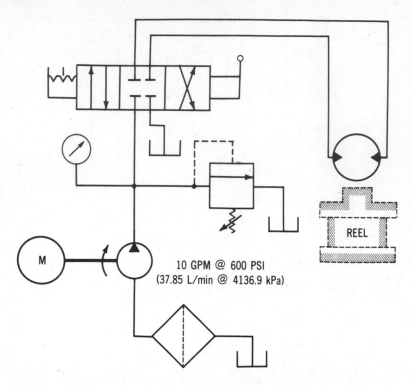

Fig. 21-6. Diagram of a basic circuit in which a closed-center three-position four-way directional control valve is used to control a hydraulic motor that drives a reel mechanism.

often provides added flexibility, since the limit switches or the limit-switch actuators usually can be repositioned more easily than the actuators and the cam valves in a system with rigid piping.

In some applications in which a piston moves from position *1* to position *3*, controlled speed during several intermediate segments of the stroke may be desirable. This can be accomplished by positioning several cams on a machine table or slide. The cams then depress the roller of a cam-operated flow-control valve. Typical applications of this arrangement can be found on milling machines or similar machines on which controlled machining is performed in several different locations during travel of the piston in the cylinder. If gradual deceleration of a piston in a cylinder is required in its travel from position *2* to position *3*, a deceleration valve, which is a version of a flow control, can be used in conjunction with a long-tapered cam.

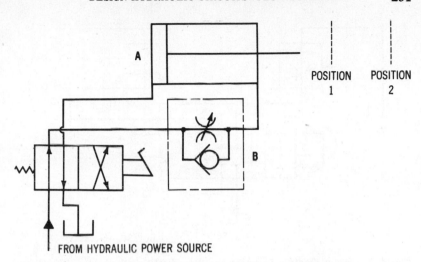

Fig. 21-7. Circuit diagram in which the forward movement of the piston is controlled by the adjustable orifice in the flow control, which controls the volume of oil exhausted from the cylinder on its forward stroke.

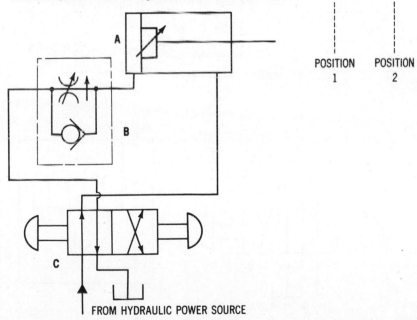

Fig. 21-8. Circuit diagram in which the orifice in the flow control controls the inlet flow to the cylinder, which, in turn, controls the forward movement of the piston.

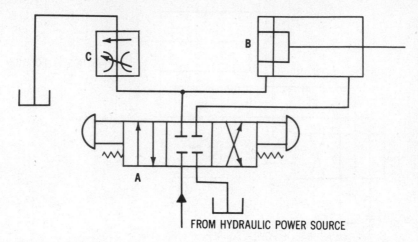

Fig. 21-9. Circuit diagram in which the "bleed-off" method is used to control piston movement in a cylinder.

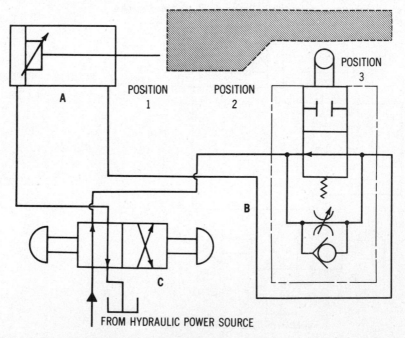

Fig. 21-10. Circuit diagram in which a cam-operated flow control is used to control the speed of a piston during a portion of its travel.

Chapter 22

Pressure Boosters or Intensifiers—Design

In the fluid power industry, the terms *pressure booster* and *pressure intensifier* are used interchangeably. Although one manufacturer may label his product a pressure booster, the next manufacturer may label the device a pressure intensifier. These components perform the function their name implies—they increase fluid pressure. The function performed by a pressure booster is similar to that performed by a fluid pump. However, the basic operation of a pump results from a revolving motion. This is not true in the basic operation of pressure boosters or intensifiers.

The basic operating principle of the pressure booster can be compared to that of the automobile tire pump. A huge, brawny man pushing downward on the handle of the tire pump can cause the air acting on the small plunger in the tire pump to exert a considerable amount of pressure within the tire. However, a small, weak lad cannot apply enough force to the handle of the tire pump to inflate the tire satisfactorily. The basic operating principle of the pressure booster is shown in the diagram in Fig. 22-1. Compare the larger diameter of the primary piston with the much smaller diameter of the

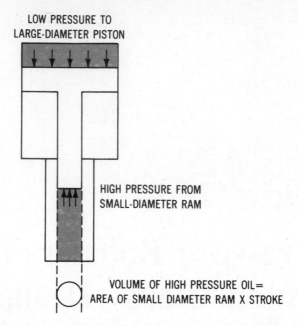

**LOW PRESSURE TO
LARGE-DIAMETER PISTON**

**HIGH PRESSURE FROM
SMALL-DIAMETER RAM**

**VOLUME OF HIGH PRESSURE OIL=
AREA OF SMALL DIAMETER RAM X STROKE**

Fig. 22-1. Diagram showing the basic operating principle of the pressure booster or intensifier.

secondary piston. For example, if the primary-piston area is 50 square inches (322.6 cm²) and the secondary-piston or ram area is 5 square inches (32.3 cm²) when the operating pressure is 100 psi (689.5 kPa), the output pressure is (50 × 100) ÷ 5, or 1000 psi [(322.6 cm³ × 689.5 kPa) ÷ 32.3 cm³ = 6886.0 kPa.]

OPERATION

The pressure boosters can be either single-acting or double-acting. In a single-acting hydraulic pressure booster (Figs. 22-2 and 22-3), intensified fluid is delivered on only one stroke of the main piston. No fluid is delivered on the reverse stroke of the piston. The main piston of the pressure booster is returned either by the line pressure of the system or by spring pressure. The rear or primary section of a pressure booster is, in reality, either a single-acting or a double-acting cylinder.

A double-acting pressure booster (Figs. 22-4 and 22-5) provides

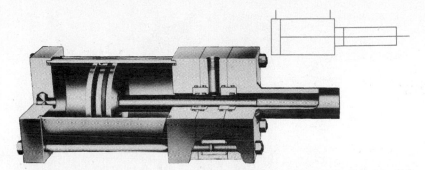

Fig. 22-2. Cutaway view of a single-acting hydraulic pressure booster. (This illustration is the courtesy of Parker Hannifin Corporation)

intensified pressure in both directions of movement of the main piston. The construction of a double-acting pressure booster (Fig. 22-3) is similar to that of a double-end double-acting cylinder in which a receiver is fastened to each cover.

In a single-acting pressure booster, the work is performed before the piston reaches the end of the forward stroke. Otherwise, the intensified pressure drops suddenly. In a double-acting pressure booster, the intensified pressure is available at all times.

Various fluid media can be used in pressure boosters. Compressed air can be used to operate the main cylinder. The fluid to be intensified can be hydraulic oil, water, or a gas. For another variation, hydraulic oil or water can be used. The fluid to be intensified can be oil, water, or other fluid. The material used in the construction of the intensifier should be compatible with the fluid used. For example, if water is the fluid to be intensified, the intensified portion of the booster should be made from bronze or stainless steel. The ram should be made from chrome-plated stainless steel. Suitable packings are also required, because the intensified full pressure works against the packings. If the intensified pressures are extremely high, such as 30,000 (206,842.5 kPa) or 40,000 psi (275,790.0 kPa), special connectors are required to pipe the fluid into the system.

The advantages of pressure boosters are as follows:

1. Elimination of a heat problem often caused when high-pressure pumps operate against the relief valve. With the pressure

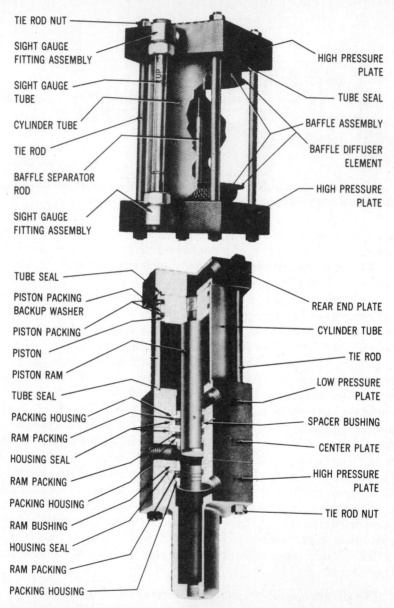

TIE ROD NUT

SIGHT GAUGE
FITTING ASSEMBLY

SIGHT GAUGE
TUBE

CYLINDER TUBE

TIE ROD

BAFFLE SEPARATOR
ROD

SIGHT GAUGE
FITTING ASSEMBLY

HIGH PRESSURE
PLATE

TUBE SEAL

BAFFLE ASSEMBLY

BAFFLE DIFFUSER
ELEMENT

HIGH PRESSURE
PLATE

TUBE SEAL

PISTON PACKING
BACKUP WASHER

PISTON PACKING

PISTON

PISTON RAM

TUBE SEAL

PACKING HOUSING

RAM PACKING

HOUSING SEAL

RAM PACKING

PACKING HOUSING

RAM BUSHING

HOUSING SEAL

RAM PACKING

PACKING HOUSING

REAR END PLATE

CYLINDER TUBE

TIE ROD

LOW PRESSURE
PLATE

SPACER BUSHING

CENTER PLATE

HIGH PRESSURE
PLATE

TIE ROD NUT

Fig. 22-3. Cutaway views of an air-to-oil pressure booster and oil supply tank. Note the baffle arrangement employed to minimize agitation of the liquid in entering and leaving the tank. (Courtesy The S-P Manufacturing Corporation)

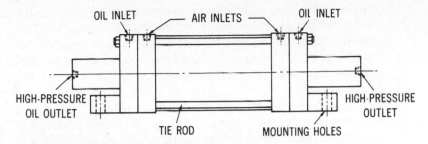

Fig. 22-4. Diagram showing the basic construction of a double-acting double-end air-hydraulic pressure booster.

booster, the high pressure can be held for longer periods of time without a heat problem.

2. If compressed air is used to actuate the main cylinder and hydraulic oil is used in the intensified section, the intensified pressures [2000 psi (13,789.5 kPa), for example] can be obtained more easily. This eliminates the need for a hydraulic power device.

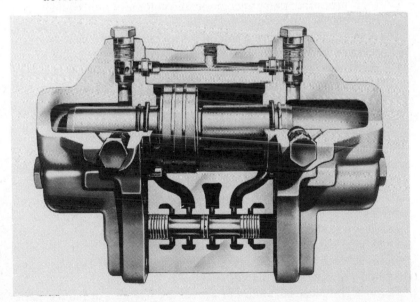

Fig. 22-5. Cutaway view of a double-acting hydraulic pressure booster. (Courtesy Rexnord Inc., Cyl. Div—HANNA)

3. The air-operated oil-intensified pressure booster eliminates the need for electrical equipment, which makes the booster ideal for explosive atmospheres and for excessively damp atmospheres.
4. The air-operated oil-intensified pressure booster is compact and portable. The small cylinders supplied with the booster also can be moved easily.
5. The pressure boosters can be used in place of expensive high-pressure hydraulic control valves. When air pressure is used in the main cylinder, only the relatively simple three-way or four-way directional control valves are required. When hydraulic fluid is used in the main cylinder, only the hydraulic control valves that are compatible with the primary-system pressure are needed.
6. The small cylinders supplied with the high-pressure fluid from pressure boosters can be positioned advantageously on the fixtures, etc., because they are compact.
7. If a high-pressure line from the single-acting pressure booster to a cylinder should rupture, only a small volume of fluid would be lost. However, if the same small cylinder were supplied by a high-pressure pump, a considerable volume of fluid could be lost before the pump was stopped.
8. A two-pressure system can be set up easily by using a pressure booster and a single pumping unit. For example, one of the pressures can be 500 psi (3447.4 kPa) and the other pressure can be intensified to as high as 50,000 psi (344,737.5 kPa).

INSTALLATION

The installation of pressure boosters is similar to the installation of cylinders. The air-operated pressure boosters should be provided with lubrication and with clean, filtered air. The pressure booster should be mounted solidly. Pressure boosters should be protected from high ambient temperatures, because they use several seals. If possible, they should be located where they are readily available for servicing. Pressure boosters should be equipped with pressure gauges at both the input and the output ports, so that accurate readings are always available.

Some of the causes of failure in a pressure booster, or failure of the booster to meet the requirements of the system are:

1. Failure in seals, resulting from excessive wear, excessive pressures, or from deterioration.
2. Excessive ambient temperatures can cause the seals to deteriorate.
3. Fluids are not compatible with the seals.
4. Dirt particles in the input system can damage the seals, cylinder walls, and pistons.
5. Broken springs in single-acting boosters do not permit the booster to make a complete stroke.
6. Failure of the check valve in a line to the oil reservoir can permit the high-pressure fluid to bleed off.
7. In a single-acting booster, the intensified fluid may be used before the work is completed.
8. The booster may be inadequate in capacity (both pressure and volume) for the job.

Chapter 23

Pressure Boosters or Intensifiers—Application

Pressure boosters or intensifiers have many uses in industry. They are used on welding machines, test apparatus, assembling machines, riveting machines, presses, forming machines, compactors, etc. The type of pressure booster or intensifier selected depends largely on the application and on the operating medium to be used.

In this chapter, several different circuits that include pressure boosters are discussed. An *air-to-water* pressure-booster circuit used in forming machine applications is diagrammed in Fig. 23-1. Some of the softer metals available in tubular form can be expanded to different shapes by using intensified water pressure. Since extremely high pressure is not required for this job, air can be used as the primary fluid medium. An example of this type of application is the expansion of copper tubing to construct the various sections of heating and cooling equipment. The use of fluid power eliminates the need for expensive mandrels and similar equipment used for expanding the tubing. Water can be used as the expanding fluid, which eliminates fire hazards and reduces cleanup time. If water is the only fluid medium used in the system, the cost of the pumping

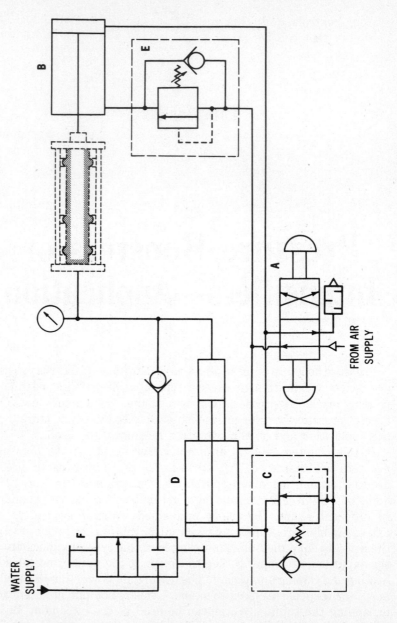

Fig. 23-1. Diagram of circuit in which an air-to-water pressure booster is utilized in forming applications.

system, controls, and other equipment required is quite high. The flexibility that can be gained from the use of two different fluid mediums is lost.

TYPICAL CIRCUITS

The inexpensive air-to-water pressure booster circuit diagrammed in Fig. 23-1 is used widely in industry. The circuit functions as follows:

1. The operator places the workpiece in the fixture, and shifts the handle of the two-position four-way air directional control valve *A*.
2. The flow director is shifted. Air pressure is directed to the blind end of the clamping cylinder *B*, where the piston clamps the workpiece.
3. When the air pressure builds up in the line to cylinder *B*, the orifice in the sequence valve *C* opens. Air pressure is directed to the blind end of the air-water intensifier *D*. The piston of the pressure booster *D* advances. The workpiece is expanded by the intensified water pressure.
4. Then the operator shifts the handle of the air valve *A* to its original position. The piston of the pressure booster retracts rapidly, releasing the intensified water.
5. When the piston of the pressure booster reaches its retracted position, air pressure builds up and opens the orifice in the sequence valve *E*. Air pressure is directed to the rod end of cylinder *B*. Its piston retracts.
6. This completes the cycle. The workpiece is removed from the fixture.

The circuit diagrammed in Fig. 23-1 is only a basic circuit. However, additional controls and a feeder mechanism can be used to develop a circuit that cycles automatically. A circuit that can be used in test applications is diagrammed in Fig. 23-2. The input medium can be air. The output medium can be either water or hydraulic oil. The pressure gauge indicates the pressure applied to the workpiece. The circuit functions as follows:

1. The operator places the workpiece in the fixture, and locks it against the pressure seals X and Y.
2. At low pressure, the water fills the workpiece. The air bleed needle valve A is shut off.
3. The operator shifts the handle of the two-position four-way directional control valve B. Air pressure at 50 psi (344.7 kPa) is directed to the blind end of the pressure booster C. The piston of the booster advances. Pressure for the primary test is recorded by the pressure gauge D.
4. After the test is recorded, the operator shifts the handle of the directional control valve to its original position. The piston of the pressure booster retracts.

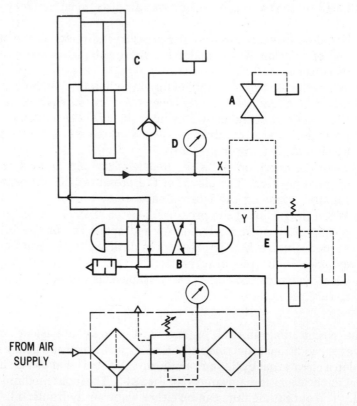

Fig. 23-2. Diagram of a circuit in which a pressure booster is used for test applications.

5. The shutoff valve E is then opened to drain the workpiece. The workpiece is removed from the fixture.

A selector valve can be installed in the line to the four-way directional control valve. Two pressure regulators can be used to provide the two operating pressures without changing the regulator setting. A pressure booster is used in the circuit to provide the hydraulic high pressure for spot-welding applications, as illustrated in the diagram in Fig. 23-3. The circuit operates as follows:

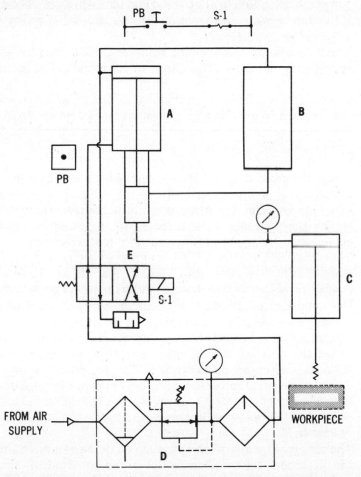

Fig. 23-3. Diagram of a spot-welding circuit that utilizes a pressure booster.

1. The workpiece is placed beneath the spot-welding head. The push button *PB* is depressed. This causes the solenoid *S-1* to energize. Air pressure is directed to the blind end of the pressure booster *A* and to the storage tank *B*.
2. The piston of the pressure booster advances. High-pressure hydraulic oil is directed to the single-acting spring-return spot-welding cylinder *C*.
3. Contact is performed under high pressure. The push button is released when the operation is completed.
4. Air pressure is directed to the rod end of the pressure booster. Its piston retracts. Air pressure to the booster is set by the air-pressure regulator *D*.
5. The piston in the spot-welding cylinder *C* then retracts, since spring pressure forces the piston to its retracted position.

The use of an air cylinder to provide quick clamping action and a hydraulic cylinder to provide high pressure and controlled feed for a deep-drawing operation is shown in the circuit diagram in Fig. 23-4. The circuit functions as follows:

1. When the push button *PB-1* is depressed momentarily, the solenoid *S-1* of the directional control valve *A* is energized momentarily. The flow director shifts. Air pressure, which is set by the regulator *B*, is directed to the blind end of the clamping cylinder *C* to clamp the workpiece securely.
2. Then the push button *PB-2* is depressed momentarily. This momentarily energizes the solenoid *S-3* of the directional control valve *D*. The flow director shifts to direct air pressure to the blind end of the pressure booster *E*. The piston of the booster advances, moving the piston of the draw cylinder *F* at a rate of speed controlled by the flow-control valve *G*.
3. When the drawing operation is completed, the push button *PB-3* is depressed momentarily. The solenoid *S-4* on the directional control valve *D* is energized momentarily to shift the flow director of the valve. Air pressure is then directed to the rod end of the pressure booster *E* and to the top of the air–oil reservoir *H*.
4. The air pressure above the hydraulic oil in the reservoir forces the oil through the free-flow section of the flow-control valve *G* and onward to the rod end of the draw cylinder *F*. The piston in

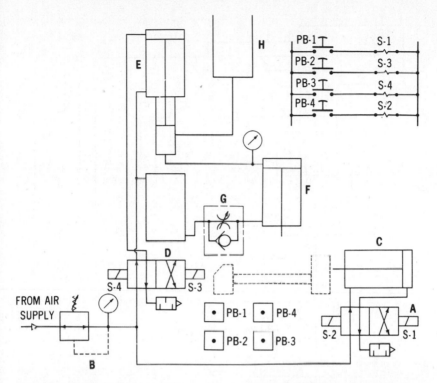

Fig. 23-4 Diagram of a circuit for a deep-drawing application in which an air cylinder provides quick clamping action and a hydraulic cylinder provides high pressure and controlled feed.

the cylinder retracts at the same time the piston of the pressure booster retracts.

5. The push button *PB-4* is depressed to energize momentarily the solenoid *S-2* of the directional control valve *A*. The flow director of the valve shifts. Air pressure is directed to the rod end of the clamping cylinder *C*. The piston of the clamping cylinder retracts to complete the cycle.

The diagram in Fig. 23-5 shows an *oil-to-oil* pressure-booster circuit. The oil pressure to the booster is controlled by the power unit. The circuit functions as follows:

1. The handle of the two-position four-way directional control

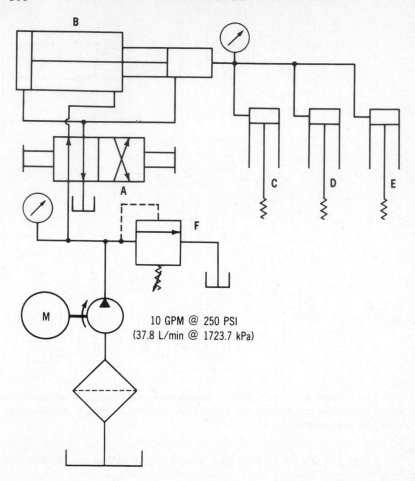

Fig. 23-5. Diagram of a hydraulic circuit in which a pressure booster is used to provide high-pressure hydraulic oil for riveting operations.

valve *A* is shifted. Oil pressure is directed to the blind end of the pressure booster *B*.

2. The piston of the booster begins to advance. High-pressure oil in the high-pressure side of the pressure booster is directed to the blind ends of the three single-acting riveting cylinders *C*, *D*, and *E*.

3. The pistons in these cylinders advance. The riveting operation is performed at high pressure.

4. When the riveting operation is completed, the handle of the directional control valve is shifted to its original position.
5. The oil pressure is then directed to the rod end of the pressure booster. The piston retracts.
6. At the same time, the springs in the three riveting cylinders retract the pistons to complete the cycle.

If the spring-return feature in the riveting cylinders (see Fig. 23-5) is undesirable, a pressure line can be extended to the rod ends of the riveting cylinders from any point in the line that connects the pump and the four-way directional control valve. This converts the spring-return cylinders to double-acting cylinders. Constant pressure can be applied on the lower sides of the cylinders. A pressure reducing valve can be placed in the line, if a pressure lower than the main operating pressure is desired. It is possible also to extend an oil line to the rod end of the pressure booster to retract the riveting cylinders when the booster piston retracts.

The use of a sequence valve in the pressure-booster circuit is shown in the diagram in Fig. 23-6. The circuit functions as follows:

1. When the push button PB-1 is depressed momentarily, the solenoid S-1 of the directional control valve A is energized momentarily. The flow director is shifted. Oil is directed to port W in the pressure booster B and then to the blind end of the press cylinder C. The piston advances until it encounters resistance.
2. The pressure builds up. The orifice through the sequence valve D is opened. Oil pressure is directed to port X of the pressure booster.
3. The piston in the pressure booster advances. High-pressure oil is forced from the port Y to the blind end of the press cylinder C. The piston of the cylinder completes the work stroke at high pressure.
4. When the work stroke is completed, the push button PB-2 is pressed momentarily. The solenoid S-2 is energized momentarily. The flow director in the directional control valve shifts to its original position. Oil pressure is directed to port Z of the pressure booster and to the rod end of the press cylinder. The piston in the pressure booster and the piston in the press cylinder retract to complete the cycle.

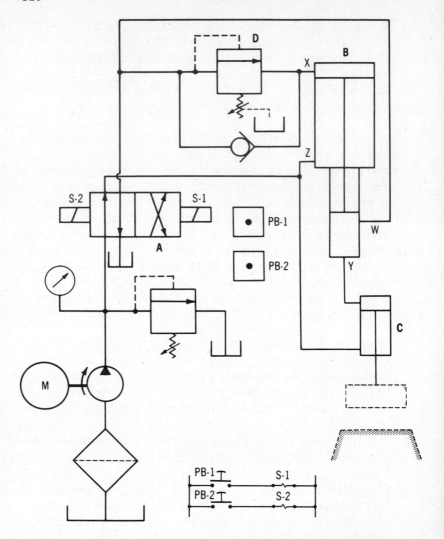

Fig. 23-6. A pressure booster circuit in which a sequence valve is used.

OTHER APPLICATIONS

A *reciprocating-type* pressure booster is shown in the circuit diagram in Fig. 23-7. The pressure booster continues to reciprocate until the full output pressure has been attained. When the full output

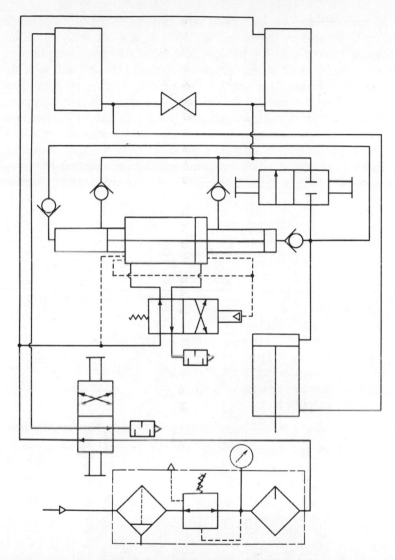

Fig. 23-7. Diagram of circuit that includes a double-acting pressure booster which reciprocates until full output pressure is reached.

pressure is reached, the pressure booster stalls. The reciprocating action resumes if any of the pressure bleeds. The chief advantage of the reciprocating-type pressure booster is that it permits the use of longer high-pressure cylinders with longer-stroke pistons.

A four-column bench press powered with an *air-hydraulic*
pressure booster is shown in Fig. 23-8. In a plant with an airline
pressure of 80 psi (551.6 kPa), a 25:1 ratio air-hydraulic pressure
booster produces an output pressure of 2000 psi (13,789.5 kPa) for
operation of the high-pressure working stroke of a heavy-duty press
cylinder. The "approach" stroke on the press cylinder is accomplished at low pressure (approximately 80 psi, or 551.6 kPa) from the
"advance" air-on-oil reservoir. The pressure booster can maintain
the high pressure for lengthy periods of time without requiring
additional power and without generating undesirable heat. The press
is equipped with a circuit that includes controls which require

Fig. 23-8. A four-column bench press powered with an air-hydraulic pressure
booster. (Courtesy Miller Fluid Power Corp.)

two-handed operation—for safety of the operator. The controls are located on the crown head of the press.

The air-hydraulic pressure booster can be located on a panel which includes a complete power package, as shown in Fig. 23-9. In addition to the booster and the panel, the package includes the air–oil tanks, the controls, and the piping required for connecting to the air line in the plant.

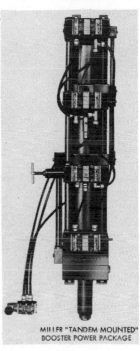

MILLER "PANEL MOUNTED" BOOSTER POWER PACKAGE

MILLER "TANDEM MOUNTED" BOOSTER POWER PACKAGE

Fig. 23-9. The air-hydraulic pressure booster can be mounted on a panel (left) which includes a complete power package. A "tandem mounted" booster power pack is shown (right). (Courtesy Miller Fluid Power Corp.)

Chapter 24

Torque Devices—Design

Torque is a rotational force, as described previously. Therefore, a device that produces a rotational force is a torque device or torque generator. Torque forces are similar to straight-line forces generated by cylinders. However, the torque force is in a rotational direction, or follows the path of an arc around a pivot point. Electric motors, air motors, hydraulic motors, and internal-combustion engines are devices that produce torque. Since, however, these units are usually involved with rotational speed, as well as force measured at a specific rotational speed, their capacities are usually expressed in terms of horsepower. In this discussion, this type of unit is normally referred to as a torque generator, or, more commonly, a rotary actuator.

Generally speaking, the rotary actuator is a torque device with an external shaft for producing rotary force through a limited arc, usually less than one complete revolution. Four basic types of rotary actuators, with variations in each design, are currently available. Actuators can be purchased in a variety of shapes, sizes, and construction details. These devices are priced quite competitively. In selecting a rotary actuator, size, package configuration, controllability, and adaptability to a specific application may be the determining factors in a specific instance.

VANE-TYPE ROTARY ACTUATOR

One of the basic designs is the *vane-type* rotary actuator (Fig. 24-1). This design consists of a cylindrical chamber that is divided by one or more barriers or "shoes" that bear a seal against the central rotating shaft. A fixed vane is attached to, and is a part of, the rotating shaft. The barrier, the seals, and the rotating vanes are enclosed by a cylinder end cap, with the rotating shaft extending through the end cap. One of the optional designs permits the rotating shaft to extend through each of the end caps, thereby offering double attachment points. The vane in the vane-type rotary actuator bears a molded rubber seal similar in cross section to an "O" ring. The seal is made with square corners that fit the contour of the vane, as well as the inside wall of the cylinder. This seals the inside diameter of the cylinder chamber, as well as the end caps.

The vane-type rotary actuator is available in either double-vane or single-vane units. In the double-vane actuator, two barriers or shoes are used. This reduces the total available rotational arc to less than one-half the arc available in the single-vane unit. The double-vane unit is approximately twice as powerful as the single-vane unit. The single-vane unit provides approximately 280 degrees of rotational arc. The double-vane unit is limited to approximately 100 degrees of rotational arc.

(A) Cutaway of hydraulic rotary vane-type actuator with spline shaft.

(B) Hydraulic actuator with hollow shaft and internal keyway.

Fig. 24-1. Hydraulic actuator. (Courtesy Bird-Johnson Co.)

Fluid pressure enters the chamber on one side of the vane or vanes, creating a rotational force in either a clockwise or a counterclockwise direction. The fluid on the opposite side of the vane, or vanes, is permitted to exhaust from the other port. The stationary barrier separates the two chambers and the two ports. In the double-vane unit, the incoming fluid reaches the second vane through either internal passages within the shaft or internal porting in the end caps. The rotational force in the opposite direction is accomplished by admitting pressurized oil to the opposite port and to the opposite side of the vane, or vanes. If both ports are pressurized, a stop-motion condition results, which locks the rotary shaft in an intermediate position. When both ports are depressurized, a free or "floating" condition of the rotor and output shaft is created.

External design details in the vane-type rotary actuator are usually the same in both the single-vane and the double-vane versions of the actuator. The actuators are available in a variety of mounting styles, including flange, foot, and end mounting styles. Normally, actuators can be operated on either air or oil. Their operating pressure usually depends on the type of bearings, the clearances permitted within the chamber, and the type of shaft seal used. Normally, two different pieces of optional equipment are available for attachment to the external torque shaft. One of these options is an *SAE* spline. The alternate option is a keyed shaft.

The vane-type actuator is capable of rotating less than the total rotational limit of the unit. It is readily adapted to a continual oscillating motion. Deceleration within the vane-type unit is accomplished in some designs by permitting the vane to close off the port orifice gradually as it approaches its rotational limits, which gradually reduces the flow rate through the exhaust port.

The torque produced by the vane-type unit can be calculated by multiplying the area of the vane, or vanes, by the distance from the center of the shaft to the center of the vane times the pressure of the incoming fluid. One manufacturer's rating chart indicates that single-vane actuators which can produce from 1200 to 84,000 inch-pounds (135.6 to 9490.7N·m) of torque at 1000 psi (6894.8 kPa) are available. Double-vane units can produce upward to 220,000 inch-pounds (24,856.7N·m) of torque. Special actuators have been constructed to produce several million inch-pounds (newton meters) of torque.

PISTON-TYPE ROTARY ACTUATOR

Another type of rotary actuator resembles a cylinder (Fig. 24-2) in external appearance, and uses a piston in the cylinder. In this unit, the piston is guided to prevent internal rotation of the piston. The piston is driven back and forth inside the cylinder, actuating a helical rod that passes through the piston and extends through the cylinder cover for external attachment. When facing the driver end of the cylinder, counterclockwise rotation is obtained by applying pressure to the driver-end port. Clockwise rotation is obtained when pressure is applied to the idler-end port. The piston can be stopped at any intermediate point in the rotation cycle by equalizing the pressure to the two cylinder ports. The angle of the helix that passes through the piston permits the work load on the shaft extension to be held firmly—even with a complete loss of hydraulic power. The locking action of the piston on the helical rod or shaft prevents the free or "floating" characteristic that can be obtained with the vane-type unit.

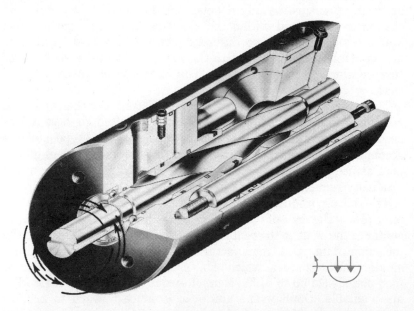

Fig. 24-2. Cutaway view of a cylinder-type rotary torque actuator. This actuator can produce up to 15,000 inch-pounds (1695 N·m) of torque, and as much as 370 degrees of rotation on standard units. (Courtesy Carter Controls, Inc.)

The construction of the *piston-type* unit with a helical shaft closely resembles that of the hydraulic cylinder. The unit contains many features of the hydraulic cylinder, including the use of cushions to decelerate the stroke of the piston and thereby decelerate the rotational speed of the external shaft. The deceleration cushions are adjustable in the same manner cushions of the hydraulic cylinder are adjustable. The seals used in this unit are standard seals. Only the seal between the piston and helical shaft is specially designed to fit the opening through the piston and the nonsymmetrical rod.

The torques delivered range from nearly zero upward to 15,000 inch-pounds (1694.8N·m) of torque, with maximum pressure ratings at 300 psi (2068.4 kPa). The actuator shown in Fig. 24-2 is available from the manufacturer with built-in valving, attached pumps, motor drive for the pump, and makeup reservoir chamber. If the unit is furnished with valves only, the valves are the pilot-operated type, and can be controlled by remote poppets or solenoids. A variety of mountings are available, including foot mounting, front or rear flange mounting, and special mountings made to order to fit a specific mounting detail.

Motion occurs when the patented helical piston applies more torque to the shaft than the external torque load and the inherent sliding or static friction between the piston and shaft provide in resisting rotation. The cushion construction (see Fig. 24-2) is used only for air service. The effective cushioned length of rotation is through approximately 22 degrees of arc. The tapered cushion noses on the pistons in hydraulic units provide an effective cushioning length through approximately 31 degrees of arc of the output shaft. Special shaft extensions, double shaft extensions, and nonstandard rotation lengths are available from the manufacturer.

The number of degrees of rotation of this unit (see Fig. 24-2) is determined by the total length of the unit. Standard units provide for 100, 190, 280, or 370 degrees of rotation. The chief difference between the units is in overall length.

Two pistons, each containing a rack, drive the pinion in the rack and pinion-type actuator in Fig. 24-3. The actuator may be operated either by compressed air or by a hydraulic medium. To cause the pinion to rotate counterclockwise, pressure is applied to the port in each end cover at the same time. The fluid exhausts from the center section, as shown in Fig. 24-3B. To cause clockwise rotation of the pinion, pressure is applied to the port in the center section. The fluid

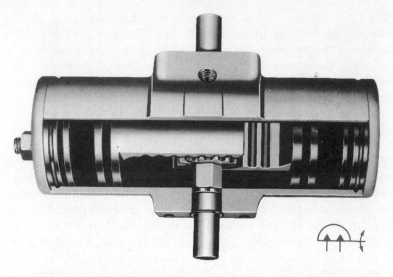

(A) Actuator with rotational stroke adjustor.

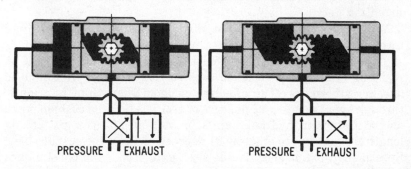

(B) Operating sequence.

Fig. 24-3. Rack-and-pinion type of actuator. (Courtesy Ohio Oscillator Co., Orrville, Ohio)

exhausts from the port on each end cover. To limit the rotation of the pinion, a stroke adjustor is built into the end cover, as shown at the left in Fig. 24-3A.

An actuator in which the rack shoe is connected to both pistons of the actuator is shown in Fig. 24-4. In this type of rack-and-pinion actuator, standard rotations of 100 degrees, 190 degrees, 280 degrees, and 370 degrees may be obtained. Greater amounts of

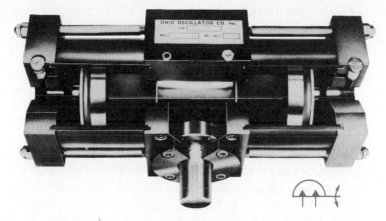

(A) Cutaway view.

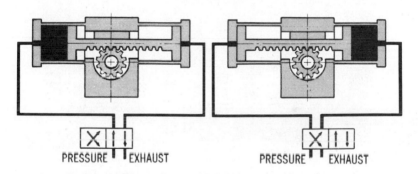

PRESSURE EXHAUST   PRESSURE EXHAUST

(B) Operating sequence.

Fig. 24-4. Rotary actuator with cushions to provide a smooth stop. (Courtesy Ohio Oscillator Co., Orrville, Ohio)

rotation are available as a special order. While this type of actuator can be obtained for either air or hydraulic service, those actuators operating with hydraulic pressure can produce as much as 200,000 inch-pounds (22,596.9 N·m) of torque. The seals shown on the piston are lip-type seals which provide zero leakage past the piston.

The pinions are designed with a high-strength tooth form and are made of high-tensile, heat-treated alloy steel. In addition to the pinion shaft extension shown in Fig. 24-4A, spline shafts are also available. Hollow shafts with an internal keyway are sometimes used. To limit the stroke of the pistons, a stroke adjustor may be included in the end covers. Cushioning is also available (see Fig. 24-4A).

The diagram in Fig. 24-4B shows how the operating sequence can be obtained. When pressure is applied to the left-hand end of the actuator, the pinion rotates clockwise. When pressure is applied to the right-hand end of the actuator, the pinion rotates counterclockwise. If a three-position valve is used in place of the two-position valve shown, the rotation can be stopped at any point.

To create a great amount of torque in a relatively small space, an actuator can be used (Fig. 24-5). This actuator has two racks which actuate one pinion (Fig. 24-5B). There are four pistons. Two pistons are connected to each rack. The actuator has stroke adjustors (see Fig. 24-5A) to accurately limit the stroke. Actuators of this type are

(A) Adjustable actuator with two racks and four pistons.

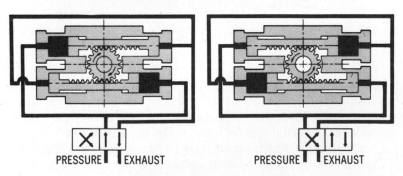

(B) Operating sequence for actuator.

Fig. 24-5. Rotational actuator. (Courtesy Ohio Oscillator Co., Orrville, Ohio)

capable of producing as much as 600,000 inch-pounds (67,790.9N·m) of torque.

In another design of actuator, two cylinders are linked together by a chain drive that passes around and imparts rotational force to a sprocket connected to a shaft.

In still another type of actuator, two opposed cylinders are used to drive a slotted lever around a shaft pivot. In comparison with other units, the degrees of rotation are limited. The torque factor changes slightly as the drive pin changes the effective radius of the oscillating lever.

In general, the actuator shaft bearings are not recommended for external radial, thrust, or bending loads, unless special bearing provisions are made. Most manfacturers recommend that a flexible coupling should be placed between the actuator and the load. Actuators that are designed to operate on petroleum-base fluids can often be modified for use with fire-resistant or synthetic fluids or water. But, the manufacturer should be contacted before using these fluids.

Chapter 25

Torque Devices—
Application

Rotary actuators or torque devices are used in many widely diversified applications and application methods. The resulting movements vary from a simple rotational movement to conversion to a straight-line movement. Rotary actuators are used in automation equipment and in machine tools which are used for indexing, clamping, feeding, locating, transferring, lifting, positioning, and turnover operations. Actuators are used to open and to close valves which sometimes require considerable force. They are used on submarines; on ships, to open and close the hatches; on conveyors, to control the movement of materials; on machinery, to bend pipe and tubing; and in vats and tanks, for mixing and agitation. Actuators are also used to open and close massive doors and to swing booms or other crane members. The list of practical applications is limited only by the imagination of the design engineer.

Actuators vary widely in shape and in package dimensions, as described in the preceding chapter. But, in general, they pack much power in a comparatively small package. The rotating motion of the output shaft often eliminates the need for complicated linkages.

Actuators can be adapted readily to double or tandem operation merely by attaching a unit to each end of a shaft which is to be rotated. Since the working mechanism of the unit is enclosed and usually sealed to prevent the entrance of foreign matter, the undesirable exposure of moving linkages can be reduced considerably.

When the torque actuator is applied to a machine application, loads other than the pure torque loads should be avoided. Overhung or radial loads applied externally create undue stresses and wear on the rotating shaft and its bearing. This decreases the normal service life of the unit. Therefore, close-coupled connections to the output shaft should be incorporated in all applications. The alignment of the output shaft to the machine member should be nearly perfect, if this is possible and practical. Most of the common applications in industry require the torque output of the actuator to be delivered to another shaft, gearing, cam, lever, or other type of mechanical linkage. The driven connections have a thrust or overhung load as a load-transmitting component in most instances. Mechanical parts should be mounted between pillow blocks, on stub shafts, or on other means if they are to receive the total torque output of the actuator. The external pillow blocks should be used in a way that permits them to absorb all the radial loads within the machine shaft, thereby isolating them from the actuator shaft assembly.

The torque device or rotary actuator can, in general, be mounted in any position—vertically, horizontally, or at an angle. In the most common mounting style, the rotary actuator unit is stationary and the shaft is rotating. However, the units can be applied with the shaft stationary and the body rotating, remembering that the limitation is in making the fluid connections to the actuator ports.

ROTARY MOTIONS

After the load requirement has been determined, it is good engineering practice to select a rotary actuator with a capacity 20 to 50 percent higher than the load requirement. Since the actuator operates on the principle of pressure differential across the moving vane or piston, the actual movement of the output shaft attached to the load depends on oversizing to create the pressure differential. Motion of the shaft and load occurs when the applied torque of the actuator exceeds the resisting load force. The velocity and acceleration provided the mass are proportional to the excess torque or force, and can be determined by the formula:

$$A = \frac{Force}{Mass}$$

After the load or mass has been set in motion, it must be stopped or decelerated, which requires an opposing force $(F = MA)$. The deceleration force usually can be obtained by restricting gradually the flow of fluid from the actuator. This permits pressure to build up on the opposite side of the actuator vane or piston.

Severe shock and possible damage to the rotary actuator and the system can be caused by abruptly restricting the flow of the outgoing fluid, which generates fluid pressure surges or shock-wave pressures within the actuator or system. The shock-type pressures easily can exceed the pressure rating of the actuator unit, unless care is taken to control them properly.

The shock and deceleration problem can be anticipated and dealt with by the use of deceleration valves in the piping system. These valves can be actuated mechanically by a cam or other means. They should be used in conjunction with pressure limiting valves to restrict the fluid flow gradually and to control the generation of excessive pressures in opposing the mass inertia. If cam-operated valves are used for deceleration, the shape and angularity of the cam should provide a gentle ramp transition to prevent the development of sudden pressure surges. A pressure-limiting valve arrangement can be used for deceleration.

The amount of energy and the rate of dissipation of energy required to stop a moving mass can be calculated if the variables, such as velocity, mass, time, pressure, viscosity, etc., can be determined (Fig. 25-1). In actual working circuits, these factors are interrelated. Their positive calculation is quite complex. Good general practice requires that more cycle time be permitted for deceleration of rotational loads than for acceleration of the same load.

Lifting a mass through an arc, moving it across the center point, and returning the mass to its original level in an alternate position is diagrammed in Fig. 25-2. The effective radius of the torque device in relation to the vertical lift varies, depending on the rotational position of the mass. It is zero at the vertical position. Therefore, the torque required to lift the weight W decreases from maximum torque

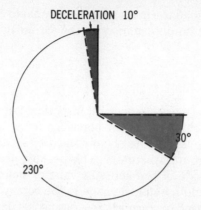

Fig. 25-1. A mass accelerated uniformly through 260 degrees of arc at 500 psi
(3447.4 kPa) generates a destructive opposing pressure of 13,000 psi (89,631.8 kPa)
to dissipate uniformly the kinetic energy within 10 degrees of arc. If the same mass
is accelerated uniformly through only 30 degrees of arc at 500 psi (3447.4 kPa) and
then moves at a constant speed through the succeeding 230 degrees of arc, it can be
decelerated uniformly in the final 10 degrees of arc by an opposing pressure of 1500
psi (10,342.1 kPa) by means of flow control valves.

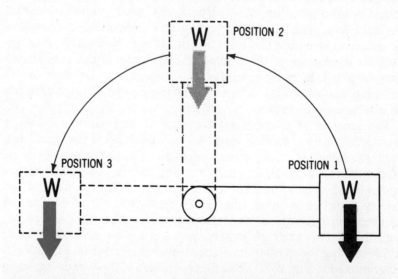

Fig. 25-2. Diagram illustrating the lifting of a mass through an arc, moving it
across the center point, and returning the mass to its original level in an alternate
position.

at position *1* to zero at position *2*. The required torque from the actuator then reverses to a point where the weight of the mass aids the rotation during its travel from position *2* to position *3*. In this instance, restriction of the flow of fluid and control of the deceleration pressure are quite essential. This illustrates what is commonly referred to as "overcentering" a load.

The rotary actuator with its inherent built-in arc limitation lends itself to rather simple methods of increasing or decreasing the moment of arc and to increasing or decreasing the torque at a given pressure. Since the movement of the output shaft is a rotational movement, one of the most common methods of altering both the arc and the output torque is to transfer the motion and the torque force to a second shaft through a meshing gear. Ignoring frictional losses in this type of gear linkage, the product of distance and force within the actuator is equal to the distance multiplied by the force of the driven member. The force decreases as the distance increases and vice versa.

STRAIGHT-LINE MOTIONS

The use of the rotational force and the movement of the rotary actuator to obtain straight-line motions can produce interesting variations in both force and speed (Fig. 25-3). The application of the rotational movement and force through a double-lever arrangement to create straight-line motion for clamping is illustrated in the diagram. As the double-lever arrangement approaches a straight line, the straight-line thrust force of the clamping block increases rapidly. Its movement decreases at the same rate that the force increases. The results of this type of movement and linkage are a rapid approach and a harmonic reduction in linear speed.

The use of a rotary actuator with a matched parallel linkage attached to a lifting platform is diagrammed in Fig. 25-4. Although

Fig. 25-3. The application of rotational movement and force through a double-lever arrangement to create straight-line motion in a torque device for a clamping action.

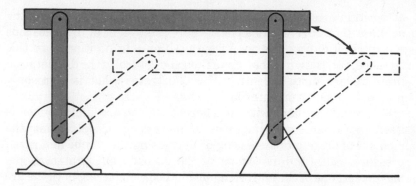

Fig. 25-4. Diagram demonstrating the use of a rotary actuator with matched parallel linkage attached to a lifting platform.

the velocity of the radial movement of the lever arm and the platform mass can be constant, the vertical movement of the platform gradually decelerates to zero as its maximum elevation is approached.

The use of a rotary actuator with an attached lever arm and ratchet arrangement to produce intermittent motion in only one direction is shown in Fig. 25-5. The fixed ratchet pawl functions as an antibackup device to hold the rotational load during the return cycle of the actuator arm. The intermittent motion can be controlled for repeating automatically. Or it can be triggered by an external signal which causes it to rotate on demand.

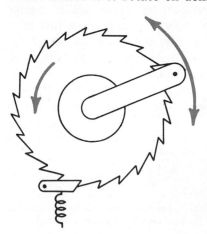

Fig. 25-5. Diagram demonstrating the use of a rotary actuator with an attached lever arm and ratchet arrangement for producing intermittent motion in only one direction.

The continuous-oscillation feature of the rotary actuator is ideal for agitating or mixing volatile fluids and other fluids in explosive atmospheres. Since both torque and the rotational speed can be controlled and varied in the rotary actuator, a definite advantage over the electric motor drive is provided, especially when viscosity changes occur in the fluid being mixed or agitated. Like the hydraulic motor, the rotary actuator can withstand sustained stalling without damage to its mechanism. In controlling all external masses—especially in slowing down or stopping a mass—resilient external bumpers should be provided at the extreme limits of rotation on rotary actuators to absorb the tremendous inertia loads.

In general, the vane-type units are more subject to internal slippage or clearance flow past the vane, because it is difficult to achieve a perfect seal within the chamber. In some applications, this slippage or "softness" can be an advantage. The piston-type unit with the helical shaft (see Fig. 24-2) does not permit slippage or overtravel of the external load. External torque loads applied to the output shaft in these units cannot produce rotation, even if a complete power failure should occur. In many applications, this is definitely an advantage. However, in other applications it can be a handicap—for example, a device that moves and positions a load that requires overriding the fluid force of the actuator. If the external overriding feature is required in the application, the vane-type unit provides a distinct advantage.

The piston-chain type of actuator is similar to the piston-rack type of unit. The operating characteristics of these units are similar to the operating characteristics of most of the common types of cylinders.

Chapter 26

Maintenance of Force Generators

The term *force generator* refers to any component that performs work as a direct result of pressurized fluid having been channeled to it. This includes the components discussed in the preceding chapters.

Each type of force component is vulnerable to misapplications and improper maintenance practices. The component parts most susceptible to wear or damage are the seals, bearings, and bearing surfaces. Although these parts vary slightly from one force component to the next, the maintenance problems are identical in both rotary motion and linear motion. Seal designs differ between linear force components and rotary force components. A wide variation in seal designs exists within each class of force components.

It is a safe assumption that wherever movement is encountered, wear and deterioration result. In the "care and feeding" of force components, the nature, degree, location, and frequency of wear are factors that indicate the probable causes of the wear. Careful examination of these wear factors often leads to simple corrective measures which can produce a pronounced increase in the service life expectancy of the component. Like the automobile, fluid power components can, with proper care, deliver many hours or years of productive service. Improper care, mistreatment, and poor maintenance can shorten appreciably the service life of a machine and result in an unsatisfactory performance.

The most frequent causes of damage and excessive wear of force components are dirt particles (internal and external), excessive heat, improper lubrication, misapplication, and misalignment. Many of the factors that cause malfunctions can be determined by careful observation of the external parts of the component. These factors should be looked for in routine inspections. Corrective measures should be undertaken immediately on their discovery.

NONROTATING

A cross-sectional diagram of a typical hydraulic nonrotating cylinder can be seen in Fig. 26-1. As in other types of force components, the cylinder contains two important types of seals which are termed *static* seals and *dynamic* seals. The static seals are used to seal mating, but nonmoving, members of a component assembly. The static seals shown in the diagram include the "O"-ring cover gasket, the "O"-ring piston seal, and the "O"-ring rod bearing seal.

The dynamic seals prevent the passage of fluid between the moving members in the component. In the diagram (see Fig. 26-1), the dynamic seals are the piston rings and the piston-rod packing. The resilient type of piston seal is shown in Fig. 26-2.

Dirt particles borne in the fluid media can collect at points of minimum clearance, such as between the piston rings and the cylinder wall. As discussed previously, these small clearances can function as filters, separating the dirt particles from the media and trapping them at the clearance points. In ring-type pistons, this results in an accumulation of dirt particles in the ring grooves. This accumulation can lock the piston rings, which prevents their expanding freely against the cylinder wall. Excessive accumulations of dirt particles in the piston-ring grooves can cause pressure loading of the rings against the cylinder wall, resulting in excessive wear of the cylinder wall.

When an accumulation of dirt particles, foreign matter, or abrasive material occurs in the cup-type piston construction, the result can be severe abrasion and wear of the cylinder wall. As deterioration of the cylinder wall progresses, further wear is contributed to—even when

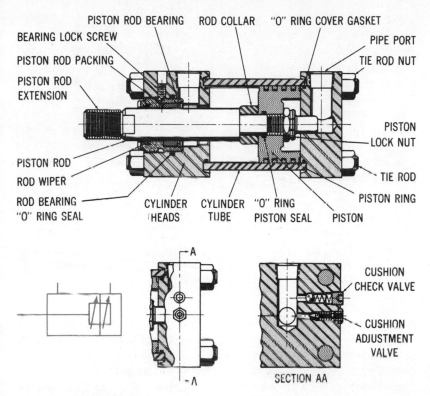

Fig. 26-1. Parts diagram for a heavy-duty hydraulic nonrotating cylinder. (Courtesy Logansport Machine Co., Inc.)

the resilient piston seals are replaced. The result is a cylinder that "chews up" soft packings with considerable frequency.

When this type of wear is discovered in time, it is sometimes possible to hone the inside surface of the cylinder tube. However, if too much wear has occurred, the amount of honing required to remove all the grooves and scratches can result in too much clearance between the piston seal and the cylinder wall. The cylinder tube should be replaced when this happens.

Normally, static seals inside the force component do not exhibit the type of wear that often occurs with the dynamic seals. If the static seals are "O" rings, damage is usually caused by excessive pressure, improper assembly of the component parts, excessive heat, or noncompatible fluids. If the static "O" rings are exposed to excessive

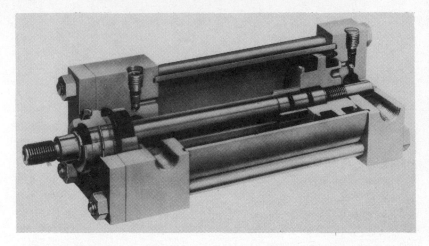

Fig. 26-2. Cutaway view of a pneumatic double-acting cylinder with resilient-type piston seals. Note that the cushion collar is sealed on the inside diameter (ID) by an O-ring. (Courtesy Schrader Fluid Power Division)

heat, a hardening or "curing out" of the rubber compound usually results. The presence or use of noncompatible fluids within the force component can cause the static seals to swell, shrink, harden, soften, or deteriorate completely. If this type of deterioration or damage is discovered within a force component, compatibility charts should be consulted to determine the type of rubber compound that should be used.

The rod wiper, or rod scraper, cleans the dirt particles and foreign matter from the piston rod to prevent their being drawn into the cylinder bearing or chamber. However, excessive accumulations of foreign matter should not be permitted to remain within the recess of the rod wiper, since their presence can result in a progressive abrading action on the piston rod.

If an examination of the piston reveals appreciable wear in an uneven or irregular pattern and if the wear has occurred primarily on one side of the piston rod, misalignment or improper mounting of the cylinder is the probable cause. If the cylinder is mounted so that the movement of the work load is not precisely parallel with the movement of the piston rod, a distorted side loading is imparted to the piston rod and its bearing.

In the hydraulic cylinder diagrammed in Fig. 26-1, all the moving parts are lubricated by the fluid medium. The cylinder shown in Fig. 26-2 is a pneumatic cylinder. It is necessary to add a lubricating agent to the air medium that powers it. The air line lubricator discussed in Chapter 6 is the device for accomplishing this type of lubrication. In the small-bore, short-stroke air cylinders, it is sometimes difficult to transfer the lubricant from the air line lubricator into the cylinder. This is a problem, because an extremely small flow of air is involved, which results in difficulty in picking up and transmitting the lubricant into the cylinder chamber. When this problem is discovered, steps should be taken to provide an injection-type lubricating device or other method of ensuring lubrication of the cylinder.

Misalignment of a cushioned cylinder is more critical than misalignment of a noncushioned cylinder, because close fits are involved in mating the cushion nose with the cushion well in the cylinder cover. Sometimes, an external examination can detect this condition. With the piston detached from its work load, excessive play of the piston in a radial direction is an indication of wear in the rod bearing. Excessive play permits misalignment of the internal cushion nose. When misalignment is detected, the piston-rod bearing should be replaced.

Some excellent suggestions for dismantling, repairing, and assembling a nonrotating cylinder are:

1. Dismantle the cylinder in a clean location. If it is a hydraulic cylinder, drain all the oil from the cylinder. Do not try to dismantle a cylinder when there is pressure to it.
2. Clean each part. If the cylinder is to be dismantled for any length of time, coat the metal parts that are to be reused with a good preservative and place in protected storage.
3. Check the piston rod for straightness. If the rod is bent, place it on vee blocks in a press, and carefully straighten it.
4. Examine the piston rod for scratches, scores, indentations, and other blemishes. If these blemishes are not too deep, remove them with a fine grade of emery cloth. If it is necessary to grind the rod, it is suggested that it be chrome plated afterward to restore the diameter to its original size.
5. Examine cover bushings and cushion bushings for wear and finish. If they are not in first-class condition, they should be replaced.
6. If the cylinder tube is damaged, either repair or replace it.

Deep scores are difficult to repair. It may be necessary to chrome plate the tube.

7. In reassembling a cylinder, it is suggested that all the seals and gaskets be replaced. When metal piston rings are used, check the cylinder manufacturer's specifications for gap clearance. If synthetic or leather seals are used on the piston, be extremely careful in placing the piston into the tube, so that sealing surfaces are not damaged. Use a light grease on packings when installing them. This makes them much easier to assemble.

8. If, for any reason, a metal piston with rings is to be replaced, be sure to grind the piston concentrically with the piston rod after it has been assembled to the rod. Grind the piston to fit the tube closely.

9. Cylinders that have foot-mounted covers should be assembled on a surface plate. The mounting pads of both covers must make full contact with the surface plate. Otherwise, a binding action may occur when the cylinder is mounted, or a mounting foot may be broken.

10. Always tighten the cover bolts evenly. If "O"-ring or quad-ring gaskets are used to seal the tube and cover, tension on the cover screws can be reduced to a minimum.

11. After a cylinder has been assembled completely, test it at low operating pressure to make certain that the piston and rod are moving freely and that they are not being scored or bound. Then increase the pressure to its full operating range. Check for both internal and external leakage. To check internal leakage, place fluid pressure on the blind-end cylinder port. Force the piston to the rod end. Then check the amount of fluid that comes out of the rod-end cover port. Next, place fluid pressure in the rod-end cylinder port, move the piston to the blind-end cover, and check leakage at this position. If leakage is to be checked at other positions of the cylinder, block the ports at this position by external means and check as above. If excessive leakage occurs, it is necessary to disassemble and make corrections. If the piston is sealed with synthetic or leather seals, the leakage should be nearly nil. If metal piston rings are used in a hydraulic cylinder, the leakage will vary with the operating pressure, the cylinder bore, and the oil temperature and viscosity.

12. When returning a cylinder to a machine or fixture, make

certain that it is mounted securely. Remember that these cylinders are capable of delivering great force.

In some pneumatic cylinders, the seal between the cylinder tube and the end cover is a flat gasket. These cylinders should be reassembled with extreme care, being careful not to draw the rod nuts down to the point where they damage or extrude the flat gaskets. In the hydraulic cylinders, especially those with the larger bores, the seal between the static cylinder tube and the end cover is provided for either on the inside diameter or on the outside diameter of the cylinder tube. In these cylinders, the tie rods are normally torqued to a prestressed tension to prevent their elongation when the cylinder assembly is subjected to pressure. Then it is good practice to follow the manufacturer's recommendations on the torque to be applied to the tie-rod nuts.

In reassembling a cylinder, a common problem is encountered in placing the rod packing over the end of the piston rod. This can be extremely critical if the diameter of a male thread on the rod end is equal to the diameter of the piston rod. When this problem is encountered, thin shim stock can be wrapped around the threads on the rod. Then the rod packing or rod packing gland can be passed over the shim stock. This prevents damage to the sealing lips of the rod packing in passing over the thread crests. Again, a lubricant applied at this point helps considerably.

For an "O" ring to effect a proper seal, even in a static sealing situation, a compression or squeezing of the "O" ring in its cross-sectional diameter is necessary. When mating parts are sealed in this manner—as in sealing the cylinder tube to the end cover, for example—extreme care should be exercised in assembly. Carelessness at this point can result in cutting or shaving the "O"-ring seal, resulting in possible leakage. A liberal application of grease and an unhurried working pace can facilitate this type of assembly. Gentle pressure of the parts against each other permits the gradual "flow" of the "O" ring into its proper position.

Occasionally, the cushioning action of a cylinder apparently ceases to function. If this occurs, the bypass ball check, which is located in the cylinder cover, and its seat and spring should be examined for breakage or damage. The cushion-adjusting needle also should be examined carefully for breakage or deep scarring that may permit a bypass of fluid (Fig. 26-3). Also check cushion collar and nose for damage due to dirt in the system.

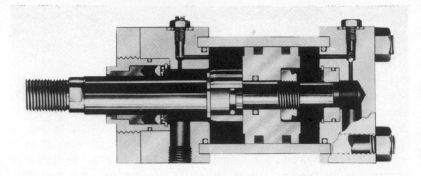

Fig. 26-3. Cutaway of double-acting cylinder showing the cushion needle and the ball check. Note the cushioning design in the rear cover. (Courtesy Carter Controls, Inc.)

ROTATING

A cross-sectional view of a rotating cylinder is shown in Fig. 26-4. The preceding observation pertaining to the nonrotating cylinders, in general, apply to the rotating cylinder. The chief difference is in the assembly details and in the distributor that delivers the fluid medium to the cylinder chambers. Misalignment or damage to a rotating cylinder can result in an out-of-balance condition which can lead to further damage to the component. In the rotating cylinder, the entire cylinder and shaft rotate with the spindle to which the cylinder is attached. Only the shaft or distributor body does not rotate. Therefore, the shaft-stem seals are rotating-type seals. The manufacturer's recommendations on lubrication of the rotating seals should be followed explicitly.

Since the design of rotary actuators varies more than the design of the other force components, careful adherence to maintenance instructions furnished by the manufacturer is required. The sealing materials used in rotary actuators are susceptible to the same types of wear and abuse that affect the seals in other force components.

In general, force components—especially rotary actuators—should be provided additional rust or corrosion protection when stored for an extended period of time. When this type of storage is anticipated, it is good practice to remove the port plugs, fill the chambers with clean mineral-base oil (or other fluid compatible with the sealing compounds), and replace the port plugs securely. All

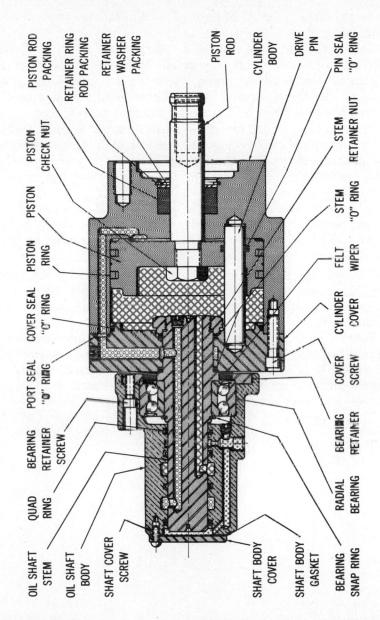

Fig. 26-4. Parts diagram for a hydraulic rotating cylinder. (Courtesy Logansport Machine Co., Inc.)

external surfaces should be covered with adequate corrosion-resistant material.

In the application of load to a rotary actuator, a semiflexible coupling and load shaft supported by separate bearings are recommended. A similar arrangement is recommended for power transmission through gears. This prevents gear load and separating forces aggravating or adding to the actuator bearing load.

If a flexible coupling cannot be used, accurate alignment of the actuator and associated equipment is essential to prevent additional actuator bearing loading. Usually, a pilot boss is provided on the face of the actuator to permit a close-fitted mating to the equipment to be driven.

End thrust or axial loading of the actuator shaft is not advisable. A thrust bearing should be provided to absorb end thrust from the load. It should be driven through a sliding spline-type coupling which minimizes internal wear of the actuator.

The service life of a force component depends on the speed, pressure, load, and motion characteristics. Proper maintenance practices result in longer and more satisfactory service with minimum repair costs.

Parts diagrams and an assembly drawing of a force component should be obtained from the manufacturer before its assembly or disassembly is attempted. Manufacturers' recommended lists of repair parts and replacement parts are usually based on actual field experience. They are usually accurate (Fig. 26-5).

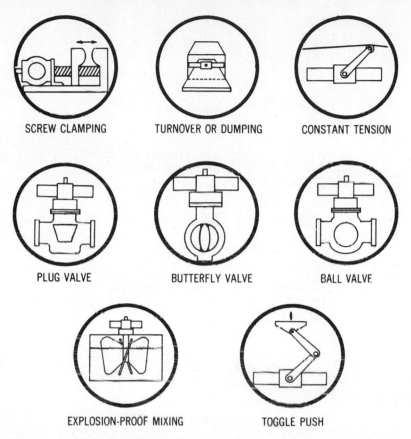

SCREW CLAMPING   TURNOVER OR DUMPING   CONSTANT TENSION

PLUG VALVE   BUTTERFLY VALVE   BALL VALVE

EXPLOSION-PROOF MIXING   TOGGLE PUSH

Fig. 26-5. Some of the practical applications in which torque is required to perform certain functions over a limited travel. Rotary actuators are ideal for this type of application. Other applications for rotary actuators are found in industry; in marine service; and on mobile equipment in such applications as door openers, torque testers, power feeding, tube bending, transferring, braking, compacting, boom swing, steering, revolving of gun turrets, and many other uses. (Courtesy Ohio Oscillator Co., Orrville, Ohio)

Chapter 27

Pumps and Motors— Design

Many different types of fluid pumps are now in use. However, relatively few of these types are commonly used to develop pressures for hydraulic power-transmission systems. Some of these pumps are: (1) Reciprocating piston; (2) Gear; (3) Vane; (4) Centrifugal; and (5) Rotary piston pumps. Hydraulic pumps also can be classified as to their type of delivery, as either constant-displacement or variable-displacement pumps.

HYDRAULIC PUMPS

The *reciprocating pumps* are characterized by a reciprocating piston action (Fig. 27-1). The diagram shows a double-acting type of piston. This type of pump can be devised easily from a double-acting hydraulic cylinder using external check valves. The inherent pulsating nature of the discharge from this type of pump is objectionable in some applications. Pumps having three or more cylinders on the same shaft are designed to reduce pulsations in flow

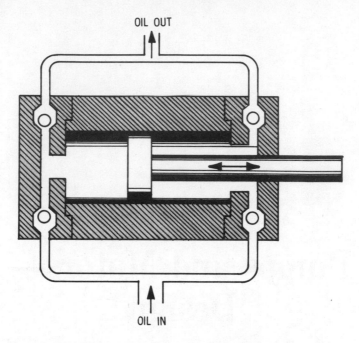

OIL OUT

OIL IN

Fig. 27-1. A reciprocating pump with a double-acting type of piston.

by overlapping the compression strokes of each cylinder. A three-piston pump is shown in Fig. 27-2.

Most reciprocating pumps are of the fixed-displacement type. One type of pump is provided with a stroke transformer which permits a variation in fluid volume by varying the length of the piston stroke. The pressure booster or intensifier is another unit with a reciprocating piston. This unit was discussed in detail in an earlier chapter.

A *gear-type* pump is a rotary pump which consists of two meshing gears. It is used widely in hydraulic applications. Gear-type pumps are constant-delivery pumps. Their delivery rate can be varied only by changing the speed of the pump. In the *external gear-type pump* (Fig. 27-3), the oil is carried in the space between the teeth and the casing to the outlet side of the pump. Then, it is forced out of the space between the teeth by the meshing action of the gears. The efficiency of this type of pump is determined largely by the accuracy with which the component parts are fitted.

Fig. 27-2. Piston-type pump in which parallel pistons combine to produce unusually high pressure (as high as 60,000 psi, or 413,685.4 kPa).

Three different types of gears are used in the external gear-type pumps. They are: (1) Spur gears (Fig. 27-4); (2) Helical gears (Fig. 27-5); and (3) Herringbone gears (Fig. 27-6). The pressure ratings of external gear-type pumps range from the lower pressures to as high as 3000 psi (20,684.3 kPa) in some models. In pumps that are rated in the higher pressure ranges, the gears are fitted to closer tolerances, resulting in less tolerance for foreign matter in the fluid being pumped. Helical gears can handle large volumes of fluid at higher speeds than the spur gears. However, they have a tendency to develop excessive end thrust. Herringbone-gear designs have overcome this disadvantage. They also deliver oil with less pulsation, since a larger quantity of oil is carried in the V-shaped center portion of the gear surfaces.

Some of the oil is trapped between the teeth of gear-type pumps at the moment the teeth begin to mesh. As the gear teeth continue to mesh, a pressure is created which tends to oppose the motion of the pump. To overcome this disadvantage, some gear-type pumps are designed with small passages extending through and between the gear teeth to permit the escape of the trapped oil. Other pumps have grooved teeth to accomplish the same purpose. Other adaptations of the gear-pump principle use either lobe-shaped or screw-form rotating elements to carry the fluids through the housing. Then the fluids are forced out of the housing by a meshing action.

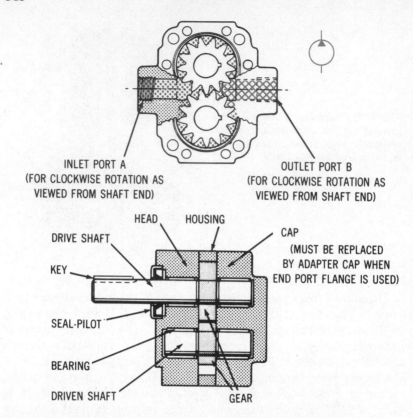

Fig. 27-3. Cutaway diagram of an external gear-type rotary pump. (Courtesy Brown & Sharpe Mfg. Co.)

Fig. 27-4. Constant-delivery positive-displacement gear-type pumps with spur gears. (Courtesy Brown & Sharpe Mfg. Co.)

Fig. 27-5. Sectional view of a rotary gear-type pump with helical gears. (Courtesy Brown & Sharpe Mfg. Co.)

The *internal gear-type pumps* are a modification of the gear-pump principle (Fig. 27-7). As in the external gear-type pump, the oil moves from the suction port to the discharge port by entrapment action between the meshed teeth of the rotating gears. Power can be applied either to the outer ring gear or to the inner gear.

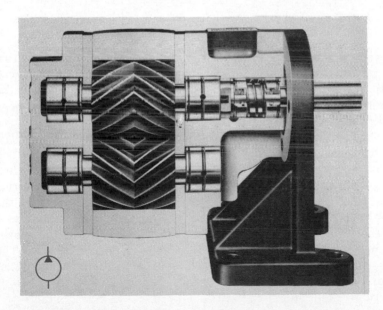

Fig. 27-6. Sectional view of a rotary gear-type pump with herringbone gears. (Courtesy Brown & Sharpe Mfg. Co.)

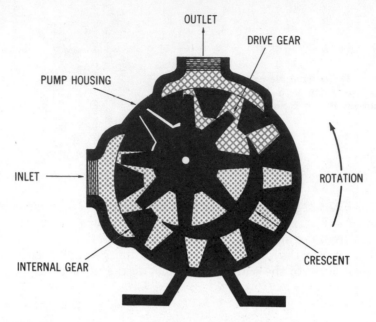

Fig. 27-7. Diagram of parts of an internal gear-type pump.

The crescent-shaped block acts as a divider between the suction and the discharge portions of the fluid. As the teeth move out of mesh near the suction port, the oil is drawn into the space between the teeth. The oil remains in the space until the teeth are again in mesh near the discharge port to force the oil outward.

The *Gerotor-type pump* shown in Fig. 27-8 is a variation of the internal gear-type pump. In this design, a small external-toothed gear is placed eccentrically within a larger internal-toothed gear—with the small gear having one less tooth than the external gear. The teeth of the smaller gear are passed continually by the teeth of the external gear, with a sealing point of contact at the crests of the teeth. Here again, the fluid is drawn inward as the spaces increase in size. It is expelled as they decrease in size. The cresent-shaped piece is not needed to separate the inlet and outlet chambers, unless the difference in the number of teeth is two or more.

In *vane-type pumps*, a cylindrical rotor with movable vanes in radial slots rotates in an oval-shaped housing. The outer edges of the

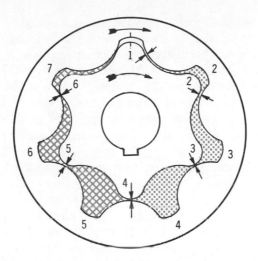

Fig. 27-8. Schematic diagram of the Gerotor mechanism. (Courtesy Double A Products Co.)

vanes are always in contact with the inner surfaces of the housing (Fig. 27-9). The vanes are usually held in contact with the housing by means of lightweight springs located behind each vane. In one design variation, a lightweight spring load prevents the vanes from

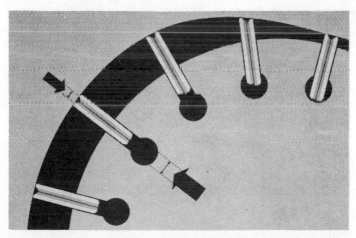

Fig. 27-9. The outer edges of the movable vanes in the radial slots are always in contact with the inner surfaces of the housing. (This illustration is the courtesy of Parker Hannifin Corporation.)

contacting the housing until a preset rpm is attained. Then the vanes are forced into a seal or contact with the housing by means of centrifugal force. This feature permits starting against minimum resistance from the pump. As the vanes move from the point where the rotor and the housing are in closest contact, the oil from the intake port is swept into an enlarging space between the rotor and the housing. As the vanes pass the point of greatest separation between the rotor and housing, they sweep and compress the oil into the outlet ports of the pump. The pump shown in Fig. 27-10 is in hydraulic balance, since the two intake and outlet ports are diametrically opposite each other.

In another design of vane-type pump (Fig. 27-11), the rotor is mounted eccentrically with respect to a circular housing. In this design, the pump is provided with an adjustable pressure chamber which can be shifted in relation to the rotor. The eccentricity of the chamber with respect to the rotor determines the volume of oil that

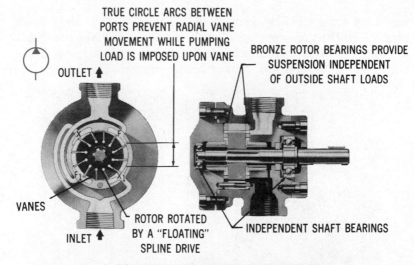

Fig. 27-10. Sectional view of a vane-type pump. (Courtesy Sperry Vickers Division of Sperry Rand Corp.)

Fig. 27-11. Variable-volume vane-type pump. (Courtesy Rexnord Inc., Hydraulics Components Div.)

is delivered. Therefore, vane-type pumps can be either constant-delivery or variable-delivery pumps.

Vane-type pumps, because of their design, can handle large volumes of oil. The average vane-type pump normally produces up to 1000 psi (6894.8 kPa). Pumps that are designed for higher pressures are usually provided with an internal means for draining slippage or leakage to the low-pressure side.

Rotary piston-type pumps are of two types: (1) Radial, and (2) Axial. Normally, they deliver pressures of 3000 psi (20,684.3 kPa) or higher.

The *radial piston-type pump* consists of a cylindrical element rotating around a fixed spindle or pintle (Fig. 27-12). The cylinder rotor is constructed with a number of radial bores that are fitted with pistons which can move inward or outward as the rotor turns. The axial holes in the pintle are provided with slots or ports which connect with the inner openings in the cylinder bores. A rotor and its support, called the slide block, move eccentrically with respect to the cylinder rotor and its spindle. The radial pistons are provided at their outer ends with shoes that bear against a freely rotating ring which is called the reaction rotor and is concentric with the circular housing.

As the cylinder rotor turns, each piston moves outward as its

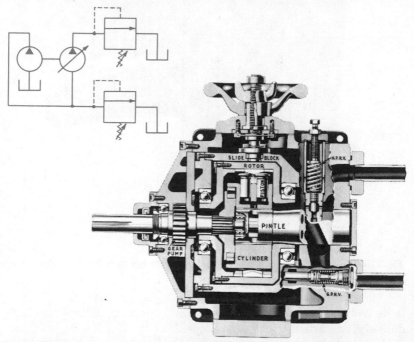

Fig. 27-12. Sectional view of a radial variable-delivery, piston-type pump. In this unit, delivery is adjusted by changing the eccentricity between the slide-block rotor assembly and the cylinder block. The gear-type pump in the end-bell housing on the left-hand side provides pressure for operating the hydraulic controls and for supercharging the system. (Courtesy The Oilgear Co.)

cylinder passes the suction ports in the spindle, drawing in oil. As the cylinder passes the discharge ports in the spindle, the piston is forced inward by the pressure of the reaction rotor on it, thereby forcing oil under pressure into the hydraulic system. The oil which leaks from the high-pressure side to the low-pressure side of the pump is drained internally by means of a drain passage connected to the suction chamber or, externally, to the oil reservoir.

The cylinder rotors are provided with an odd number of pistons. Thus, not more than one cylinder is blocked completely by the spindle at any time. This construction results in a more even delivery of oil.

Radial piston-type pumps are available in both constant-delivery and variable-delivery models. The volume and pressure of the hydraulic oil in the variable-delivery pump are controlled by means of the slide block, which increases and decreases the eccentricity of the reaction motor with respect to the cylinder rotor. This changes the length of stroke of the pistons.

The eccentricity of the cylinder rotor (which governs rate of delivery) may be controlled by different methods. Manual control by means of a handwheel is the simplest method. For automatic control of delivery to accommodate variable-volume requirements during the operating cycle, a hydraulically controlled cylinder can be used to position the slide block. A gear-type motor controlled by either a push button or a limit switch is sometimes used for this purpose. A hydraulic servomotor with a lever control can also be used in connection with a servo-operated valve.

In the operation of the *axial piston-type pump*, the cylinder block is built with a series of cylinder bores in which the pistons are free to move in and out in a plane parallel to its axis (Fig. 27-13). The driving member or drive plate, which rotates at the same speed as the cylinder block, is linked to the cylinder block, with its axis tilted at an angle. The pistons are driven through connecting rods which are attached to the drive plate either by ball-and-socket joints or by means of shoes sliding against a tilted wobble plate.

As the cylinder block rotates, each piston moves back and forth inside its cylinder, the length of travel depending on the angle at which the cylinder block is tilted. As each piston moves toward the top of the cylinder, oil is drawn in through the valving mechanism or port block. As the piston descends toward the bottom of the cylinder, oil is forced outward under pressure. As in the radial pumps, leakage oil is drained either internally or externally.

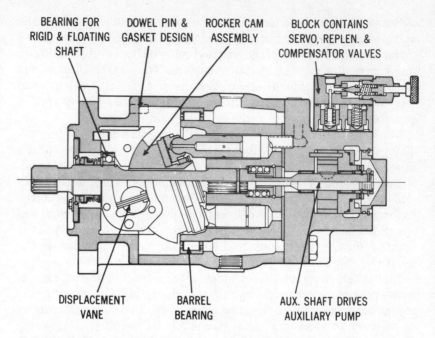

BEARING FOR RIGID & FLOATING SHAFT

DOWEL PIN & GASKET DESIGN

ROCKER CAM ASSEMBLY

BLOCK CONTAINS SERVO, REPLEN. & COMPENSATOR VALVES

DISPLACEMENT VANE

BARREL BEARING

AUX. SHAFT DRIVES AUXILIARY PUMP

Fig. 27-13. Cross-sectional view of a variable-displacement axial piston-type hydraulic pump with servo control. (Courtesy Abex Corporation, Denison Division)

The axial piston-type pumps are also available in both constant-delivery and variable-delivery designs. The variable-delivery pump is provided with a means of changing the angle between the driving element and the cylinder block, which changes the length of stroke of the pistons. The rate of delivery in variable-delivery pumps can be controlled by changing the angle of the cylinder block or wobble plate by means of a handwheel or one of the automatic devices mentioned in connection with the radial piston-type pumps.

HYDRAULIC MOTORS

Nearly all the rotary hydraulic pumps (with the exception of the centrifugal pump) can, theoretically, be used as hydraulic motors. However, some types are used more commonly because they are more efficient. Commercially available rotary motors are the gear-type, vane-type, and piston-type motors. The speed of a hydraulic motor is controlled easily, within the limits of the motor, by controlling the volume of pressurized fluid that is delivered to the motor. Some degree of speed control can be obtained by controlling the pressure of the driving fluid. However, this method is not considered to be dependable.

Gear-type motors do not operate satisfactorily, unless they are a hydraulically balanced design. An unbalanced mechanism develops both radial and axial pressures that are sufficient to lock the gears against the housing—even when the gears rotate on antifriction bearings capable of withstanding considerable loading. This is caused by the unbalanced rotating members. Manufacturers use different methods to obtain hydraulic balance. Most of the methods involve channeling the pressurized oil to counterbalancing surfaces within the motor mechanism.

The forces which act axially in gear-type pumps and motors are controlled by the wear plates. On the sides of the gears, the end plates impart a measured amount of force to control leakage. A recent innovation is to port the fluid pressure to a chamber on the side of the wear plate opposite the gear side. The fluid in this pressurized chamber exerts a force on the wear plate, pressing it against the side of the gear to prevent leakage between the side of the gear and the wear plate. In a more recent innovation, graduated chambers behind the wear plate keep the wear-plate holding forces proportional to the separating forces. The forces that tend to separate the gear and the wear plate are higher in the high-pressure areas. Metering of the fluid to the back side of the wear plate causes the wear plate to push on the side of the gear with more force in the high-pressure areas, with moderate force in the moderate-pressure areas, and with less force in the low-pressure areas.

This feature has increased the efficiency of single-direction rotary external gear-type motors to an extremely high level by placing wear-plate forces directly opposite maximum pressure-separation forces. This reduces the leakage paths to a minimum. This arrangement is not practical in reversible-type motors, since wear-plate pressure must be distributed uniformly.

The vane-type motors are of the same general construction as their pump counterparts. The cam ring is machined with a circular contour in the unbalanced-rotor designs and with an elliptical contour in the balanced-rotor designs. A rotor, rotating inside the cam ring, is provided with vanes that track inside the cam ring. A shaft attached to the rotor delivers the output power. Oil is forced into the vane-type motor, pushing against the vane. The vane begins to extend itself as it tracks inside the cam ring. Timing of fluid injection and fluid exhaust is extremely important to control leakage paths with the resultant motor slip and to make full use of the pressure acting against the vane areas for developing torque.

The vane-type motors are designed with either a balanced rotor or an unbalanced rotor. In the unbalanced-rotor design, the oil is introduced on one side of the cam ring. It is exhausted at approximately 180 degrees of arc from this point. In the balanced-rotor design, fluid is introduced at two points which are 180 degrees apart. This balances out the resultant rotor deflections that exist in the unbalanced-rotor design. The fluid is exhausted from the two ports which are 180 degrees apart. It is then combined within the cored channels for discharge from a common exhaust port. Fluid is introduced into the cartridge. It is exhausted at a point 90 degrees from the introduction point. Two vanes that are 180 degrees apart simultaneously develop torque with canceling radial bearing loads, to provide a complete dynamic and static rotor balance.

Both the radial and the axial piston-type motors (Fig. 27-14) are similar in design to the corresponding radial and axial piston-type pumps. Fluid pressure in the motor forces the pistons outward, causing the cylinder block to rotate. The rate at which the fluid is introduced to the motor determines its speed. The pump end of a pump-motor combination is often referred to as the A-end. The motor is the B-end.

Motors, like pumps, can be either variable-delivery or constant-delivery types. The axial piston-type motors are characterized by extremely low inertia. These motors accelerate quickly, and they are excellent for quick reversals.

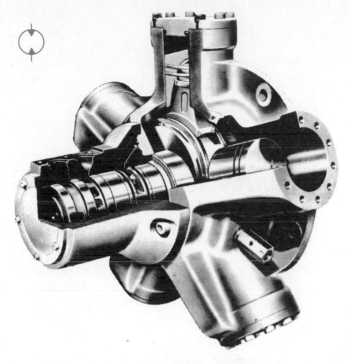

Fig. 27-14. Cutaway view of five-piston radial-type hydraulic motor. (Courtesy Double A Products Co.)

Chapter 28

Pumps and Motors—
Application

By definition, a *hydraulic pump* is a device which converts mechanical energy to fluid energy. A *fluid motor*, by definition, is a rotary device for converting fluid energy to mechanical energy. Both the hydraulic pump and the fluid motor are comparatively simple in construction. The application of these components, either as individual units or as combination units, can be simple or complex, depending on the application and the ingenuity of the application engineer.

The simplest method of supplying pressurized fluid on demand is to start and to stop the hydraulic pump, operating it at a speed that delivers the volume required by the circuit. This method is not practical, especially when the hydraulic pump is driven by an electric motor. Although the method can be successful with the smaller hydraulic units that are driven by synchronous motors, the cost of starting and stopping the electric motor is prohibitive when the larger or more powerful units are used.

BASIC CIRCUITS

A simple basic circuit for a hydraulic pump (Fig. 28-1), uses a constant-speed electric-motor drive and a relief valve to establish the operating pressure. Also, it provides a bypass return to the reservoir for the unused pump volume. This is considered to be the most basic motor drive—hydraulic pump-relief-valve combination. Variations of this basic circuit are used to effect a variety of functions. The use of the basic circuit to produce more than one operating pressure is treated later as a dual-pressure circuit.

Continual functioning of this basic circuit during periods in which the volume produced by the pump is in excess of the volume required by the circuit results in considerable heat being generated within the hydraulic fluid. It is sometimes desirable to effect a means of reducing this pressure during the idle periods in a machine cycle. With a fixed-volume pump, two basic methods are used to accomplish the unloading function. One method involves the use of the directional control valve, in which the idle position permits the oil to return to the tank under a minimum of pressure. This method is shown in the circuit in Fig. 28-2.

The second method of accomplishing the unloading function is to use a pilot-operated relief valve with a provision for venting the pilot section of the relief valve to the tank. This, in turn, drops the bypass pressure of the relief valve, permitting the pump to bypass the

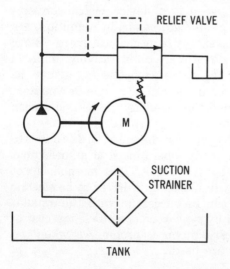

Fig. 28-1. Schematic diagram of basic circuit for a hydraulic pump.

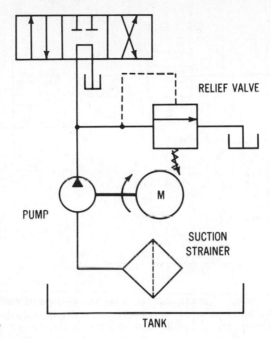

Fig. 28-2. Schematic diagram of a circuit used to reduce the oil pressure during the idle periods in a machine cycle. The idle position of the directional control valve permits the oil to return to the tank under minimum pressure.

output volume directly to the reservoir, short-circuiting the system at a nominal pressure. This nominal pressure value usually ranges from 65 to 100 psi (448.2 to 689.5 kPa), which is the pressure required for operating the pilot-operated valves in the system. This principle is shown in Fig. 28-3.

The solenoid-operated two-way valve in the circuit (see Fig. 28-3) can be opened by a limit switch or switches to unload the pump automatically at one or more of the idle positions in the machine cycle. If the idle period is lengthy in relation to the working time, the two-way solenoid-operated valve should be a "normally open" valve. The terms *normally open* (N.O.) and *normally closed* (N.C.) refer to the position of the valve when the solenoid or actuating mechanism is de-energized. The choice between a "normally open" and a "normally closed" valve depends on which valve requires the solenoid to be energized for the shortest length of time.

If a circuit is required that includes lengthy holding cycles at the

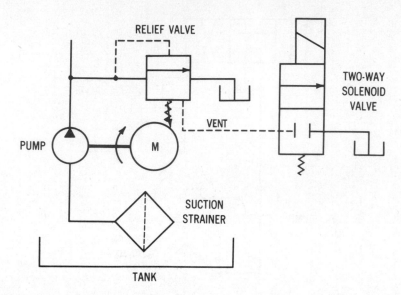

Fig. 28-3. Schematic diagram showing a second method of reducing the oil pressure during idle periods of the machine cycle. A pilot-operated relief valve with a provision for venting the pilot section of the relief valve to the tank is used.

maximum established pressure, but little or no volume is required, the circuit in Fig. 28-3 is unsatisfactory. This requirement is typical of a lengthy curing cycle for a molding press.

Two methods are commonly used to accomplish this type of full-pressure holding cycle. One of these methods is diagrammed in Fig. 28-4. Two pumps, one having a comparatively small volume and the other a larger volume, are used. In the circuit, the high-volume pump usually functions at a comparatively low pressure range, serving to supply a larger volume of oil for closing or moving the press member to the point where compression is required. At this point in the circuit, the high-volume, low-pressure pump is unloaded to the tank through an unloading valve that receives its operating signal from the pressure rise in the main pressure line to the press. The low-volume, high-pressure pump continues to operate at the pressure value established by the relief valve in the circuit. Most of the oil bypassed to the tank has been submitted to very little work, resulting in a minimum of heat generation within the fluid. The high-pressure, low-volume pump maintains the pressure, returning a comparatively small volume of oil to the tank at the higher pressure

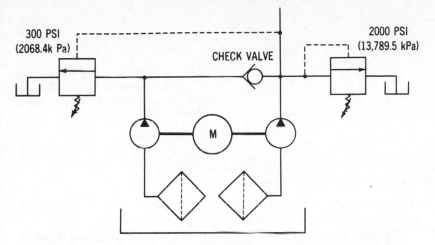

Fig. 28-4. Schematic diagram of a hydraulic circuit in which two pumps are used to accomplish a lengthy holding cycle at maximum established hydraulic pressure.

setting. Two advantages of this type of circuit are: (1) less horsepower is required to maintain the pressure at the smaller volume, and (2) less heat is generated within the fluid.

The second approach to the lengthy holding cycle is to use a pressure-compensated variable-volume pump (Fig. 28-5). In this circuit, the pump delivers maximum volume at the lower pressure requirement. It is stroked to a near-zero volumetric output as it reaches the higher pressure level. Here again, less horsepower is required and less heat is generated within the fluid. At the point in the machine cycle where a larger volume is required and the pressure demand is reduced, the pump automatically increases its stroke and volumetric output to its maximum setting.

HYDRAULIC TRANSMISSION

The use of a hydraulic pump to drive a hydraulic motor is usually referred to as a *hydraulic transmission*. There are many variations and adaptations of this comparatively simple arrangement, with both torque and speed being controlled by a variety of methods. A hydraulic motor-pump combination in its simplest form includes a fixed-displacement pump and a fixed-displacement motor. This combination is shown in Fig. 28-6, with no provision for regulating

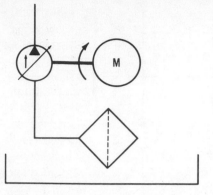

Fig. 28-5. Schematic diagram in which a pressure-compensated variable-volume pump is used to achieve a lengthy holding cycle.

the speed of the motor other than the volumetric output of the pump and its relation to the volumetric displacement of the motor.

The volume of oil delivered by the hydraulic pump varies inversely with the pressure resistance at the discharge side of the pump. The speed of the hydraulic motor is affected by the slippage rate within the motor element. The slippage rate is determined in the same way that the volumetric output efficiency of the hydraulic pump is determined, with the net result being a function of the pressure differential across the hydraulic motor. The maximum pressure differential across the hydraulic motor is, of course, determined by the maximum input pressure driving the motor. The actual pressure drop across the motor is determined by the resistance of the motor and the motor load to the volumetric flow of the delivered fluid. Since the pressure drop and slippage rate vary, depending on the load, the

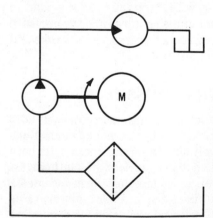

Fig. 28-6. Schematic diagram of the basic circuit for a hydraulic transmission in which a fixed-displacement pump is used to drive a fixed-displacement unidirectional motor.

net result is fluctuation in the speed of the hydraulic motor with variations in the load.

The maximum speed for a hydraulic motor is usually indicated specifically in the manufacturer's performance data. Another limitation that the application engineer should be aware of is the minimum effective speed of the motor unit. Although the minimum effective speed tends to vary among the different motor designs, a hydraulic motor, generally, is not highly effective below approximately 150 rpm. Below the minimum effective speed, intermittency in the rotational speed of the motor is encountered. This phenomenon is sometimes referred to as "gallop." The varying slippage rate and the inefficiencies encountered in the hydraulic motor at minimum speeds are the reasons for the low starting torques provided by these motors. Performance charts furnished by the manufacturer should be consulted before these units are used for the low-rpm applications that require high torques.

Theoretically, the hydraulic motor is capable of infinitely variable speeds within its speed range. Precise control of these speeds, however, is a different matter (Fig. 28-7). When variable load resistances are anticipated, it is wise to use an over-size hydraulic motor, because a reserve of power is needed for control purposes. To maintain a more consistent pressure drop across the hydraulic motor, the most common method for controlling the speed of revolution is the use of a throttle or speed-control valve to control the oil exhausting from the downstream port of the motor. When a downstream flow control is employed, care should be taken to provide adequate external drainage.

If a given hydraulic motor is built to rotate in only one direction, internal drainage of the internal leakage is usually provided by channeling the leakage internally to the downstream side of the element. When internal porting is present, throttling of the downstream porting of the motor is unsatisfactory. This deflects the internal leakage toward the seals, resulting in failure of the shaft seal and in possible internal damage to the motor. A bidirectional hydraulic motor usually provides for an external drain connection, since the two ports are pressurized alternately.

Another method that provides for speed-variation requirements on a hydraulic-motor application is to power the hydraulic motor with a variable-volume pump. This type of circuit is diagrammed in Fig. 28-8. The variable control of the pump output can be effected by

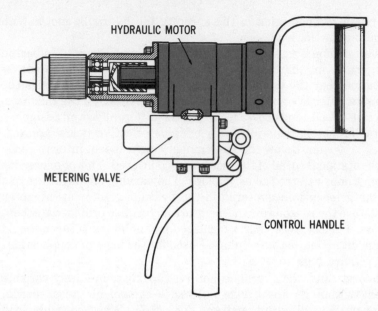

Fig. 28-7. Diagram of a hand drill with a ½-inch chuck and spindle adaptation driven by a high-torque hydraulic motor. The metering valve on the control handle controls the motor input volume to control the spindle speed. This drill is designed for use with hydraulic-powered mobile equipment.

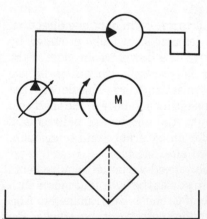

Fig. 28-8. Schematic diagram of a hydraulic circuit in which a variable-volume pump is used to provide for speed-variation requirements on a hydraulic motor application.

electrical, mechanical, or manual means, depending on the requirements of the control feature.

In some hydraulic-motor applications, the motor itself decelerates or stops the load by hydraulic means. It is necessary to provide for the control of this type of rotational inertia if damage or destruction of the hydraulic motor is to be avoided. If the hydraulic motor in Figs. 28-7 and 28-8 were blocked suddenly by an intermediate valve, the rotational inertia could generate extremely high pressures inside the hydraulic motor. A common method of applying the potential braking power of the hydraulic motor to the rotational load is diagrammed in Fig. 28-9.

In the circuit (see Fig. 28-9), commonly known as a braking and replenishing circuit, two relief valves (one valve on each side of the motor) are installed to apply maximum back pressure to the motor elements, thereby retarding the rotation of the motor. The check valves may or may not be necessary, depending on the type of relief valve employed. This system provides pressure braking, rather than volume braking, as in either retarding or stopping the flow. The presence of the two relief valves in this circuit indicates that the hydraulic motor is bidirectional. If the motor is unidirectional, but with an external drain connection, only one relief valve is required.

The hydraulic motor also lends itself to applications other than those requiring torque. One of these alternate applications involves the use of a double-element motor, or two motor elements operating on a common shaft. A hydraulic motor with one inlet port and two outlet ports (Fig. 28-10) can be used quite effectively as a flow divider. The function of this type of flow divider is to meter a single source of hydraulic fluid through the two separate motor elements, thereby delivering equal amounts of fluid to the two outlets. If it is desired to synchronize two equal, but separate, hydraulic cylinders, this method is an effective means of accomplishing the task.

Since the two motor elements cannot be matched perfectly and since a wide difference in the loads of the two cylinders being controlled can create different slippage rates within the two motor elements, some external means should be provided in the circuit to permit re-equalizing the system when the elements reach the point of being out of phase.

The external shaft shown in Fig. 28-10 is not used when the unit is employed as a flow divider. Therefore, the manufacturer also provides the unit with a stub shaft and two blank end covers. The same effect can be created by coupling two matched motors together on a common shaft. This can be expanded to three or more units by using double-shaft motors in the interior positions.

Several manufacturers offer hydraulic power packages, including any combination of components and accessories desired. Custom-built packages are shown in Figs. 28-11, 28-12, and 28-13. A real test of your ability to identify fluid power equipment is offered in Fig.

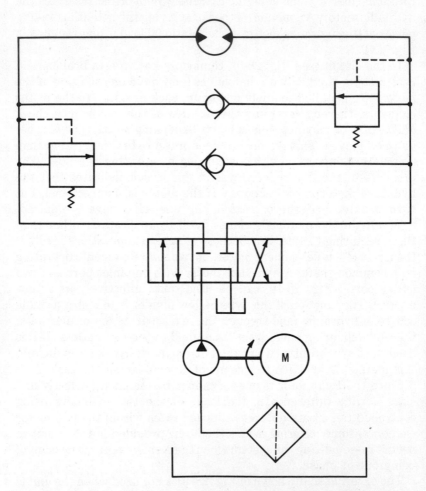

Fig. 28-9. Schematic diagram of a hydraulic braking and replenishing type of circuit.

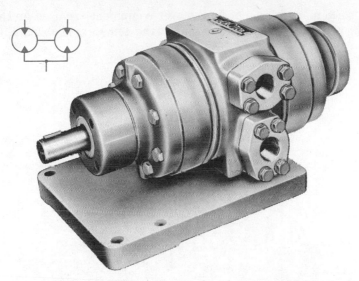

Fig. 28-10. A hydraulic motor with one inlet port and two outlet ports. (Courtesy Double A Products Co.)

Fig. 28-11. Oilgear *Power Pak* including the following: three 150-hp two-way variable-delivery pumps; three 60-hp constant-delivery pumps; three 200-hp electric motors; a 1500-gallon reservoir; and system valves for a 350,000-lb drawbench. (Courtesy The Oilgear Co.)

28-14, since many pieces of fluid power equipment are used on the printing press. A closeup view of some of the items is available in Fig. 28-15.

Fig. 28-12. Oilgear *Power Pak,* including a two-way variable-delivery pump, an electric motor, a 430-gallon (1.628 kiloliters) reservoir, and the system valves for a hydraulic press. (Courtesy The Oilgear Co.)

Fig. 28-13. *Power Package,* incorporating two electric motors and two hydraulic pumps complete with automatic valving plumbed and mounted to fit in a special housing on the machine.

Fig. 28-14. Printing press employing a wide variety of fluid power components.
(Photo courtesy of the Packaging Machinery Div. of FMC Corporation)

Fig. 28-15. A *Char-Lynn* hydraulic motor is used to drive the rubber ink roll on a 706-A FMC flexographic press. The motor is speed controlled by a color-coded needle valve on the return line of the motor. The other needle valve is the shutoff valve which is used to stop the motor. This particular drive is used as an ink drive motor when the press is not printing, to keep two rolls turning in the ink. As the press speed increases, this motor acts as a brake and allows the rubber roll to run at a much slower speed. (Photo courtesy of the Packaging Machinery Div. of FMC Corporation)

Chapter 29

Pumps and Motors— Maintenance

As emphasized earlier in Chapter 27, hydraulic pumps and hydraulic motors are precision pieces of equipment manufactured and assembled to close tolerances. Generally, they are lubricated by the fluid being pumped. Exceptions are those units that are used to pump nonlubricating fluids. Then, the pumps are provided with grease fittings or some other means of lubricating the bearings and moving parts.

CAUSES OF FAILURE

Tremendous forces and thrusts are generated within a hydraulic pump, causing the wearing members or sliding points of contact to absorb these tremendous forces. When hydraulic pumps are operated above their rated pressures, the most common cause of pump failure or seizure is the breaking of the oil film that lubricates the sliding friction members. The result is seizing and breakage of the finely machined parts within the pump.

Probably the most common cause of pump failure is contaminated or dirty fluid. Since the fluid being pumped is also used to lubricate the surfaces that bear the high-thrust loads within the pump, the presence of grit and abrasive matter within the fluid contributes to extremely rapid wear and deterioration of these surfaces.

The various pumps are provided with different tolerances for contaminants within the fluid stream. The pumps most sensitive to contamination are, of course, the high-pressure pumps, because the tolerances of the mating parts are closer. The piston-type pumps, especially the variable-volume, servo-controlled piston-type pumps, are the most sensitive. In contrast, the external gear-type pumps can operate quite satisfactorily for a lengthy period of time with contaminated fluid. It is characteristic of the gear-type pumps to decline gradually in efficiency, rather than to fail suddenly, which is commonly experienced with the piston-type pumps.

The failure of a pump to deliver fluid may occur instantly. The failure may develop slowly over a long period of time during which the problem seems to be only that the equipment is slightly slower, slightly less powerful, and slightly less responsive to control than usual. The more obvious the symptoms are, the worse the problem is. If the pump stops suddenly and if the shaft cannot be rotated either by hand or with a short-handled wrench, the cause is nearly certain to be a major mechanical failure of an internal pump part. Then, the solution is to dismantle the pump and replace the failed component. At this point, a specification sheet from the manufacturer of the pump is needed. These specification sheets are available with a parts breakdown and nomenclature that permit ordering the needed replacement parts (Fig. 29-1).

After the pump has been disassembled, the problem located, and the failed part replaced, the machine should not be operated, under any circumstance, until the system has been drained, flushed, and refilled with hydraulic oil. The purpose of this action is to make sure that debris from the failed part does not remain in the system. Otherwise, there is a serious risk involved in the form of scored cylinders and jammed valves, which may require another replacement unit in a short time.

Gradual loss of power is more difficult to discover in a pump, because it is spread over a longer period of time. However, once a loss in power is suspected, the flow rate and pressure delivered by the pump should be checked immediately. Some operators and designers

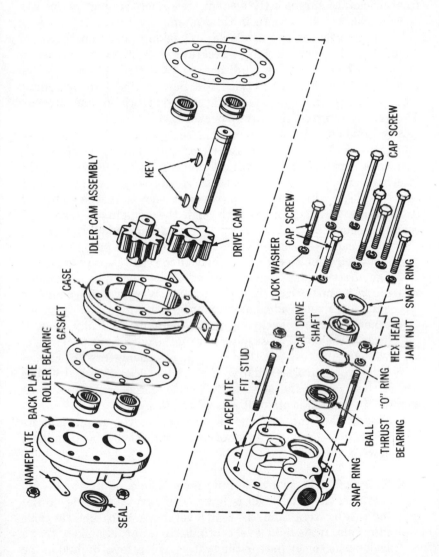

Fig. 29-1. Parts diagram for a gear-type pump. (Courtesy Roper Pump Company)

install a pressure gauge in the pressure line for routine observations. Since the instrument is extremely sensitive, it is economical to provide a shutoff valve to the gauge, thereby removing the gauge from shocks and surges in the system, except for the short period it is used to check the pressure in the system.

If a flow meter is not available for making precision checks of volumetric delivery, a fairly accurate determination of pump performance can be made by discharging the flow into a barrel or container through a restriction. The flow is timed to determine whether the specified gallonage is delivered in a given time interval. This test is referred to as the "bucket test."

Any loud or unusual pump noise should be investigated immediately. This indicates restricted inlet supply (starvation), excessive aeration of the hydraulic oil, or excessive wear of the internal parts. Any one of these three problems can result in a "gravel effect" within the pump, which is a sound that is similar to sand or gravel being pumped through the equipment. The most common cause of gravel effect is cavitation inside the pump. Cavitation can be caused either by starvation of the pump or by the presence of air bubbles within the fluid. In the vane-type pumps (Fig. 29-2), an effect similar to gravel effect can be caused by an uneven wearing action on the cam ring. This uneven, wavy wearing action on the cam ring is referred to as "washboard effect."

Another important checkpoint in the system is for leaky seals. If the power of the pump is escaping around the gaskets or packings in the cylinders, valves, or other components, pump repairs cannot solve the problem. Particularly, if a gradual loss of speed occurs within the system, the seals in the other components should be examined for an indication of loss of efficiency.

Assuming that the problem is localized in the pump and its immediate accessories, a number of checkpoints should be examined, as follows:

1. *Reservoir.* The hydraulic fluid level should be up to the full mark. If the fluid level is low and if service records do not indicate a recent refill, the fluid level should be checked at close intervals. If the fluid level continues to lower, it is likely that a leak exists somewhere in the system. A low level of fluid in the reservoir may expose the suction strainer, permitting air to be drawn into the pump with the fluid.

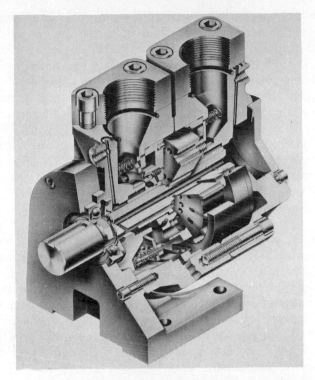

Fig. 29-2. A vane-type pump having two vanes per slot. (This illustration is the courtesy of Parker Hannifin Corporation.)

2. *Intake piping.* A cloth or scrap of paper may have been drawn into the piping, plugging the intake so that the pump cannot receive enough oil. Kinks and frays in the suction lines also can restrict the oil supply.

3. *Suction-line filter.* If the filter is plugged, it should be cleaned or replaced as the manufacturer recommends. A clogged filter can shut off the oil supply to the pump. If the equipment is operating under extremely dusty conditions, it is worthwhile to check the filter more frequently. If the filter requires cleaning or replacing more often than normal or if frequent checking is inconvenient, it may be economical to install a larger filter and filter housing. However, the manufacturer's representative should be consulted before making a permanent mechanical change of this type.

4. *Air leaks in the suction line.* If the pump picks up air along with the oil, there is additional risk in quicker oxidation, in foaming, and in spongy operation. This action can be detrimental, since aerated oil loses some of its lubricity. The oxidized oil forms sludges and deposits at points where they are least desired. Repair the leaks and replace the lines and fittings if necessary. Be certain that all bolts and fittings are tight, because old leaks may return if the pump vibrates too much.

5. *Oil viscosity.* If the hydraulic oil thickens too much, the pump cannot prime itself. Either the incorrect grade of oil for the operating temperature is being used or the oil has become thickened from oxidation or contamination. The oil should be replaced before more damage is done.

Oxidation is nearly inevitable in hydraulic oil. Oil changing recommendations are within the service life of an oil operating under average operating conditions. Therefore, additional heat, higher pressure, or the presence of contaminants, such as metal particles or water, can speed the rate of oxidation. The end result is gum, sludge, plugged valves, and rapid wear on the pump, cylinder, and valve parts. The solution to the oxidation problem is to make sure that the correct grade of oil is used and to make sure the equipment receives regular preventive maintenance.

Failure of the shaft seal is a common source of oil aeration. This type of failure may not be noticeable while the pump is running. However, if a deposit of oil is found below the shaft of the pump after an extended shutdown period, the shaft seal should be examined. If the shaft seal is in a condition that permits oil to drain from the pump while it is not in operation, the seal is probably permitting air to enter the pump during operation.

If excessive wear occurs at the shaft seal and if both the shaft and the shaft bearing are involved, the alignment of the pump with the driving motor should be gauged carefully. Flexible-shaft couplings are designed to operate under misaligned conditions. However, they can subject the pump shaft assembly to tremendous punishment and unwarranted side thrusts. The driving motor shaft, the flexible-shaft coupling, and the pump shaft should be in perfect alignment, if possible. The flexible-shaft coupling should be inspected carefully. Some couplings, particularly the rubber-insert type of coupling, wear unevenly. They can subject the pump shaft assembly to damaging end thrust and/or side thrust, thereby causing excessive and premature wear in the assembly. The coupling manufacturer's recommendations for clearances within the coupling assembly should be followed carefully.

Hydraulic oil contains some dissolved air, usually about 10 percent by volume. Under increased pressure, however, the oil can absorb much more air. If an air leak exists on the suction side of the pump through a fitting, a shaft seal, or the filter case, air is added to the oil in the system. When the oil is returned to the reservoir, the entrapped air causes a froth of bubbles.

The presence of foam in a hydraulic system creates a problem involving jerky, uneven power and accelerated oxidation. Uneven power results when the foaming hydraulic oil is trapped in a hydraulic cylinder. When hydraulic pressure is applied, the normal steady pushing action of the oil against the piston is impossible, because the air in the oil is compressed. As a result, the piston moves with a jerky and jolting movement. If this type of hydraulic-cylinder movement is observed, checking for air leaks in and around the pump and the related components may reveal the source of the problem.

Entrapped air promotes accelerated oxidation, because the air temperature increases as the air is compressed. The temperature of the air bubbles may reach 2000° F (1093° C) at a pressure of 2000 psi (13,789.5 kPa), which is high enough to actually scorch the hydraulic oil in the area of the foam pocket. The scorching action oxidizes the oil and leads to sludge and varnish formation. If a hydraulic system that contains foam is kept in operation, the entire oil supply may be oxidized. Eventually, the sludge and varnish can ruin every hydraulic component in the system.

The basic remedy for foaming is to eliminate all air leaks. With the system in operation, check the pump intake system for leaks. Pour a quantity of oil on each connection. At the same time, listen closely to its operating sound—with a listening device, if necessary. If a distinct difference in the sound of operation occurs when the oil is poured on the suspected leakage spot, it is an indication that the oil has temporarily sealed a leak. This also can be used to detect leaks at the pump shaft seal.

Inside the reservoir, it is good practice to locate the return line at least 4 inches below the oil level in the reservoir. This avoids the surface turbulence that can add to foaming and aeration of the reservoir fluid.

A gradual loss of pump output volume may be caused by deterioration and wearing action of the mating pumping elements. When this occurs, the clearances between the pumping elements are increased, resulting in an increase in slippage within the pump. Slippage usually can be detected by an increase in the operating temperature of the pump. The pump then becomes the hot spot of the entire system. As the slippage flow increases and as the oil temperature increases, the viscosity of the oil decreases, contributing even more to the slippage rate within the pump. If this occurs, the best remedy is to replace either the pump elements or the entire pump assembly.

Inside a hydraulic pump or motor the points of wear action vary, depending on the design of the unit. The external gear-type pump probably provides the smallest number of critical points for wear (Fig. 29-3). The areas that should be checked for wear are, of course, the surfaces of the gear teeth, the end surfaces of the gears, and the wear surfaces of the body cavity. In external gear-type pumps that are provided with balancing wear plates, the condition of the wear plates should be examined carefully. Occasionally, the shaft and the shaft bearing become unstable within the body housing. This is especially true if misalignment of the drive shaft has prevailed for a lengthy period of time.

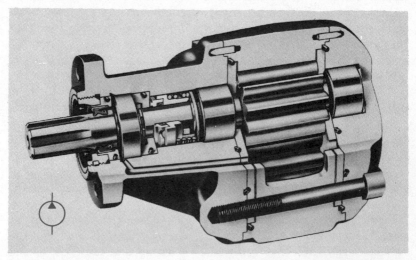

Fig. 29-3. Cutaway view of an external gear-type pump. (Courtesy Hydreco)

The internal gear-type pump or motor provides approximately the same points of wear as does the external type, with the exception of the external wear surface of the outer gear (Fig. 29-4). This external surface is a bearing surface which rides on a film of oil that separates it from the inside diameter of the pump housing. Severe scoring of this surface is an indication of an extremely dirty fluid and/or overpressuring of the unit. Under normal operating conditions and with proper filtration of the fluid, some internal gear units have operated for more than ten years in the field with so little wear that the part number rubber stamped on the bearing surface was not removed. Improper filtration in conjunction with overpressuring of the units can shorten the life of the pumping elements to days or hours.

The vane-type pumps and motors normally receive the most wear on the surfaces against which the ends of the vanes slide, which increases the clearance. Another spot that receives much wear in these units is the inside diameter (ID) of the oval shaped cam ring. Here, the wearing action previously referred to as "washboard effect" can be accelerated rapidly if the pump is permitted to cavitate or to draw in air on the suction side. Most of the vane-type pumps and motors with a worn cam ring can be repaired easily by a simple replacement of the cam ring. In vane-type units that have been abused, broken vanes may be found. In disassembling these units, extreme care should be exercised in removing the rotor from the unit to be sure that the vanes are not lost, reversed, or mismatched.

The piston-type pumps and motors, as mentioned previously, can tolerate less dirt than any other pump type. The pistons and the cylinder bores should be examined carefully for excessive scoring, since scoring considerably increases the slippage flow within the pump. Also, it can detract from the efficiency of the unit. If the piston-type unit is provided with check valves that are built into the porting, the check-valve seats should be examined for scoring, hammering, and fractures. A problem can be expected to appear on the surfaces that bear the largest thrust forces within the unit, such as the shoes or bearing surfaces where the wobble plate or thrust ring bears against the piston or piston linkage. Other points that should be examined carefully are the bearing surfaces and the shaft bearing and seal. On tearing down the piston-type unit, it can be seen readily that it contains many more intricate and precision parts than the less sensitive gear-type units. Therefore, it is wise to consult a complete

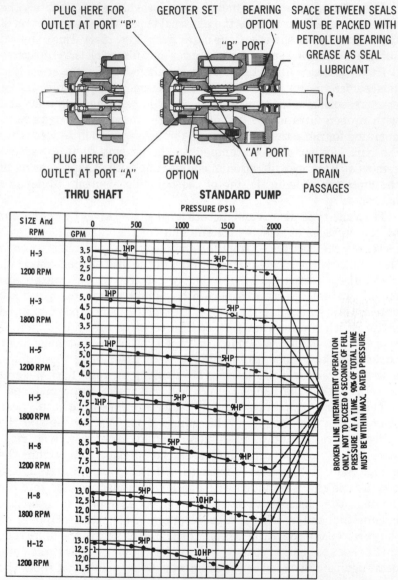

Fig. 29-4. Cross-sectional diagram of pump assembly and performance chart for a Gerotor-type hydraulic pump.

set of manufacturer's specifications before endeavoring to dismantle the piston-type pump or motor.

All hydraulic pumps that incorporate the servo-type compensating control (Fig. 29-5) usually contain fine orifices that can become blocked or sludged. These pieces of control equipment are finely tuned. They are designed to respond to minute pressure variations. Therefore, they must be kept clean and free from scoring or sludging. Usually, when this type of pump fails to deliver pressure, the cause is a silting action or blocking of the fine orifices that transmit the pressure-demand signals to the compensator unit. For all hydraulic pumps and motors, it is good practice to keep the system clean, to stay within the manufacturer's specifications regarding pressure ratings, and to maintain proper temperature control of the system, thereby avoiding unnecessary downtime and corrective mainte-

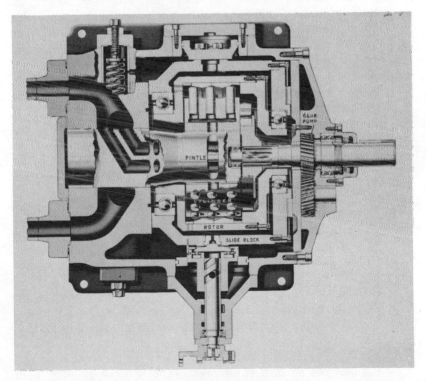

Fig. 29-5. Cross-sectional view of a variable-delivery two-way flow pump with a hydraulic servomotor lever control. (Courtesy The Oilgear Co.)

nance. Again, preventive maintenance in a hydraulic system is more economical than corrective maintenance.

DISASSEMBLY AND REPAIR

A typical pump assembly with parts identification is shown in Fig. 29-6. Sectional diagrams of two different shaft assemblies for the same pump are shown in Fig. 29-7.

Disassembly instructions for the pump (see Figs. 29-6 and 29-7) are as follows:

1. Pump should be detached from plumbing.
2. After disengaging the pump from its shaft coupling or other drive connection, detach the base or flange and shaft key. Place the pump cap or head end upward in a vise which has protective jaws. Remove the four cap screws, cap, and gasket. The bearing should slip from the shaft as soon as the snap ring is off.
3. The 12 body screws then are loosened to permit disassembly of the head and shaft. Remove rotor set, locating pin, and key.
4. Before pulling the body, bearing, and seals, make certain the shaft key is removed.
5. Remove retainer rings, if necessary. Remove the bearings and two shaft seals from their locations in the body.

Repair and overhaul instructions for the pump (see Figs. 29-6 and 29-7) are as follows:

1. Wash metal parts in clean mineral solvent. Blow them dry with filtered compressed air. Place them on a clean surface for inspection.
2. Replace shaft seals, "O" rings, and gasket exposed for repair. Replace completely at each overhaul.
3. Bearings should be rotated while applying pressure to check for wear, cracked or pitted races, or pitted balls. Replace if defective. Otherwise, replace all bearings and bushings at each overhaul.
4. Shaft journal for the seals should be scanned for wear or scores. Replace the shaft if the scores cannot be removed by light polishing. A smooth groove not exceeding 0.005 inch (0.127mm) is acceptable.

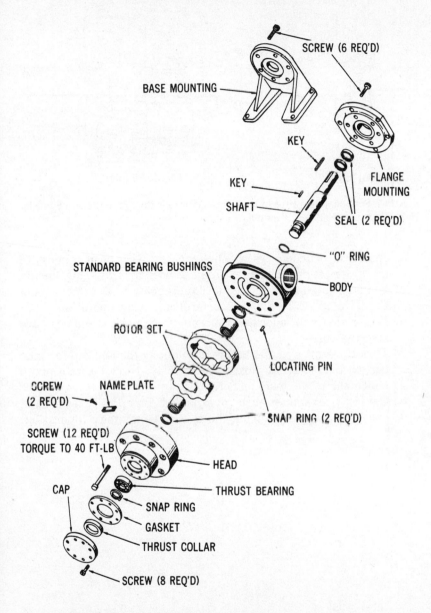

Fig. 29-6. Exploded view of a typical hydraulic pump assembly with parts identification. (Courtesy Double A Products Co.)

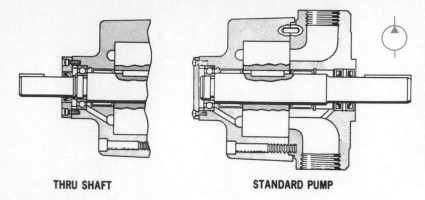

THRU SHAFT   STANDARD PUMP

Fig. 29-7. Sectional diagrams of shaft assemblies for the hydraulic pump in Fig. 29-6. (Courtesy Double A Products Co.)

5. Install the bearings against the snap ring in the head or body. In a 0-40 series, an arbor press may be used for this. They should be installed about 1/16 inch (1.6 mm) below the surface facing the *Gerotor* set to avoid interference during operation.
6. Be certain that the locating pins aligning the head and body are inserted correctly.
7. Turn the shaft by hand before assembling the end cap to make certain that internal parts rotate freely. Never start a pump which shows evidence of binding.
8. Carefully examine each overhauled pump for leaks and excessive vibration during the first five minutes of operation.

Chapter 30

Fundamentals of Reservoir and Plumbing Design

The reservoir serves primarily as a storage and supply point for the hydraulic system fluid. However, if it is designed properly, it can contribute to the proper functioning of the entire system. Also, it can reduce the normal maintenance requirements. A well-designed reservoir can assist in the separation of air and contaminants from the fluid, can help dissipate heat generated within the system, can protect the system from undue contamination, and can provide easy access to the system and its components for mounting, cleaning, and servicing.

FUNCTION OF THE RESERVOIR

The reservoir should, primarily, be large enough to store more than the largest volume of fluid the system will require, and it should be as large as possible, within space limitations. The oil level should never fall below a reasonable working level under any operating condition, with additional capacity for oil that drains from the system during shutdown. If the system uses single-acting ram-type cylinders or accumulators, the accompanying fluctuation of the fluid level should be provided for in the capacity of the reservoir.

Heat is radiated or transferred from the hydraulic fluid in a system by means of all the surfaces contacted by the fluid. The reservoir provides at least five surfaces whose total radiating surface is determined by the size of the reservoir. The risers or legs of the reservoir should provide 6 inches (15.2 cm) of minimum height above the floor. This permits proper air circulation across the bottom surface of the vessel. It also eases the housekeeping efforts. It is strongly recommended that a reservoir should be adequate in size to contain a fluid volume at least three times the gallon-per-minute (or volume) rating of the pump. Although increasing the size of the reservoir increases the surface area only arithmetically, its volume of fluid capacity increases geometrically. The fluid thereby is permitted a longer rest period or idle time for the dissipation of heat through the walls of the reservoir. Therefore, it is advantageous to design the reservoir oversize when considering its size.

Above the minimum needs, an ample supply of oil permits the oil to remain in the reservoir long enough for cooling and for the dirt and water entrained in the fluid to separate. Most of the heat is dissipated by radiation and convection from the walls of the reservoir. Whenever possible, the hydraulic unit should be placed in a location that permits free circulation of the cooling air across the entire system. Connections should be provided in the original design for the subsequent addition of an oil cooler, if the temperature cannot be controlled by another means.

RESERVOIR DESIGN

If the reservoir is too shallow, the wall area may not be large enough for effective cooling. Also, if the reservoir is too deep and too narrow, enough surface for proper separation of the entrained air bubbles or foam may not be provided.

A vertical baffle plate rising to approximately two-thirds the height of the fluid level should traverse the longitudinal axis of the reservoir to maintain a separation of the return oil from the oil entering the pump. The baffle permits slower and more complete oil circulation, which helps to release the air and the other contaminants from the return oil before it is recirculated through the system. It also aids in preventing stratification of the oil and short-circuiting of the hot oil from the return lines to the pump suction.

The bottom of the reservoir should be sloped or dished, with a drain valve at the lowest point for easy and complete draining of the vessel. The previously mentioned 6-inch (15.2 cm) clearance for the bottom also permits the use of drain pans or hose for capturing the drained oil without spillage.

The reservoir cover may be welded to the vessel. It may be bolted in place with gaskets provided to prevent the entrance of contaminants. It should be of adequate thickness and strength to support the pump and driving motor without undue flexing or distortion. An alternate method of mounting the pump and motor units requires a separate motor–pump mounting plate sturdy enough to provide rigidity. It is attached to the top surface of the reservoir cover with a three-point mounting. This alternate method also provides better insulation and prevention of heat transfer between the electric motor and the oil reservoir. The mounting plate should cover only the portion of the top area of the reservoir that is necessary to provide for the mounting and alignment of the pump, coupling, and motor.

All fluid lines to the reservoir should enter through the cover. The return or tank lines should extend to a point that is always lower than the fluid level within the reservoir—to reduce misting and aeration of the fluid. Provision should be made for the drain lines from working components that are external to the reservoir to return the draining fluid and empty into the area above the fluid level of the reservoir. This reduces resistance to the flow within the drain lines to a minimum.

Large, gasketed cleanout doors should be provided in the end of the reservoir, to provide access to both sides of the baffle plate. Space should be provided for a filter or strainer on the suction line, with the minimum fluid level always higher than the filter or strainer, to prevent drawing air into the suction line. The intake or suction-line opening through the reservoir cover should be large enough for inserting and removing the filter unit. All openings should be sealed adequately to prevent the entry of foreign material into the vessel.

A flush-mounted and protected fluid-level indicator should be installed near the filler opening. The indicator should be marked for high and low levels. If the small circular porthole-type indicators are used, they should be used in pairs with one indicator at the high limit and the other at the low limit of fluid level. A float-actuated switch is sometimes used in conjunction with a warning light or the motor control to guard against excessive reduction in fluid level.

The inside surfaces of the reservoir vessel should be painted with a sealer material to prevent condensation and the subsequent oxidation that can result from the presence of the condensate. Care should be taken to select a sealer that is compatible with the type of fluid being used in the system. Satisfactory sealers usually can be recommended by a supplier.

The reservoir cover should include an air-breather unit that can protect the reservoir from contaminant-laden air. The capacity of the unit should be large enough to maintain an approximate atmospheric pressure during maximum demands of the hydraulic system. The movement of the air through the breather unit is determined by the rate of change in the oil level inside the reservoir during cycling of the system. A system employing single-acting cylinders, oversize rod cylinders, or accumulators develops the largest fluctuations in the fluid level.

The fill connection should be easily removable. However, it should be retained by a chain or similar device to prevent misplacement or to prevent its contact with surrounding contamination. The fill opening should be provided with a fine screen to remove contaminants in the fluid and to prevent the entrance of foreign matter when the fill cap is removed. Combination filler and breather caps are sometimes used, with an air-filter element built into the fill cap.

Several pipe couplings should be welded flush with the inside surface of the bottom of the reservoir for the installation of pipe plugs with magnets attached. The magnets attract the ferrous particles that either enter the system or are generated within the system components through wearing action. A complete hydraulic power unit is shown in Fig. 30-1.

Additional equipment that is sometimes a part of the reservoir package includes the heat exchanger, pressure gauge, filter, thermometer, unloading valve, relief valve, motor controls, and other control valves that are panel mounted to the reservoir. Most of these items are discussed elsewhere in this text.

Fig. 30-1. Complete hydraulic power unit, including a 35-gallon (132.5 liters) reservoir and a 15-gpm (56.8 liters/min) hydraulic pump. Note the sight-level gauge and thermometer on the side of the reservoir. (Courtesy Continental Hydraulics, Inc.)

DESIGN VARIATIONS

The design of a hydraulic reservoir that conforms to standard specifications is shown in Fig. 30-2. However, a number of design variations can be used to provide specific advantages. One of these variations is the L-shaped unit (Fig. 30-3). As noted in the illustration, the motor and pump assembly are mounted on a platform at the side of the reservoir. The platform is an extension of the base of the reservoir. The column pressure of the oil is a distinct advantage in this design. Since the oil level in the reservoir is higher than the intake port of the pump, a "flooded suction" is provided for the pump. Both atmospheric pressure and the column pressure of the oil above the intake port are available to force the oil into the intake port of the pump.

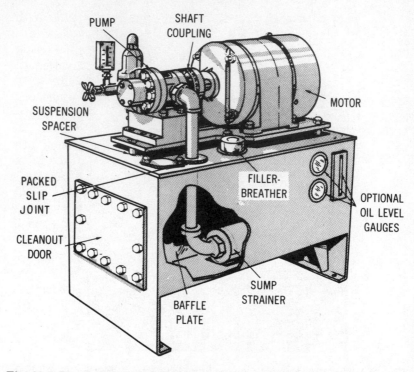

Fig. 30-2. The hydraulic reservoir should be designed to conform to standard specifications utilizing adequate materials.

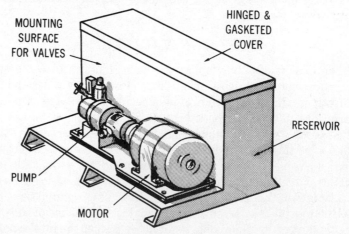

Fig. 30-3. An L-shaped hydraulic reservoir.

Additional advantages of this design include mounting of the control components on the side wall of the reservoir and placing the tank and drain lines within the reservoir. This minimizes the leakage problem in the lines, and permits compact circuitry within the confines of the reservoir. Another advantage is accessibility to the reservoir, since there is no encumbrance on the lid or top surface of the vessel. This facilitates checking the relief-valve return lines, valve return lines, and the drain lines when checking the system. The suction strainer of the pump is also readily accessible through the open top of the reservoir. The reservoir top can be locked in the closed position, and it can be gasketed to prevent entry of contaminants.

Another design variation is shown in Fig. 30-4. An overhead tank, with the pump motor unit mounted directly below the reservoir, is shown in the diagram. Column pressure of the elevated oil also provides flooded suction for the pump in this design. If floor space is a problem, this overhead reservoir construction may be more advantageous than the L-shaped construction. A possible disadvantage with this design may lie in the installation of drain lines from

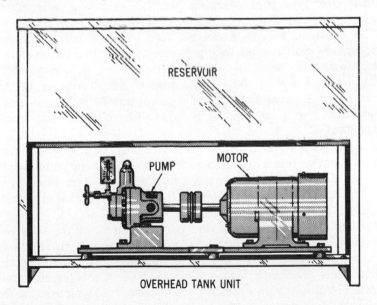

Fig. 30-4. Hydraulic unit in which the pump-motor assembly is mounted directly below the reservoir.

components, such as valves and hydraulic motors situated at a level lower than that of the elevated reservoir.

Both the L-shaped unit and the overhead-reservoir unit offer a possible advantage in eliminating the need for supercharging the pump. This depends entirely on the characteristics of the pump and the operating conditions. The manufacturer's operating instructions should be consulted and the suction characteristics of the pump should be known before a decision is made on the supercharging pressure. Some pumps require supercharging either by a centrifugal pump or by a low-pressure pump in addition to the combined atmospheric and column pressure that can be obtained from the overhead-reservoir design.

If a supercharging pump is required, it can be attached to the opposite end of a double-shafted motor. Or it can be driven by a belt from the single-end motor that is also coupled to the pressure-type pump—to which the supercharging pump can be attached. If the supercharging pump is a positive-displacement type of pump, a low-pressure relief valve is provided between the output port of the supercharging pump and the intake port of the pressure pump. A built-in gear-type supercharging pump is shown as part of the piston-type pump assembly in Fig. 29-4 in Chapter 29.

The hydraulic reservoir can be used to serve more than one independent system—or a dual-pressure system. In an independent multiple system, the double-shaft electric motor may be used with a pump on each end of the electric motor. An alternative to this arrangement is to use a compound-element pump that operates on a single shaft and is attached to and driven by a single-shaft motor (Fig. 30-5).

A multiple-system hydraulic unit is shown in Fig. 30-6. Each independent system is provided with its own motor and pump, which are mounted atop the reservoir cover. An independent suction line and suction strainer are provided each unit, but they share the same reservoir and hydraulic fluid.

Two rather sophisticated power packages in which the reservoir design provides for the mounting of many hydraulic components, as well as providing a control station and accessibility for service and maintenance are shown in Figs. 30-7 and 30-8. A contrasting small, compact hydraulic power package with a lucite reservoir used for the purpose of demonstration of design features can be seen in Fig. 30-9.

In a multiple- or dual-pressure system, independent pumps can be used on a double-shaft electric motor. Also, a multiple-element pump with a single or common shaft driven by a single-shaft electric motor can be used. The dual-pressure system usually is provided with an unloading valve to permit returning the low-pressure pump volume to the tank at minimum pressure.

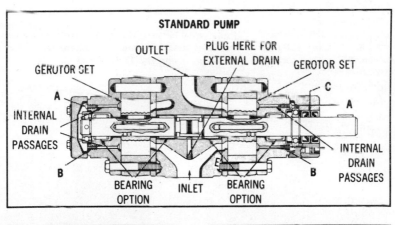

Fig. 30-5. The hydraulic reservoir can be used to serve a compound-element pump that operates on a single shaft and is attached to and driven by a single-shaft motor. (Courtesy Double A Products Co.)

Fig. 30-6. The 180-hp Oilgear *Power Pak* consists of three two-way variable-delivery pumps, a 680-gallon (2.6 kiloliters) reservoir, and the system valves for a 300-ton (272 metric tons) ingot stripper. (Courtesy The Oilgear Co.)

Fig. 30-7. Oilgear *Power Pak* includes two constant-delivery pumps capable of delivering 100 gpm (378.5 liters/min) at 5000 psi (34,473.8 kPa) and system valves for a large metal shear. Reservoir capacity is 600 gallons (2271.2 liters). (Courtesy The Oilgear Co.)

Fig. 30-8. Oilgear *Power Pak* is 19 feet (5.79 m) high, 15 feet (4.57 m) wide and 18 feet (5.49 m) long and includes six electrohydraulically controlled variable-delivery pumps supplying up to 306 gpm (1158.3 liters/min) at 5000 psi (34,473.8 kPa), all the control valves, and auxiliary components necessary to power and operate a 25,000,000-lb (11,339, 750 kg) aluminum plate stretcher. Reservoir capacity is 4000 gallons (15,141.7 liters). (Courtesy The Oilgear Co.)

Fig. 30-9. *Circuit Pak* hydraulic power package incorporating three control valves, liquid-level gauge, thermometer, suction strainer, submerged pump, pressure gauge, and internal cooling coil for water cooling. (Courtesy Double A Products Co.)

Chapter 31

Design of Fluid Power Systems for Actuating More Than One Device

Since some machines require fluid systems with circuits that are more complex than the basic circuits, the designer may be required to provide for the actuation of more than one force component, either in sequence or in unison. Some fluid systems can become quite complex. They can present problems not apparent to beginners who are using fluid power devices for the first time. Even those persons who are fairly well versed in fluid power often overlook some of these problems. In this chapter, several pneumatic and hydraulic circuits are considered. Some important suggestions regarding the selection of the components are made.

ACTUATING TWO CYLINDERS

If there is no possibility or probability of attaining the peak pressure before the first cylinder that is actuated reaches the end of its forward stroke, a sequence valve can be used to trigger the action of a second cylinder, as shown in the diagram in Fig. 31-1. In this circuit, the push button *PB-1* is depressed momentarily, which energizes solenoid *S-1* of the control valve *A* momentarily. The flow director shifts, directing oil pressure to the blind end of the clamping

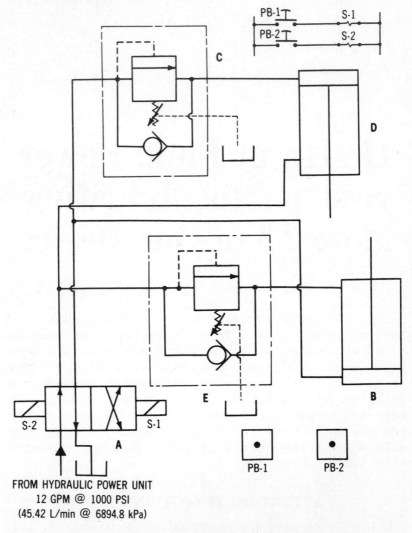

FROM HYDRAULIC POWER UNIT
12 GPM @ 1000 PSI
(45.42 L/min @ 6894.8 kPa)

Fig. 31-1. Diagram of a circuit used for operating two cylinders in sequence.

cylinder B. The piston of cylinder B advances, locking the clamp mechanism. The oil pressure builds up, opening the passage through the sequence valve C. Oil under pressure is directed to the blind end of the working cylinder D. The piston of cylinder D advances to perform the desired operation. When the piston of cylinder D reaches the end of its stroke, the push button PB-2 is contacted momentarily. The solenoid S-2 is energized momentarily. Oil pressure is directed to the rod end of the working cylinder D. The piston of cylinder D retracts. The pressure builds up. The orifice through the sequence valve E opens, directing oil pressure to the rod end of the clamping cylinder B. The piston of cylinder B retracts to release the workpiece.

By using sequence valves, a four-way directional control valve can be eliminated. However, if there is a possibility that the sequence valve may be triggered prematurely, two four-way directional control valves are more desirable. If a flow control valve is required to control the forward stroke of the first cylinder, a sequence valve may cause a problem. A circuit using two four-way directional control valves to perform the sequencing operation is diagrammed in Fig. 31-2. In the circuit, the piston of cylinder A advances. A swing-type cam on the piston rod contacts the roller of the limit switch LS-1—on the return stroke. It swings over the roller without depressing it. When the roller of the limit switch LS-1 is depressed, the solenoid S-1 of the directional control valve B is energized momentarily. The flow director shifts to direct oil pressure to the blind end of the work cylinder C. The end of the piston rod of the working cylinder is equipped with a swing-type cam which swings over the roller of the limit switch LS-2 as the piston rod advances. When the piston is retracted, the cam momentarily depresses the roller of the limit switch LS-2. The solenoid S-2 of the directional control valve D is energized momentarily, shifting the flow director to its original position. The oil pressure is directed to the rod end of cylinder A. The piston retracts.

In some instances, it is desirable to operate a cylinder from one pump and a second cylinder from another pump, as shown in the diagram in Fig. 31-3. The flow director of the directional control valve A is shifted to direct the oil pressure from the pump B to the blind end of the clamping cylinder C. The piston advances. The workpiece is clamped. The flow director in the directional control valve D is then shifted. Oil pressure from pump E is directed to the

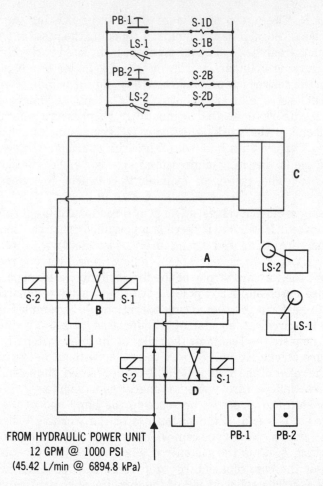

Fig. 31-2. Diagram in which limit switches are used for actuating two directional control valves which control two cylinders for performing a sequencing operation.

blind end of the working cylinder F. The piston in cylinder F advances at a speed controlled by the orifice in the flow control valve G. When the work has been completed, the flow director in the directional control valve D is returned to its original position. The oil pressure is directed to the rod end of the working cylinder F. The piston retracts. The flow director in the directional control valve A is returned to its original position. Oil pressure is directed to the rod

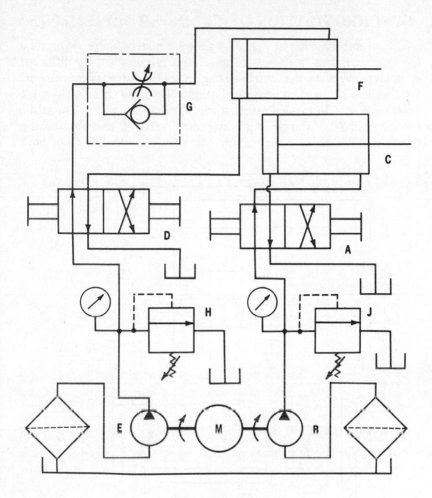

Fig. 31-3. Diagram of a circuit in which a cylinder is operated from one pump, and a second cylinder is operated from a second pump.

end of the clamping cylinder C. The piston of cylinder C retracts, releasing the workpiece.

If it is desirable to interlock the two cylinders, so that the clamping cylinder cannot release the workpiece before the piston in the working cylinder has retracted, solenoid valves can be used on the directional control valves A and D. An electrical interlock can be installed.

SYNCHRONIZATION OF CYLINDER MOVEMENTS

The circuit diagram in Fig. 31-4 shows a nearly foolproof method for synchronizing the movements of the pistons in two different cylinders. Air is the medium used to provide the force for movement. Hydraulic oil is used to provide the control. In the circuit, the push button *PB-1* is depressed momentarily. This momentarily energizes the solenoid *S-1* of the air four-way control valve *A*, thereby shifting the flow director to direct air pressure to the blind ends of both

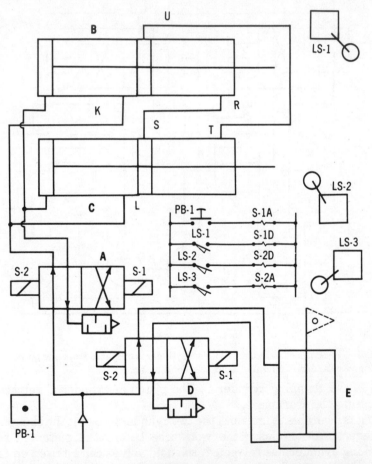

Fig. 31-4. Diagram of a circuit utilizing an air cylinder and two air-hydraulic cylinders to synchronize the movements of the pistons in the two cylinders.

cylinders B and C. The pistons of both cylinders advance in unison, as a result of the oil flow in the closed hydraulic circuit in cylinders B and C. As the two pistons advance, the oil flows from port R of cylinder B to port S of cylinder C and from port T of cylinder C to port U of cylinder B. As the pistons of cylinders B and C approach the end of their forward stroke, the limit switch LS-1 is contacted momentarily. This causes the solenoid S-1 of the control valve D to be energized momentarily. The flow director in control valve D shifts. Air pressure is directed to the blind end of cylinder E. The piston of the push-off cylinder E advances, ejecting the workpiece. Then the cam on the piston rod contacts the limit switch LS-2. This momentarily energizes the solenoid S-2 of the control valve D. Air pressure is then directed to the rod end of cylinder E. The piston of cylinder E retracts. As it approaches the retracted position, the swing-type cam on the end of the ejection mechanism trips the limit switch LS-3. This momentarily energizes the solenoid S-2 of the control valve A. The flow director shifts to its original position. Air pressure is directed to the ports K and L of cylinders B and C, respectively. The two pistons in cylinders B and C retract in unison to complete the cycle.

Any number of air-hydraulic cylinders can be synchronized by piping the hydraulic oil from the oil port in the center cover of a cylinder to the oil port in the rod end of another cylinder. Cylinders with identical bore sizes and lengths of stroke must be used. Uneven loading does not affect the movement of the pistons in the cylinders.

An all-hydraulic circuit for synchronizing the movements of the pistons in two different hydraulic cylinders is diagrammed in Fig. 31-5. This method is not so accurate as the method illustrated in Fig. 31-4. However, higher pressures can be used. In the circuit, flow dividers are used. The flow divider can be either a valve or a pair of matched hydraulic motors. In the diagram in Fig. 31-5, the flow director of the two-position four-way directional control valve A is shifted, directing the oil flow to the inlets of the two matched hydraulic motors B, which are coupled together. The same volume of oil passes from the outlet of each motor to the blind ends of cylinders C and D. The pistons of the two cylinders advance in unison to the end of their forward stroke. The flow director of the directional control valve A is then shifted to its original position. Oil pressure is directed to the rod ends of cylinders C and D. The pistons in the cylinders retract. The oil exhausting from the two cylinders passes through the check valves E and F to the control valve A and onward to the

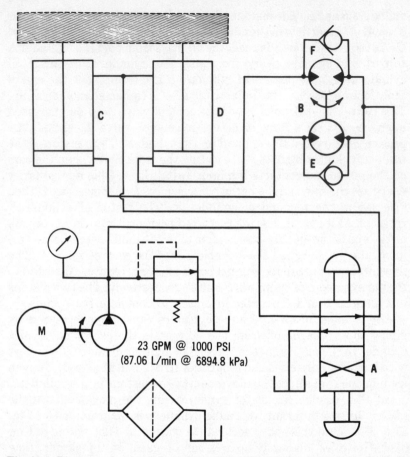

Fig. 31-5. Diagram of a circuit in which a pair of hydraulic motors with matched output volumes are used as flow dividers to achieve synchronization of piston movements.

reservoir of the power unit. The output volumes of the two fluid motors can be matched very closely.

Cylinder movements are often synchronized by mechanical means. In one method, a rack is placed on the end of each piston rod and the two gears are connected by a common shaft. Although this mechanism is somewhat bulky, it effects a close synchronization of the two cylinders.

In another method of synchronizing piston movements, the piston rods of several cylinders are attached to a heavy movable structural member. The ends of the member are placed in close-fitted machine ways with long bearings. The cylinders are anchored on another heavy structural member which is stationary. By containing the cylinders closely, synchronization is rather easy to obtain.

In a pneumatic circuit, it is sometimes necessary to actuate an air cylinder and an air motor to obtain both linear and rotary motion. For example, the air cylinder may be used to move a brushing head or a polishing head which is actuated by an air motor. This type of circuit is illustrated in the diagram in Fig. 31-6. The push button *PB-1* is depressed momentarily. This causes the solenoid *S-1* of the four-way control valve *A* and the solenoid *S-1* of the two-way control valve *B* to be energized momentarily. The flow directors in both control valves shift. Air pressure at 90 psi (620.5 kPa), which is set by the pressure regulator in the assembly *C*, is directed to the blind end of carrier cylinder *D*. The air pressure at 60 psi (413.7 kPa), which is set by the pressure regulator *E*, is directed to the air motor *F*. The piston in the cylinder *D* advances at a speed controlled by the speed control *G*. As the piston reaches the end of the stroke, the cam on the end of the piston rod contacts the limit switch *LS-1*. This energizes the solenoid *S-2* of the control valve *A* and solenoid *S-2* of control valve *B*. The flow directors of both valves shift to their original positions. The flow to the air motor *F* is stopped. Air pressure is directed to the rod end of the cylinder *D* through the speed control *G*. The piston in the cylinder retracts to complete the cycle. If it is not practical to trip the limit switch with a cam on the end of the piston rod, it may be advantageous to select a cylinder with a switch built into the front head. Then the switch is actuated by the piston.

INTERLOCKING HYDRAULIC AND PNEUMATIC SYSTEMS

A complete hydraulic system and a complete pneumatic system can be interlocked to accomplish three different actions (Fig. 31-7). The three motions accomplished are a rotary action, a clamping action, and a reciprocating action. The two fluid systems are interlocked by means of solenoid-operated directional control valves and electrical relays.

The operator loads the workpiece onto the spindle, and energizes the solenoid *S-1* of the directional control valve *A*. Air pressure is

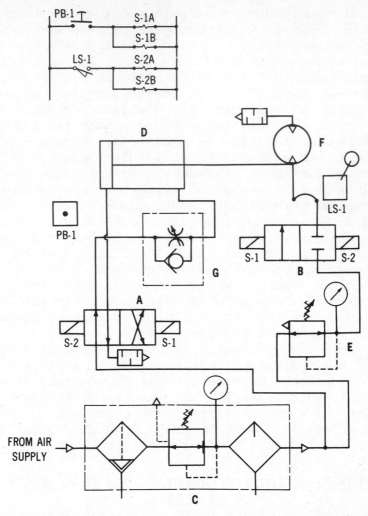

Fig. 31-6. Diagram of a pneumatic circuit in which an air motor is used in conjunction with an air cylinder to obtain both linear and rotary motion.

directed to the rod end of the rotating cylinder B. The piston of the cylinder retracts, gripping the workpiece. The solenoid S-3 of the directional control valve C is energized. The valve spool shifts, directing oil pressure to port 1 of the fluid motor D. The speed of rotation of the fluid motor is controlled by the flow control E.

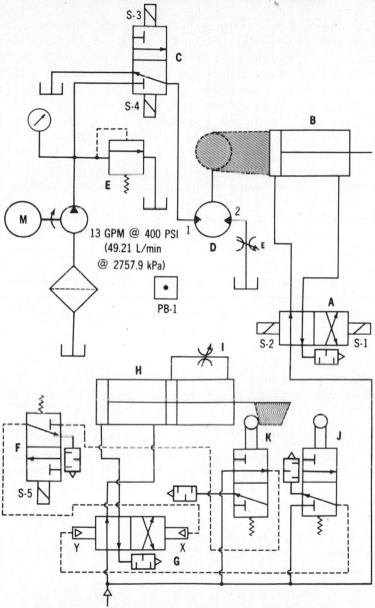

Fig. 31-7. Diagram of a circuit in which a complete hydraulic system and a complete pneumatic system are interlocked to accomplish a rotary action, a clamping action, and a reciprocating action.

The solenoid S-5 is energized, permitting air pressure to flow through the three-way control valve F to the pilot connection X of the control valve G. The spool of valve G shifts. Air pressure is directed to the blind end of the air-hydraulic cylinder H. The piston of the cylinder advances at a speed controlled by the orifice in the valve I in the oil section of the air-hydraulic cylinder.

As the piston in cylinder H approaches the end of its stroke, the cam on the worktable contacts the roller on the three-way control valve J. Air pressure is directed to the pilot connection Y of the control valve G. The spool shifts to its original position. Air pressure is directed to the rod end of the air section of the air-hydraulic cylinder H. The piston then retracts.

The feed mechanism continues to reciprocate until the final work is performed on the workpiece. The solenoid S-5 is released. The piston in cylinder H stops in the retracted position. The solenoid S-4 is energized. The spool in the control valve C shifts to its original position. The oil flow to the hydraulic motor D is stopped. The spindle stops. The solenoid S-2 of the control valve A is energized. Air pressure is directed to the blind end of the clamping cylinder B. The piston advances, releasing the workpiece.

Hydraulic circuits are especially adaptable for controlling tool movements on lathes, grinders, and other precision-type machines. A control-type circuit is diagrammed in Fig. 31-8.

The circuit shown (see Fig. 31-8) is a feed arrangement for a lathe. It is controlled by cylinders A and B. Cylinder B moves the toolholder into the cutting area. Cylinder A moves the toolholder for the cross-feed cutting action.

Both cylinders approach the cutting area at a rapid rate, until the cam roller on the cam-operated flow control valve C is depressed. Then the movement of the cylinders is subjected to feed control.

While the cam follower is against the template, cylinder A holds its position. The piston in cylinder B advances. When the piston of cylinder A is advancing, cylinder B holds its position. This is a result of reduced pressure to cylinder B, which is provided by the pressure reducing valve D.

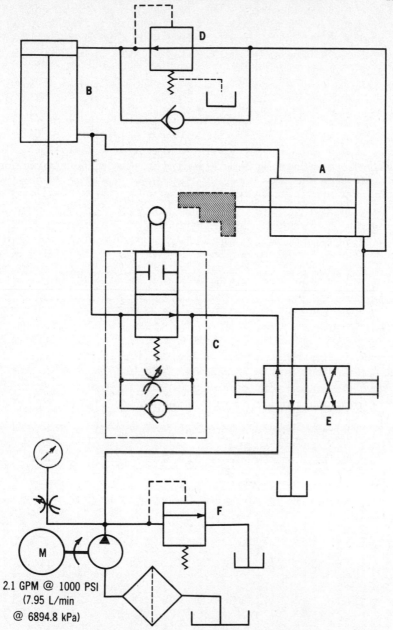

Fig. 31-8. A hydraulic circuit used to control tool movements on a lathe.

Chapter 32

Fluid Power Circuits—
Multiple Pressures

In both pneumatic and hydraulic circuits, it is often desirable to operate one portion of the system at a different pressure from another portion of the same system. Requirements for several different operating pressures may exist in the same fluid system. For example, it may be necessary to operate portion A at 1500 psi (10,342.1 kPa), portion B at 1000 psi (6894.8 kPa), and portion C at 750 psi (5171.1 kPa). Also, in many test applications it is necessary to operate the entire system at several different pressures to gather the required test data. Several circuits that can be used to obtain multiple pressures are discussed in the following paragraphs.

MULTIPLE-PRESSURE PNEUMATIC CIRCUITS

A pneumatic circuit that uses several different pressure regulators to obtain several different test pressures is diagrammed in Fig. 32-1. The pressure regulator A is set at 15 psi (103.4 kPa), regulator B is set at 25 psi (172.4 kPa), regulator C is set at 35 psi (241.3 kPa), and regulator D is set at 50 psi (344.7 kPa). A three-way directional

419

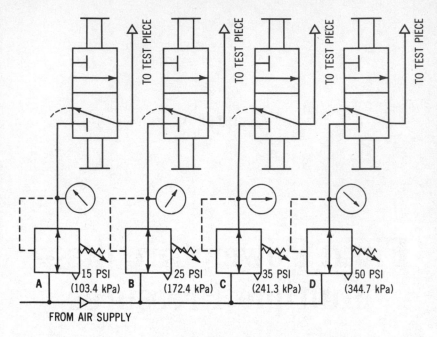

Fig. 32-1. Diagram of a pneumatic circuit in which several different pressure regulators are used to obtain several different test pressures.

control valve that is manually operated is used at each testing station. The three-way control valve is a "normally closed" valve. When air pressure is directed to the test piece, the flow director in the three-way directional control valve is shifted to direct the pressure from the pressure regulator to the test piece. When the test has been completed, the flow director in the valve is returned to its normal position. The air pressure is permitted to exhaust from the test piece. By using the proper controls, the movement of the test piece to each test station can be automatic. The testing operation also can be automatic.

Another circuit diagram in which three different pressures are accomplished by the use of a single pressure regulator is shown in Fig. 32-2. The regulator spring is depressed by a cam roller mechanism. The cam is designed for three different pressure settings, the lowest pressure being obtained first. The test piece is placed beneath the ram of the cylinder A. When the cam is in the retracted position, the operating pressure of the system is 30 psi

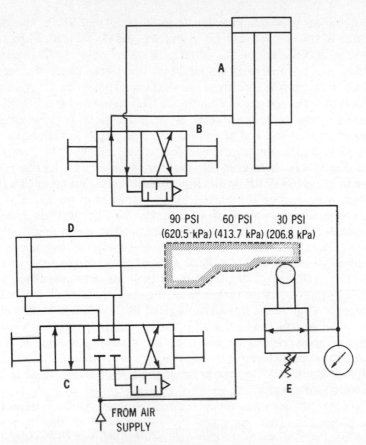

Fig. 32-2. Diagram of a pneumatic circuit which can be used to obtain three different pressures from a single pressure regulator.

(206.8 kPa). The flow director in the directional control valve B is shifted to direct air pressure to the blind end of cylinder A. The piston advances rapidly, forcing the test mechanism against the test piece. The flow director in the control valve B is then shifted to its original position. Air pressure is directed to the rod end of cylinder A. The piston retracts. The flow director in the four-way valve C is shifted to direct air pressure to the blind end of cylinder D. The piston of cylinder D advances the cam to the second lobe of the cam. The flow director is returned to the neutral position. This causes the cam roller on the regulating valve E to be depressed further, which increases the operating pressure of valve B to 60 psi (413.7 kPa). The flow director of the control valve B is shifted, and 60-psi (413.7 kPa) air pressure is directed to the blind end of cylinder A. The piston advances the test mechanism against the test piece. The flow director in the control valve B is returned to the normal position. Air pressure is directed to the rod end of cylinder A. The piston retracts. The flow director of the control valve C is again shifted. Air pressure is directed to the blind end of cylinder D. The cam is advanced to the third lobe, which further depresses the cam roller of the pressure regulator E. This increases the operating pressure of the control B valve to 90 psi (620.5 kPa). The flow director of the control valve B is shifted. Air pressure is directed to the blind end of the cylinder A. The piston advances at 90 psi (620.5 kPa), and the test mechanism meets the test piece. The flow directors of the directional control valves B and C are shifted. Air pressure is directed to the rod ends of cylinders A and D. Their pistons retract to complete the cycle.

A pneumatic circuit designed for a two-pressure operation, and used in conjunction with power chucking equipment, is shown in the diagram in Fig. 32-3. The advantage of this system is that a high clamping pressure can be used for the roughing operation. A lower clamping pressure can be used for the finishing operation. This reduces the probability of distorting the workpiece during the finishing operation. The operation of the circuit is discussed in the following paragraphs.

The operator loads the workpiece into the chuck, and depresses the hold button X. This energizes the solenoid S-1 of the four-way directional control valve A. The spool shifts to direct air pressure to port 1 of the rotating cylinder B. The piston of cylinder B retracts under the high pressure. The jaws of the chuck close on the workpiece—gripping it securely. The roughing operation is performed as the spindle revolves. The operator then depresses the hold button Y. The solenoid S-2 of the four-way control valve C is energized. The spool of the valve C shifts. Low-pressure air passing through regulating valve D is directed to port 2 of the rotating cylinder B. This applies low-pressure air on the blind end of the piston and high-pressure air on the rod end. The clamping force on the workpiece is reduced for the finishing operation, because pressure is applied to both sides of the piston. This is accomplished without stopping the spindle of the machine.

When the finishing operation has been completed, the operator

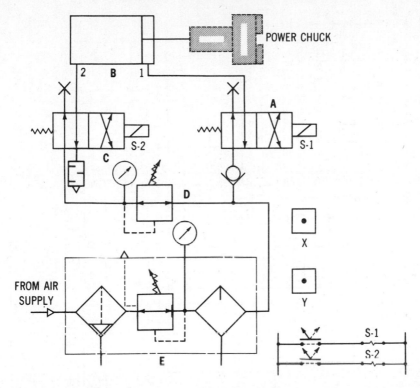

Fig. 32-3. Diagram of a pneumatic circuit designed for two-pressure operation in conjunction with power chucking equipment.

stops the spindle and releases the hold button X. This de-energizes the solenoid S-1 on the control valve A. The spring shifts the spool to its original position. The inlet of the control valve A is blocked. The air in the blind end of the cyclinder B advances the piston. The chuck jaws are opened to release the workpiece. The hold button Y is then released. This de-energizes the solenoid S-2 on the control valve C. The spring shifts the spool to its original position. The inlet of valve C is blocked. Air in the blind end of cylinder B is directed to exhaust. The cycle is completed.

In some applications, several different forces are required to complete an operation (Fig. 32-4). These different forces can be obtained by using cylinders that differ in size. However, this may be impossible, since the force required for a job may be so small that a

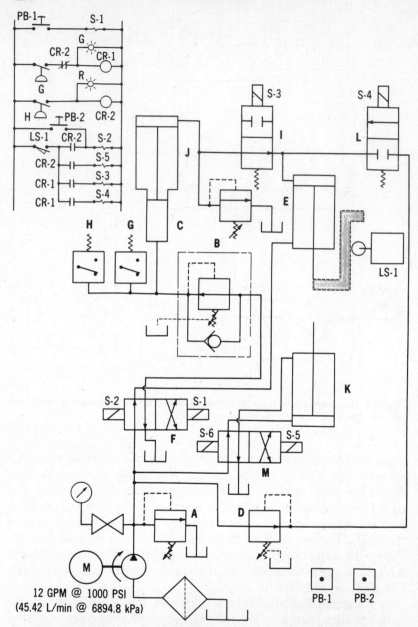

Fig. 32-4. Diagram of a hydraulic circuit that provides multiple operating pressures for completing an operation.

cylinder of suitable bore size is not available. A change in the operating pressures available in various portions of the system can accomplish this, as outlined in the following paragraph. This system is used to press fit (light press) the cylinder liners into the cylinder bodies.

The main operating pressure of the system (see Fig. 32-4) is controlled by the relief valve A on the power unit. The pressure reducing valve B controls the pressure to cylinder C. The pressure reducing valve D controls the pressure to the top or blind end of cylinder E. To operate the circuit, the operator loads the liner into the fixture and momentarily depresses the push button $PB-1$. This momentarily energizes the solenoid $S-1$ of the directional control valve F. The spool of valve F shifts, directing oil pressure to the pressure reducing valve B, and onward to the blind end of cylinder C. The pressure switches G and H are also in this line. A green indicator light is connected to the pressure switch G. A red indicator light is connected to the pressure switch H. The green light on switch G indicates a satisfactory press fit. The red light on switch H indicates a press fit that is too tight. The passage in the two-way control valve I is "normally open." As the piston in cylinder C begins to move low-pressure oil in cylinder J and since a large difference in piston area exists between cylinders C and J, the oil pressure is directed to the top or blind end of the press cylinder E. The piston of cylinder E advances. If the press fit is too tight and the red indicator light has been on when the limit switch $LS-1$ is contacted, the solenoids $S-2$ and $S-5$ are energized. The piston of cylinder E is retracted. The piston of cylinder K advances at the pressure set by the relief valve A. If the green indicator light is on when the limit switch $LS-1$ is contacted, the solenoid $S-3$ closes the control valve I. The solenoid $S-4$ opens the control valve L. The pressure set by the pressure reducing valve D is applied to the top or blind end of cylinder E to finish the operation. The operator then momentarily depresses the push button $PB-2$. This energizes the solenoid $S-2$. The spool of the control valve F is returned to its normal position. The oil pressure is directed to the rod end of cylinder E. The piston of cylinder E retracts, completing the cycle.

MULTIPLE-PRESSURE HYDRAULIC SYSTEMS

A rather ingenious circuit diagram is shown in Fig. 32-5. Air, a

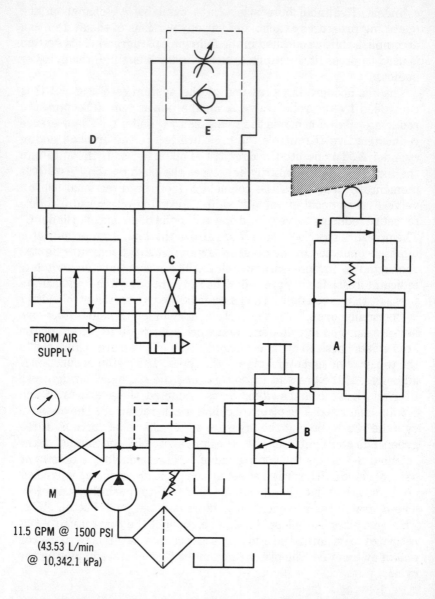

Fig. 32-5. Diagram of a setup in which air, a combination of air and hydraulics, and hydraulics are used to provide a multiple-pressure hydraulic system for test operations and press applications.

combination of air and hydraulics, and hydraulics are used to set up a multiple-pressure hydraulic system which can be used for test operations and press applications in which the pressure is varied several times during each cycle. The most essential component in the circuit is a cam-operated relief valve that can handle a wide range of pressures. By means of a long cam with a small rise connected to the end of the combination air and hydraulic cylinder, the stem of the relief valve is depressed gradually, until the maximum pressure is reached in the hydraulic system. The combination cylinder can be stopped at any position during the stroke by merely returning a three-position air directional control valve to the neutral position, thereby blocking both cylinder ports. It has been found that if a small-volume pump is used on the hydraulic power device, a slight fluctuation which seems to be caused by the pulsations of the pump appears on the pressure gauge. This fluctuation is not so noticeable on the large-volume pumps. This condition can be overcome by the addition of a bladder-type accumulator to the circuit to serve as a dampener. The functioning of the circuit is described in the following paragraph.

The operator places the work beneath the die on the end of the ram cylinder A, and then shifts the handle of the hydraulic two-position four-way directional control valve B. Oil flows to the blind end of the ram cylinder A at low pressure. The operator shifts the handle of the three-position four-way air control valve C. The air flows to the blind end of the air hydraulic cylinder D. The piston of the air-hydraulic cylinder D starts its forward stroke at a speed set by the flow control E on the closed hydraulic circuit. On this cylinder, the cam on the end of the piston rod contacts the cam roller on the hydraulic relief valve F. Pressure begins to build up in the hydraulic circuit. The ram of cylinder A moves forward at a pressure set by the cam-operated relief valve F. When the air-hydraulic cylinder reaches the end of its stroke, full pressure is built up through the relief valve F. The operator can control this pressure by merely shifting the four-way valve C into the neutral position at any time during the outward stroke of cylinder D. The operator shifts the handles of the air four-way control valve C and the hydraulic four-way control valve B, causing the pistons of cylinders D and A to retract. The minute feed of the air-hydraulic cylinder C permits any pressure value within the entire range of the relief valve to be obtained. This is an important feature in many applications.

In many hydraulic applications, it is advantageous to use a low-pressure hydraulic system and still be able to produce a higher pressure in a portion of the system for performing certain functions. This type of system is diagrammed in Fig. 32-6. When the four-way control valve A is actuated, the piston of the ram cylinder B advances rapidly at low pressure. The intensifier C is also prefilled. As the piston of cylinder B contacts the workpiece, pressure builds up sufficiently to open the bypass valve D (external pilot). The full pump flow is directed to the large ram of the intensifier C. Intensified pressure is directed to the blind end of cylinder B. The check valve E prevents high-pressure fluid from reaching the control valve A. The high-pressure fluid completes the work cycle. The spool in the control valve A is shifted. The pump output at low pressure is directed to the rod end of cylinder B. The low-pressure oil is also directed to the pilot connection of the check valve F. This opens the decompression piston of the check valve F. This permits the slow decompression of the high-pressure fluid from the lower side of the piston of cylinder B. When the pressure is decompressed to the proper value, the piston of the check valve F opens and permits the rapid return of the piston of cylinder B. The low-pressure input side of the intensifier exhausts through the return free flow check of valve D. The intensifier is reversed by the pressure built up previously on the ram cylinder and the return flow, as the ram retracts. The high-pressure line from the intensifier output port to the cylinder should be kept at a minimum length.

A circuit that includes a high-pressure cylinder and pump and a low-pressure four-way directional control valve and pump is used in some important applications (Fig. 32-7). To operate the circuit, the operator momentarily depresses the electric push button $PB\text{-}1$, which energizes the electric relay and the solenoid of the solenoid-operated four-way control valve A. It is necessary that the valve size is large enough to deliver the volume of oil produced by the pump B. The spool of valve A shifts. The oil is delivered to the blind end of cylinder C by pump B, until a resistance higher than 250 psi (1723.7 kPa) is met. Then pump D delivers at 10,000 psi (68,947.6 kPa) to finish the advance stroke. The contacts of the pressure switch E then open. The relay drops out, de-energizing the solenoid of the control valve A. The spool of the control valve shifts to its original position. Oil from pump B is directed to the rod end of cylinder C and to the pilot chamber of the check valve F, which has a 40-to-1 ratio.

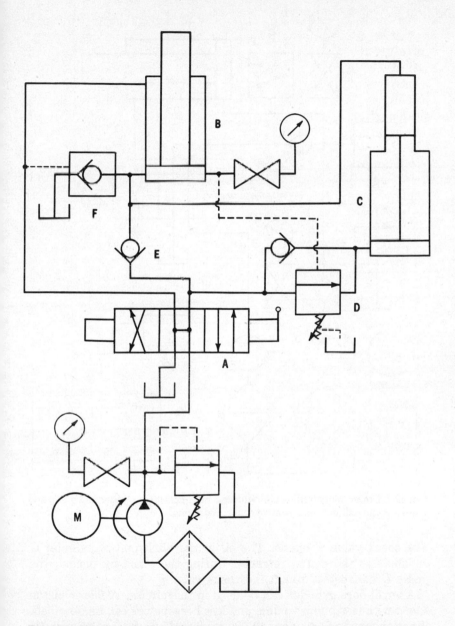

Fig. 32-6. Diagram of a hydraulic circuit in which a low-pressure power supply is used to obtain a high-pressure work cycle.

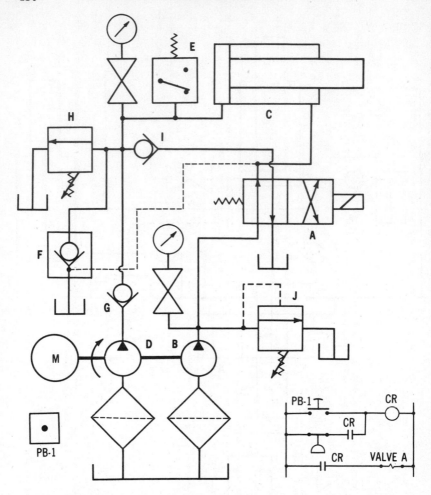

Fig. 32-7. Diagram of circuit that includes a high-pressure cylinder and pump and a low-pressure directional control valve and pump.

The check valve F opens. The oil in the blind end of cylinder C exhausts as the piston retracts. During the stand-by period, the pump D exhausts through the check valve F.

A small-bore cylinder can be used in this circuit. A faster piston speed with less pump volume and less horsepower can be obtained. When the pump D delivers 10,000 psi (68,947.6 kPa) to the cylinder C, the necessary tonnage is applied by the cylinder.

The use of a hydraulic pressure reducing valve to provide a two-pressure circuit is shown in the diagram in Fig. 32-8. The flow director in the control valve A is shifted to direct full pressure to the blind end of the clamping cylinder B. The piston advances to the end of its stroke. The flow director of the control valve C is shifted. The oil passing through the pressure reducing valve D is directed through the control valve C to the blind end of the working cylinder E. The

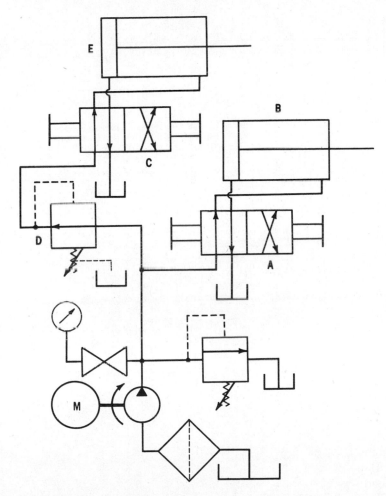

Fig. 32-8. Diagram of a hydraulic circuit showing how to use a pressure reducing valve to provide two different pressures.

piston advances to the end of its stroke. The flow director of the control valve C is returned to its original position. Oil at reduced pressure flows to the rod end of cylinder E. The piston retracts. The flow director of the control valve A is shifted to its original position. Oil pressure is directed to the rod end of cylinder B. The piston retracts to unlock the clamp and to complete the cycle.

Chapter 33

Accumulators—Design

The hydraulic accumulator is primarily a device for storing pressurized hydraulic fluid. The three basic designs or types of accumulators are: (1) weight or gravity type; (2) spring-loaded type; and (3) gas type.

WEIGHT-TYPE ACCUMULATOR

The weight-type or gravity-type accumulator consists of a cylinder, a movable piston, and a mass or weight applied to the piston (Fig. 33-1). The deadweight can be concrete, iron, steel, water, or other heavy material. The weight is sometimes placed in a container. This permits either increasing or decreasing the weight against the piston of the accumulator. One design provides a container for water, which is easily adjusted for weight.

A precision fit is required for the piston within a honed cylinder wall. An effective seal is necessary to prevent leakage past the piston. This seal withstands the full pressure of the hydraulic fluid. As the hydraulic oil is pumped into the cylinder chamber below the piston, the weighted piston rises to provide a larger chamber for holding the stored oil. In the stored condition, the energy placed in

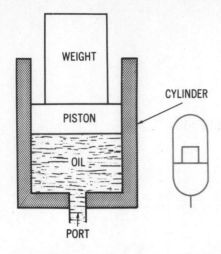

Fig. 33-1. Weight-type accumulator.

the hydraulic fluid by the hydraulic pump is stored in the mass or weight of the piston. This energy is released in the downward motion of the piston in quantities required by the demands of the hydraulic system.

This type of accumulator is often custom-built to fit a given installation. Since the total pressure that can be created or stored within the accumulator is equal to the weight applied to the piston area divided by the piston area, the pressure value can be increased or decreased by varying the weight of the piston.

An advantage of this type of accumulator lies in the nearly constant oil pressure provided by the gravity force of the piston throughout its entire stroke. By providing an adequate piston area and an ample piston stroke, a large volume of fluid can be supplied at any desired pressure value. One large gravity-type accumulator often provides service for several different machines, especially in a centralized hydraulic system. The accumulator can be mounted only in a vertical position. The gravity-type accumulator has declined in popularity in recent years, because of the large mass of the unit and the lack of flexibility in locating and mounting the unit.

SPRING-LOADED ACCUMULATOR

The spring-loaded accumulator (Fig. 33-2) is similar in construction to the weight-type accumulator. However, the piston "weight"

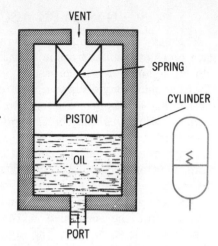

Fig. 33-2. Spring-loaded type of accumulator.

is provided by a spring located behind the piston. The chief advantage of the spring-loaded accumulator over the weight-type accumulator is that simple adjustments in pressure settings can be made by varying the spring tension by means of screw-type controls. While the oil is being pumped into the accumulator, the piston moves against the spring, compresses it, and stores the pressure energy in the compressed spring. Then, the stored energy is released on demands for oil. The spring is permitted to extend and relax.

One distinct difference between the spring-type accumulator and the weight-type accumulator is the lack of constant pressure at all positions of the piston. While more oil is being forced into the storage chamber of the spring-loaded accumulator, the spring's compression rate increases. The pressure of the stored hydraulic fluid also increases. Therefore, the piston seals, like the seals in the weight-type accumulator, must be capable of withstanding the full pressure of the fluid.

The spring-loaded accumulator is much more limited in size and volumetric capacity, because it is limited in spring closure. However, it is slightly more flexible than the weight-type accumulator, because it can be mounted in positions other than the vertical position. Since the spring-loaded unit can be enclosed completely, it is slightly easier to prevent contaminants from entering the unit, but another limitation lies in maintenance of the spring.

GAS-TYPE ACCUMULATOR

The gas-type accumulator differs in construction from the other two types of accumulators in that the pressurized hydraulic fluid is stored against a gas pressure (usually dry-pumped nitrogen), thereby using Boyle's law in regard to the relations of volume and pressure. The gas-type accumulators can be divided further into the *nonseparator-type* and the *separator-type* accumulators.

In the nonseparator type of accumulator, the oil inside the accumulator is in direct contact with the gas (Fig. 33-3). The nonseparator accumulator consists of a fully enclosed cylinder, adequate ports, and a charging valve. A so-called "free surface" exists between the hydraulic oil and the gas. While more oil is being pumped into the accumulator, the gas above the hydraulic oil is compressed even further. The energy of the compressed hydraulic fluid is stored in the compressed gas. It is released as it is required by the demands on the system.

Since hydraulic oil under compression possesses an affinity for air or gas, its capacity for absorbing and holding the gas increases as the pressure increases in the hydraulic fluid. This results in a gradual loss of gas pressure in the nonseparator-type accumulator. It must be recharged periodically. A portion of the hydraulic oil must remain in the bottom of the cylinder to prevent the gas escaping into the hydraulic circuit. Usually, a mechanical float-type device is used to

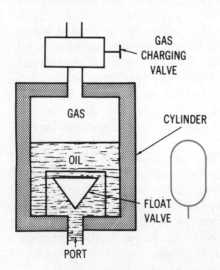

Fig. 33-3. Nonseparator gas-type accumulator.

limit the discharge of oil, thereby providing a minimum fluid level within the vessel.

The nonseparator-type accumulator can be mounted only in a vertical position. This prevents the escape of the gas into the system. When the gas volume is reduced appreciably, the effectiveness of the accumulator is also reduced.

The most common and by far the most widely accepted accumulator is the separator-type accumulator. Two types of separator accumulators are the free-floating *piston-type* accumulator and the flexible-barrier or *diaphragm-type* accumulator. In the piston-type accumulator, the piston is a barrier between the gas and the hydraulic oil. High-pressure gas is charged into the space on one side of the piston. Hydraulic oil is pumped into the opposite side. The packing on the piston separates the oil and the gas. This packing usually consists of two "O" rings, with pressure balancing ports to equalize the pressures on both sides of the "O" rings. This eliminates pressure lock between the "O" rings. A dashpot decelerates the piston at the end of its stroke. The entire volume of the unit can then be used without any hammering. The piston-type accumulator is similar to a hydraulic cylinder without a piston rod.

Since the piston is free-floating, the differential pressure across the piston is reduced to the frictional drag of the piston against the accumulator walls. The only position in which the differential pressure becomes appreciable is when the piston is extended. Then the compressed gas is expanded, and the oil is discharged completely from the oil section of the accumulator. The inner surface of the accumulator wall should be machined and honed with extreme care and precision to minimize frictional drag and to effect the best possible seal.

The piston-type accumulator can be installed in any position—either vertical or horizontal. The most desirable position, however, is with the piston moving in a vertical direction, with the gas connection at the top. A double-shell construction of a piston-type accumulator is shown in Fig. 33-4. This construction provides a pressure-balanced inner shell which contains the piston and serves as a separator between the gas and the hydraulic fluid. The outer shell serves as a gas container and as a protective cover to prevent small dents interfering with the free movement of the piston. The outer shell absorbs the total pressure of the unit. Rapid decompression of the precharged gas resulting from a rapid discharge of the working

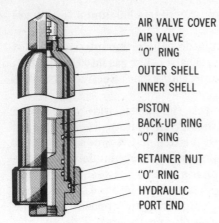

AIR VALVE COVER
AIR VALVE
"O" RING
OUTER SHELL
INNER SHELL

PISTON
BACK-UP RING
"O" RING

RETAINER NUT
"O" RING
HYDRAULIC
PORT END

Fig. 33-4. Piston-type accumulator.

hydraulic fluid provides a cooling effect for the entire working area of the inner shell. This can be especially important during rapid cycling.

A diaphragm-type accumulator with a flexible barrier is shown in Fig. 33-5. The accumulator may be of double-hemisphere construction, with the two hemispheres bolted together and a rubber diaphragm clamped between them. While the gas is being compressed, the pressure increases. The compressed gas functions like a spring against the diaphragm and the confined hydraulic fluid.

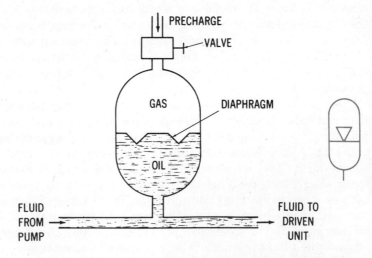

PRECHARGE
VALVE
GAS
DIAPHRAGM
OIL
FLUID
FROM
PUMP
FLUID TO
DRIVEN
UNIT

Fig. 33-5. Diaphragm-type accumulator.

Since the separating member is flexible, the oil pressure and the gas pressure within the vessel are equal.

The bag-type or bladder-type accumulator (Fig. 33-6) is usually a seamless steel shell that is cylindrical in shape and spherical at both ends. A gas valve is located at one end of the shell and opens into the shell. A large opening for insertion of the bag is located at the hydraulic fluid end. The bladder is made from synthetic rubber, usually *Buna-N*, or a compound compatible with the fluid used in the system. The enclosed bladder, including a molded air stem, is fastened by means of a lock nut to the upper end of the shell. At the opposite end of the shell, a plug assembly, including the oil port and a poppet valve, is mounted. The purpose of the poppet valve is to prevent the bladder extruding into the hydraulic oil passage. An additional safety feature is provided in this type of accumulator in that it cannot be disassembled while the gas charge is inside the bag.

The manufacturers of bladder-type accumulators specify that they should be mounted in a vertical position with the gas valve at the top of the unit. The flexibility of the piston-type accumulator is provided by this unit. However, caution should be exercised in subjecting the bladder or diaphragm-type accumulator to high oil temperatures. The piston-type accumulator with "O" ring seals can be adapted more easily and more economically to high-temperature applications, because high-temperature "O" rings are more readily available.

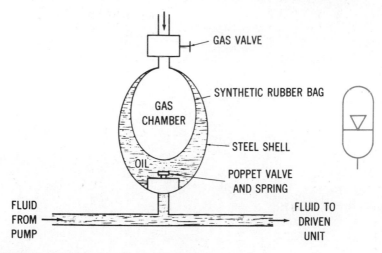

Fig. 33-6. Bag or bladder type of accumulator.

Chapter 34

Accumulators—
Application

In regard to the history of fluid power, the hydraulic accumulator was used prior to the development of the hydraulic pump and many of the other hydraulic components. In its original design, the accumulator was primarily a storage chamber for accumulating a mass of fluid for performing work through the accumulated weight of the fluid. By directing a stream of falling water into a container, the accumulating weight of the container and the contained water were used to perform work.

The modern widespread use of hydraulic equipment and its components probably originated with the development of the hydraulic press. From the beginning of fluid power development, hydraulic presses have been the chief application for fluid energy. In many of the early hydraulic presses, an accumulator was the sole source of fluid force. As the industry developed, motor-driven pumps were developed. The hydraulic accumulator became even more essential. The use of the accumulator in conjunction with the rather basic motor-driven pump sources created a practical efficiency for multiple-press installations.

The most common hydraulic systems have used a single pump as a source of power, with the individual pump sources for powering individual machines being a comparatively recent innovation. Many multiple-press installations still use a single power unit, along with one or more accumulators for smoothing the fluctuating demand from various presses or machines. Any operation that requires a considerable, but intermittent, flow of pressurized fluid is readily adaptable to the intermediate storage facilities provided by an accumulator. Thus, a rather large installation with appreciable idle time can be powered by a comparatively small, but efficient, power unit whose intermittent capacity is stretched by means of the accumulator.

Normally, hydraulic fluid is considered incompressible, but a compressibility factor is involved. For example, at 3000 psi (20,684.3 kPa), the total compression of oil is approximately 1.2 percent of the volume. If 1 gallon (3.785 liters), or 231 cubic inches (3785.4 cm³), of hydraulic oil is compressed to 3000 psi (20,684.3 kPa), the release of only 2.8 cubic inches (45.9 cm³) from that volume results in a pressure drop from 3000 psi (20,684.3 kPa) to zero pressure. This indicates the need for a compressible gas to store the oil against, inside the accumulator. Since hydraulic fluid is a liquid, its dynamic power storage quality is low. On the other hand, gases possess a high degree of compressibility. Fluid energy can be stored readily against the compressibility of the gas.

FUNCTIONS

When the accumulator is used for storage under pressure, it functions as an auxiliary source of energy in excess of the pump output during times of peak demands. Or it serves as a source of auxiliary energy when the pump is either inactive or inoperable in that portion of the system. In Fig. 34-1, the accumulator merely stores the excessive pump output during one portion of the cycle, and releases it to the cycle when the volumetric demand exceeds the pump output. The diagram in Fig. 34-2 shows a circuit in which the pump output is stored within the accumulator during the idle time of the pump. Then the stored fluid energy is available to a portion of the circuit that is not included in the main operation.

Another common function of the accumulator is to absorb pulsations and shocks within the hydraulic system. As mentioned

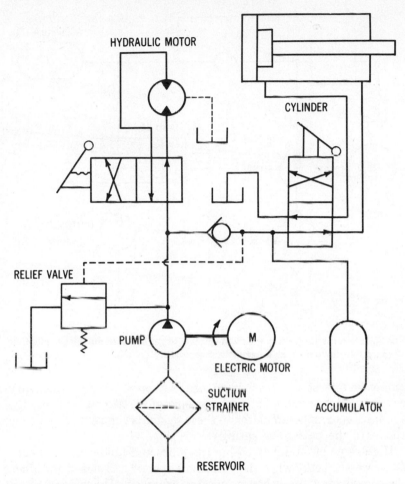

Fig. 34-1. Circuit diagram in which the excess pump output in one portion of the cycle is stored for use in another portion of the cycle.

previously in the discussion of pumps and motors, the various pumps can produce different results in regard to pulsation within the fluid delivery. The degree of pulsation varies in each type or style of pump. The pulsations, even within a given type of pump, are usually of higher magnitude at the higher pressure levels. Even in the piston-type pump, which is more often used when the higher pressures are required, the pulsations are quite prevalent. The pulsations can produce undesirable effects in hydraulic systems, and

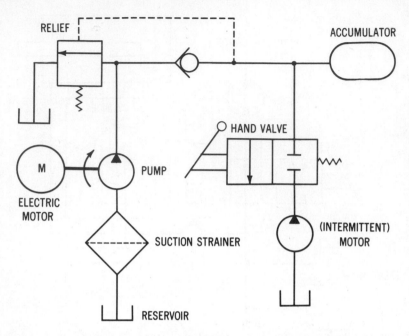

Fig. 34-2. Circuit diagram in which the pump output is stored during the idle time of the pump for use in an intermittent secondary circuit.

cannot be tolerated in some instances. Although the pulsations may be modulated to some extent by means of restrictive orifices, the application of an accumulator to these systems can reduce and even eliminate the pulsations entirely.

High rates of fluid flow within the pipes and fluid conductors can fluctuate markedly when the valves are opened and closed and when the components are either stopped or reversed. The sudden stops, reversals, and ensuing shocks can exert tremendous forces on the components, gauges, and fluid conductors and joints. These forces are often quite audible. The force transmissions are visible in the form of external movement and jerking of the fluid conductor lines. These surges and shocks can create pressure forces, although they may be only momentary, that far exceed the rated pressures of the components that contain the fluid. In some installations, the shock forces defy containment. Tremendous problems are created in eliminating leakage in a system. Repeated shock forces can result in the failure of the pipes, fittings, and even the components. Aside from failures which may occur, the vibration and the undesirable movement of the conductors and components can be objectionable. Accumulators placed at the pipe elbows and other strategic points within the system can either eliminate these shocks and surges or reduce them to acceptable levels. In this type of application, the accumulator functions only as a shock cushion, using the compressibility of the enclosed gas to modulate or dampen the undesirable and damaging surges.

Another typical application for the hydraulic accumulator is in a closed system that provides a volumetric capacity for a given volume of fluid (Fig. 34-3). The normal expansion of the fluid resulting from the changes in temperature is compensated for by the accumulator in the circuit. The accumulator also provides make-up oil to compensate for internal or external leakage.

The accumulator also can be used to store energy created by an external force or factor. In a rather unusual application, the accumulator is used as a storage chamber for the energy represented by the weight of materials lowered from a stored location to a floor level. A typical example of this application is the unloading of a truck or other carrier onto a loading dock which is several feet below the level of the carrier bed (Fig. 34-4).

As shown in Fig. 34-4, no auxiliary power unit or source of energy other than the energy of the weight represented by the material on the elevator platform is needed. A gas charge large enough to expel the oil from the accumulator chamber and large enough to raise the lift cylinder of the elevator platform is placed in the accumulator. At this point in the functioning of the circuit, a valve is closed, locking

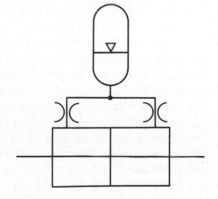

Fig. 34-3. Diagram in which a double-end-rod cylinder is used as a speed control device for a free-moving member. The two orifices establish the speed, and the accumulator creates a replenishing supply of oil to compensate for fluid that may be lost at the rod packings.

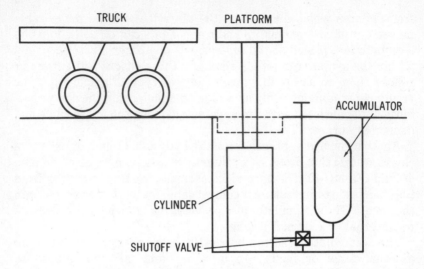

Fig. 34-4. An accumulator application in which neither an auxiliary power unit nor a source of energy other than the weight represented by the material is needed to unload a truck or carrier onto a loading dock that can be located several feet below the level of the bed of the carrier.

the fluid in the elevator cylinder and holding the elevator platform in the raised position. As the material is unloaded from the higher level onto the elevator platform, the energy represented by its weight forces the oil from the elevator cylinder into the accumulator, compressing the accumulator storage gas when the valve is opened to permit the flow. The elevator platform travels to the lower unloading level as a result of the weight of the platform and the material thereon. The valve is closed again, trapping the pressurized oil in the oil section of the accumulator.

The elevator platform remains in the lowered position, even after the weight has been removed from the elevator platform. The energy represented by the weight of the material has been stored in the accumulator against the compressed gas in the unit. The empty elevator platform can be raised again to the top position merely by opening the valve and permitting the stored pressurized fluid to flow from the accumulator into the cylinder of the elevator. Thus the energy represented by the weight of the material is transferred to the hydraulic fluid and stored within the accumulator against the compression of the enclosed gas.

CALCULATIONS

The selection of the accumulator in the preceding illustration of the elevator platform involves the principles and mathematics of Boyle's law. The weight of the elevator platform and its accessories should be determined first. Assume the weight to be 500 pounds (226.8 kg) for the following calculations. The cylinder for lifting the elevator, in this instance, is the single-acting type or uses oil only on the blind end of the cylinder piston. By calculating the volumetric displacement for the full lift stroke of the elevator cylinder, the required factors for calculating the accumulator size are known. For purposes of this illustration, it can be assumed that a 2-inch (5.1 cm) bore cylinder can provide adequate mechanical strength for raising and lowering the elevator platform and that the distance the elevator platform is to be lifted is 4 feet (1.22 meters).

The following calculations are based on the formula derived from Boyle's law:

$$P_1V_1 = P_2V_2 = P_3V_3, \text{ etc.}$$

With a 2-inch (5.08 cm) bore cylinder, approximately 160 psi (1103.2 kPa) is required to lift 500 pounds (226.8 kg) of weight. Allowing for a reasonable safety factor, 200 psi (1378.95 kPa) should be established as the minimum pressure. In this illustration, the pressure is the gas pressure inside the accumulator. The volume is the gas volume inside the accumulator. The following steps can be used in the calculations:

1. To be sure that the gas charge against which the oil is forced into the accumulator is large enough, it is good practice to begin with a 100-psi (689.5 kPa) gas precharge. Therefore, the value of P_1 is 100 psi (689.5 kPa) and V_1 is to be determined.
2. The value of P_2 is 200 psi (1378.95 kPa), which is the minimum pressure required to perform the lifting action, and V_2 is equal to V_3 + 151 cubic inches (2.47 liters), which is the displacement of the 2-inch (50.8 mm) bore, 48-inch (1219.2 mm) stroke hydraulic cylinder.
3. P_3 is the maximum pressure attained [arbitrarily selected as 2000 psi (13,789.5 kPa), a standard cylinder rating], and V_3 is equal to V_2 − 151 cubic inches (2.47 liters).

4. Substituting the known values in the formula $P_2V_2 = P_3V_3$, the calculations are:

$$200 \ (V_3 + 151) = 2000 \ V_3$$
$$200 \ V_3 + 30{,}200 = 2000 \ V_3$$
$$1800 \ V_3 = 30{,}200$$
$$V_3 = 16.8 \ \text{cubic inches}$$
$$V_2 = 16.8 + 151$$
$$= 167.8 \ \text{cubic inches}$$

Substituting the metric values into the formula $P_2V_2 = P_3V_3$, the calculations are:

$$1378.95 \ (V_3 + 2.47) = 13{,}789.5 \ V_3$$
$$1378.95 \ V_3 + 3406 = 13{,}789.5 \ V_3$$
$$12{,}410.6 \ V_3 = 3406$$
$$V_3 = 0.2744 \ \text{liters}$$
$$V_2 = 0.2744 + 2.47$$
$$= 2.7444 \ \text{liters}$$

5. Substituting the known values in the formula $P_1V_1 = P_2V_2$, the calculations are:

$$100 \ V_1 = 200 \times 167.8$$
$$100 \ V_1 = 33{,}560$$
$$V_1 = 335.6 \ \text{cubic inches}$$

Substituting the metic values into the formula $P_1V_1 = P_2V_2$, the calculations are:

$$689.5 \ V_1 = 1378.95 \times 2.75$$
$$689.5 \ V_1 = 3792.1$$
$$V_1 = 5.5 \ \text{liters}$$

6. Therefore, 335.6 cubic inches (5499.7 cm³) divided by 231 cubic inches (3785.4 cm³) per gallon is (335 ÷ 231), or (5499.7 ÷ 3785.4), or 1.45 gallons (5.5 liters) of storage capacity is required in the accumulator.

Since the above calculations indicate that 1.45 gallons (5.5 liters) of storage capacity is required, either a 1.5-gallon (5.7 liters) or a 2-gallon (7.6 liters) accumulator should be selected to allow for a leakage factor (metric sizes are not yet established). These calculations are slightly conservative, because they are based on isothermic compression and expansion of the gas. Both the 1.5-gallon (5.7 liters) accumulator and the 2-gallon (7.6 liters) accumulator are available in standard sizes from several different manufacturers (see Chart 34-1).

At this point, the "Accumulator Factor" in Chart 34-1 is important. The values in this line represent each of the ten seconds in the cycle. The plus (+) values indicate the pump volume in excess of the system demand. The minus (−) values indicate the system demand in excess of the pump output. Therefore, the plus values indicate the pump volume available for charging an accumulator. The negative values indicate the volume of fluid to be supplied the system by the accumulator.

To apply an accumulator, two additional factors need to be determined: (1) the minimum operating pressure of the system, and (2) the maximum pump output pressure that can be made available. For further calculations, it can be assumed that the system is expected to operate at a minimum of 1500 psi (10,342.1 kPa) and that the pump is capable of delivering 2000 psi (13,789.5 kPa).

A study of the chart (see Chart 34-1) reveals that the No. 1 second of the cycle represents the largest surplus of fluid volume available from the pump. The No. 2 and No. 3 seconds of cycle time each require 3.08 cubic inches (50.5 cm³) of fluid from the accumulator. In the No. 4 second, a surplus of 30.34 cubic inches (497.2 cm³) of fluid is to be stored in the accumulator. In the No. 5 second, 0.68 cubic inch (11.14 cm³) is to be stored, and in the No. 6 second, an excess of 0.66 cubic inch (10.82 cm³) is to be stored in the accumulator. The net result of the first 6 seconds of cycle time is the availability of 66.42 cubic inches (1088.4 cm³) of surplus fluid which is to be stored in the accumulator. The remaining 4 seconds of the cycle withdraw fluid from the accumulator; therefore, the volumetric requirement for the accumulator is 66.42 cubic inches (1088.4 cm³) of fluid.

The calculations for determining the accumulator size for the above application can be performed as follows:

1. The known values for the formula are:

$$P_1 = 1000 \text{ psi (6894.8 kPa)} \qquad V_1 = \text{(to be determined)}$$

Chart 34-1. Fluid Demand [in³ (cm³)] During Extension and Retraction Strokes (With Standard Oversize Rod Diameters)

Total Cycle Time: 10 Seconds

Sequence	1	2	3	4	5	6	7	8	9	10
Cyl. 1	18.85 (308.9)			9.94 (162.89)						
Cyl. 2		62.83 (1029.6)	62.83 (1029.6)		19.14 (313.6)	38.29 (627.5)				
Cyl. 3				19.47 (319)	9.73 (159.4)				9.82 (160.9)	9.82 (160.9)
Cyl. 4					11.06 (181.2)	22.12 (362.15)			10.3 (168.8)	10.3 (168.8)
Cyl. 5							84.83 (1390)	84.83 (1390)	47.1 (771.8)	47.1 (771.8)
Total in³ Volume (cm³)	18.85 (308.9)	62.83 (1029.6)	62.83 (1029.6)	29.41 (481.9)	30.2 (494.9)	60.41 (989.9)	84.83 (1390)	84.83 (1390)	67.22 (1101.5)	67.22 (1101.5)
Accumulator in³ Factor (cm³)	+40.9 (+670.2)	-3.08 (-50.5)	-3.08 (-50.5)	+30.34 (+497.2)	+0.68 (+11.14)	+0.66 (+10.82)	-25.08 (-411)	-25.08 (-411)	-7.41 (-122.4)	-7.47 (-122.4)

extension stroke ——→   retraction stroke ——→

Cyl. 1 = 2-in. bore, 6-in. stroke, w/1⅜-in. dia. piston rod. (50.8-mm bore, 152.4-mm stroke, w/34.93-mm rod)
Cyl. 2 = 4-in. bore, 10-in. stroke, w/2½-in. dia. piston rod. (101.6-mm bore, 254-mm stroke, w/63.5-mm rod)
Cyl. 3 = 1¼-in. bore, 20-in. stroke, w/1-in. dia. piston rod. (38.1-mm bore, 508-mm stroke, w/25.4-mm rod)
Cyl. 4 = 3¾-in. bore, 4-in. stroke, w/2-in. dia. piston rod. (82.55-mm bore, 101.6-mm stroke, w/50.8-mm rod)
Cyl. 5 = 6-in. bore, 6-in. stroke, w/4-in. dia. piston rod. (152.4-mm bore, 152.4-mm stroke, w/101.6-mm rod)

(arbitrary)

P_2 = 1500 psi (10,342.1 kPa) V_2 = V_3 + 66.42 cu in.
(1088.4 cm³)

P_3 = 2000 psi (13,789.5 kPa) V_3 = V_2 − 66.42 cu in.
(1088.4 cm³)

2. Substituting the known values in the formula $P_2V_2 = P_3V_3$:

$$1500 \ (V_3 + 66.42) = 2000 \ V_3$$
$$1500 \ V_3 + 99,630 = 2000 \ V_3$$
$$500 \ V_3 = 99,630$$
$$V_3 = 199.26 \text{ cubic inches}$$
$$V_2 = 199.26 + 66.42$$
$$= 265.68 \text{ cubic inches}$$

Substituting the metric values into the formula $P_2V_2 = P_3V_3$, the calculations are:

$$10,342.1 \ (V_3 + 1088.4) = 13,789.5 \ V_3$$
$$10,342.1 \ V_3 + 11,256,341 = 13,789.5 \ V_3$$
$$3447.4 \ V_3 = 11,256,341$$
$$V_3 = 3265.2 \text{ cm}^3$$
$$V_2 = 3265.2 + 1088.4$$
$$= 4353.6 \text{ cm}^3$$

3. Substituting the known values in the formula $P_1V_1 = P_2V_2$:

$$1000 \ V_1 = 1500 \times 265.68$$
$$1000 \ V_1 = 398,520$$
$$V_1 = 398.52 \text{ cubic inches}$$

Substituting the metric values into the formula $P_1V_1 = P_2V_2$, the calculations are:

$$6894.8 \ V_1 = 20,342.1 \times 4353.6$$
$$6894.8 \ V_1 = 88,561,366$$
$$V_1 = 12,844.66$$

4. Therefore, 398.52 cubic inches (6530.55 cm³) divided by 231 cubic inches (3785.43cm³) per gallon is (398.52 ÷ 231), or

(6530.55 ÷ 3785.43), or 1.725 gallons (6.53 liters) of storage capacity is required in the accumulator.

The volumetric requirement is in excess of 1.5 gallons (5.7 liters). Either a 7-quart (6.6 liters) or, preferably, a 2-gallon (7.6 liters) accumulator size should be utilized. (Metric sizes are not yet established.)

Chapter 35

Accumulators—
Maintenance

The operating principle of the hydraulic accumulator is based on the compression of a gas within the gas chamber of the accumulator. The hydraulic fluid is stored against the gas pressure. The most commonly used gas for this purpose is oil-pumped dry nitrogen which has been processed properly—all particles of moisture having been removed. Some accumulator manufacturers use special shipping provisions, whereby the accumulator is shipped with the specified precharge of gas. Some manufacturers, however, use different methods for shipment, and the accumulator is shipped without the precharge of gas. Dry nitrogen gas in high-pressure storage bottles is available from local sources in most industrial areas.

MAINTENANCE AND SERVICE

To maintain and service the hydraulic accumulator properly, one minor and rather inexpensive piece of equipment should be kept available. This piece of equipment is referred to as a charging and gauging assembly (Fig. 35-1). Although these assemblies vary in

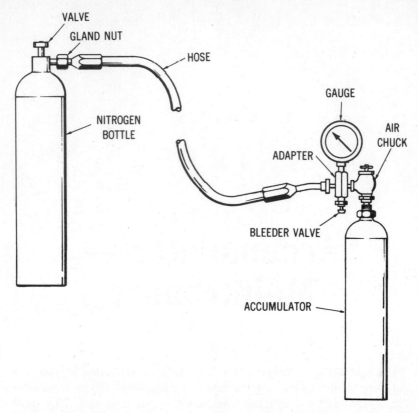

Fig. 35-1. A charging and gauging assembly is an essential piece of equipment used for properly maintaining and servicing hydraulic accumulators.

some details and in construction, they comprise basically an air chuck which attaches to the accumulator valve and a length of hose that is attached, in turn, to a combination adapter, bleeder valve, pressure gauge, and tank valve and to the nitrogen supply bottle. The high-pressure hose within the assembly can vary from 1 to 3 feet (0.3 to 0.9 m) in length. Even when an accumulator is shipped with the specified precharge of gas, the charging and gauging assembly is required to maintain the accumulator unit.

Occasionally, a permanent installation of the gauge assembly to the accumulator for periodic and frequent checking of the pressure is desirable. This installation can be convenient if access to the accumulator and the gas valve is difficult. In a permanent installation, it is advisable to install a high-pressure needle or shutoff valve to isolate the gauge from pressure fluctuations—except for the required checking of the pressure. The gauge is isolated to protect it and to avoid damage from high-frequency pressure fluctuations. Normally, the charging and gauging assembly is used only for checking and recharging the gas pressure occasionally.

Oxygen and other combustible gases should never be used for charging the gas chamber. If high-pressure air is available for the charging function, extreme care should be exercised to remove all moisture and to make sure that the air is clean and free from toxic materials.

Before proceeding to charge a hydraulic accumulator that is installed in the line and may contain some hydraulic fluid, care should be exercised to be sure that the pressure of the hydraulic fluid has been reduced to zero and that all the fluid has been forced from the accumulator. Most accumulators are provided with a guard which is placed over the gas valve. The removal of the guard exposes the small valve, which resembles the valve on an automobile tire. The air-chuck end of the charging and gauging assembly should be attached to the nitrogen cylinder. The connection should be fastened tightly.

The gauge end of the charging and gauging assembly should be attached to the accumulator gas-valve assembly, rotating only the nut with moderate torque—preferably finger-tight. The connector assembly depresses the valve core in the accumulator valve stem assembly. This registers a pressure reading on the pressure gauge.

If there is no gas charge in the accumulator or if it is necessary to increase the gas charge to meet the system requirement, the valve on the nitrogen cylinder is opened. This valve should be opened and closed slowly and periodically to receive a nonfluctuating pressure reading on the gauge. Since some charging and gauging assembly designs do not provide accurate pressure reading while there is flow through the assembly, interrupting the flow occasionally provides a more correct pressure reading.

If excessive pressure is injected into the accumulator or if it is necessary to reduce a precharge, the bleeder valve provided in the assembly can be used. After the correct precharge pressure is reached, the nitrogen supply bottle should be closed completely. The charging and gauging assembly should be removed from the accumulator. In removing the assembly from the accumulator, the connector assembly should be turned off the accumulator valve stem rapidly to permit the valve core to close quickly, thereby preventing the loss of a portion of the gas charge. Some charging and gauging assemblies provide for a screw-type engagement with the valve core after the attachment to the unit has been made.

When the attachment has been accomplished, it is wise to check for gas leakages around the accumulator valve stem and the valve core. A simple method is to apply soapy water or oil. If a leak is present, it can be detected by the presence of bubbles around the valve stem or valve core. If leaks occur around the valve core, it should be depressed and released quickly with a small blunt tool to reset it. The gas valve guard should be replaced and turned down tightly to prevent damage to the gas valve stem and to prevent dirt and contamination around the valve core.

Since the proper functioning of the accumulator and its contribution to the hydraulic system depend on the correct precharge pressure, a regular periodic check should be made on the precharge pressure within the accumulator. This check should be performed with the hydraulic pressure completely removed and with the hydraulic fluid forced from the fluid end of the accumulator. Gas pressure readings taken while the hydraulic fluid pressure is available to the accumulator merely result in a reading that includes the operating pressure of the system. This reading is, of course, in excess of the precharge pressure, because the forcing of hydraulic fluid into the fluid end of the accumulator results in further compression of the gas in the charged end of the accumulator. Therefore, the reading does not indicate the correct precharge gas pressure within the accumulator.

The functional uses of the hydraulic accumulator were discussed in the previous chapter. The success of the accumulator in performing these functions depends on the separation barrier maintaining its efficiency in functioning as a separator and on the maintenance of the correct gas volume and pressure in the gas end of the accumulator. Failure of the accumulator can result from a rupture of the flexible barrier in the diaphragm-type or bladder-type units. Failure of the nonflexible or piston-type barrier can result from damage to, or loss of, the seals on the floating piston.

When the barrier fails to perform its separating function, the gas charge eventually enters the hydraulic fluid and is dissipated— usually into the reservoir of the system. Then the gas is replaced by hydraulic fluid. The accumulator loses its storage or resilient function. In a hydraulic circuit in which the accumulator is used to store excessive pump volume for speeding up one portion of the machine cycle, a loss in speed of the machine function results in this faster portion of the cycle.

Before proceeding to dismantle or repair the accumulator, the charging and gauging assembly should be used to determine whether any of the gas charge remains within the accumulator unit. A quick check should be made at this point to be sure that no fluid is coming from the gas valve assembly of the accumulator. Also, at this point it should be noted that the accumulator cannot perform its storage function if the precharge pressure of the gas is too high or in excess of the system fluid pressure. The latter situation is created if the precharge pressure is too high originally or if the operating pressure of the system is reduced by either wear or adjustment of the pumping or pressure control components.

A loss of the gas charge usually occurs when all the hydraulic fluid is forced from the fluid end of the accumulator and returned to the system. For this reason, it is good practice to check the precharge pressure while the hydraulic pressure in the oil end of the accumulator is reduced to zero.

DISASSEMBLY AND REPAIR

In servicing and repairing the bladder-type accumulators, the manufacturer recommends the following procedure. As in any component, and especially in accumulators, some specific designs require special tools. The original purchase of these components should be checked. Any special tools that may be needed for servicing the unit should be secured.

In removing the accumulator from the system, first exhaust all pressure from the hydraulic system by opening the bypass or control valve to reduce the pressure within the accumulator to zero. Remove the valve guard and loosen the valve cap. Use a valve core tool to release the gas pressure (if any) from the gas chamber. Remove the gas valve core to expedite the removal of the gas. Then, and only then, should the accumulator be removed from the system.

To disassemble the bladder-type unit, clamp the unit, preferably in a chain vise, making sure that the accumulator shell is protected by metal strips on the vise jaw. Remove the pipe plug. Then remove the plug lock nut, using a spanner wrench and an adjustable wrench—one wrench for torque and the other for countertorque. Remove the spacer ring, and with the palm of the hand, push the plug into the shell. Insert the hand into the shell and lift out the plug, the "O" ring, the washer, and the split retaining ring. The split retaining ring is removed by twisting it on its pivot pin. Remove the plug assembly from the shell. Using a suitable open-end wrench, remove the valve-stem lock nut. Insert one hand into the shell. Depress the bag with closed fist to eliminate all gas pressure. Then grasp the heel of the bag and withdraw it from the shell.

After disassembly of the unit, clean all the parts with an approved cleaning solution. Use air pressure to dry all the parts. Lay them on a bench free of foreign matter. Replace all parts that show signs of damage, paying special attention to the condition of the threaded parts.

To reassemble the unit, place the shell in a vise, as described in the disassembly procedure. Pour a liberal quantity of "cushion fluid" into the shell, using a clean supply of the same fluid used in the system. Attach the bag to the pull rod furnished by the manufacturer. Insert the pull rod into the shell mouth and through the opening for the valve stem. Fold the bag longitudinally, so that it enters the shell mouth with a minimum of friction. Pull the rod from the shell with one hand. At the same time, feed the bag into the shell with the other hand. Remove the pull rod. Install the valve nut securely. Insert a hand into the shell. Depress the bag to expel the air. Install the valve cap before releasing the bag. This provides adequate working space for installing the plug. Grasp the threaded section of the plug. Insert the shoulder end into the shell mouth. Place the split retaining nut on the plug with the small shoulder toward the shell mouth. Then follow with the washer. Install the plug "O" ring on the plug. With the fingers, push the "O" ring onto the plug as far as possible. Withdraw the threaded end of the plug through the shell mouth. Using the hands, pull until it seats solidly in position, squeezing the "O" ring between the plug and the shell mouth. Install the spacer on the plug with the smaller shoulder diameter toward the shell. Install the lock nut on the plug and tighten securely. Replace the pipe plug. Remove the cap from the gas valve. Install the valve core, using the core tool. Precharge the bladder to approximately 10 psi (68.9 kPa) before completing the tightening of the valve stem nut. Inflate the bladder with nitrogen gas to the proper precharge pressure. Replace the valve cover, turning it one-half turn beyond hand tightness. Then replace the valve guard.

The disassembly of piston-barrier accumulators is similar to the servicing of the bladder-barrier designs. The removal of the unit from the system also follows the same procedure. Before removing the unit, however, the condition of the piston seals can be tested. To perform the test, the accumulator is shut off from the system. The piston is permitted to bottom. Then the gas pressure is measured against the specified precharge pressure. An increase in precharge pressure indicates leakage of the oil into the gas side. A decrease indicates leakage of the gas into the oil side. A substantial difference between actual pressure and the specified pressure indicates the need for new piston seals.

To disassemble a piston-barrier accumulator, clamp the unit, preferably in a chain vise, protecting the shell with metal strips placed on the vise jaws. Remove the snap ring with special pliers. With a soft mallet, tap the plug until it extends approximately 1 inch (2.54 cm) into the shell. With the fingers or a suitable tool, remove the threaded ring from the shell. Also remove the plug assembly from the shell. To remove the piston assembly, turn the threaded ring into the shell for approximately three turns. Then, using shop air, gently blow the floating piston against the threaded ring. This brings the piston to a position where it can be removed. Remove the threaded ring. Then remove the piston.

All the accumulator parts should be inspected and cleaned. Check for and replace damaged parts, paying special attention to the condition of the threaded parts and "O" rings. To reassemble the unit, adequately cover the internal surface of the shell with the system oil to facilitate reinsertion of the piston. Reinstall the piston, making sure the seal rings are lubricated properly. The hollow side of the piston should enter the accumulator shell first. Replace the oil plug assembly, making sure the "O" rings are lubricated with the system oil. Push the oil plug inward, until it is flush with the top of the shell. Replace the threaded ring. Tighten it against the shoulder. Pull on the oil plug, until it contacts the shoulder of the threaded ring. Replace the retainer plate, and lock with the bolts. Replace the gas valve core, using the core tool. Replace the cap, turning it one-half turn beyond hand tightness. Then replace the cover plate.

Chapter 36

Fluid Conductors—
Design, Applications,
and Limitations

Fluid conductors are the lines of communication between the fluid power source and the point of application of that force within a component. These lines should be capable of withstanding both the working pressures and the shock pressures created within the system. The fluid conductors should be designed properly. Their capacities should be adequate for transmitting the required volumetric flow rate demanded by the system. A practical facility for assembly and disassembly should be provided.

TYPES OF CONDUCTORS

Fluid conductors or carriers can be divided into three different classes: (1) rigid, or piping; (2) semirigid, or tubing; and (3) flexible, or hose and conduit. The type of conductor to be used for a specific application or installation is usually dictated by the requirements of

461

the installation. If movement is required, such as that for fluid lines to pivoting or swiveling cylinders, flexible carriers are necessary. If movement of the component parts containing the fluid connections is not required, rigid or semirigid conductors can be used successfully. Also, if the installation requires occasional assembling and disassembling that involves the fluid lines, semirigid conductors or tubing can facilitate the operations. When space is important, semirigid lines usually provide the greatest savings. The proper selection of piping to fit the pressure, flow, and assembly requirements of a fluid power system is essential to providing maximum efficiency and proper functioning of the system.

The air line installations in a plant are usually made from galvanized pipe. This avoids the formation of rust inside the carrier lines. Hydraulic installations often use rigid piping. However, galvanized pipe should never be used for carrying a hydraulic fluid. The presence of the zinc coating rapidly increases the oxidation rate of many hydraulic fluids. The reaction of the hydraulic oil with the zinc can cause peeling or flaking of the coating on the inside surface of the galvanized pipe.

Rigid

Rigid steel pipe is available in a number of different weights, depending on the service pressure and the shock loading. The pipe weights are classified as follows:

1. Standard (STD), or Schedule 40.
2. Extra strong (XS), or Schedule 80, for 1000 psi (6894.8 kPa).
3. Schedule 160, for 3000 psi (20,684.3 kPa).
4. Double extra strong (XXS).

Pipe fittings, such as elbows, street elbows, crosses, tees, and unions, are all available within the above weight classifications. The sizes of the fittings correspond to the pipe sizes.

Pipe sizes are determined and specified by the nominal inside diameter (ID) of the pipe. The actual inside diameter of a given pipe size varies, depending on the weight. This results in a heavier wall and a smaller inside diameter for the high-pressure pipe. The pipe in a given pipe size is made with the same outside diameter (OD), since it is the outside diameter of the pipe that is threaded to mate with the corresponding pipe fittings.

Hydraulic installations that use rigid pipe can create problems occasionally, since shock or vibration is transmitted throughout the entire length of the rigid system. Since oil is comparatively noncompressible, it is sometimes advisable to install an accumulator for absorbing these shocks and vibrations, as was pointed out in a previous chapter.

Semirigid

Semirigid conductors, or tubing, are available primarily in four different types of materials. Steel, aluminum, copper, and plastic tubing are available in a wide range of sizes and wall thicknesses. These materials can be grouped as follows:

1. Seamless steel (*SAE* 1010), fully annealed
2. Stainless steel, seamless (18-8), fully annealed, suitable for bending and flaring
3. Aluminum, seamless (*B50S-0*)
4. Copper, seamless, fully annealed
5. Plastic
6. Welded steel, cold drawn and annealed

Tubing, unlike rigid piping, is specified by its outside diameter and its wall thickness. As in rigid pipe, however, the outside diameter is critical, since it fits inside the sleeve and the nut of the tube fittings. Steel tubing, stainless-steel tubing, and aluminum are all used in hydraulic oil systems. Copper tubing is not recommended for hydraulic oils, since it tends to act as a catalyst in breaking down some of the oil additives. Copper tubing also tends to work-harden as a result of vibration. This causes fatigue and breakage. Stainless-steel tubing, aluminum tubing, copper tubing, and plastic tubing are often used for carrying pressurized air. The commonly available tubing sizes, wall thicknesses, and working pressures for steel tubing can be found in Table 36-1.

Plastic tubing is widely used with air tools, as well as in air systems. It is easily applied and is rather low in cost. Several types of available plastic tubing often used for air service and other low-pressure applications are shown in Fig. 36-1. A self-storing nylon tubing that can be used for low-pressure hydraulic service or for compressed air service is shown in Fig. 36-2.

Fittings that are used in conjunction with tubing are similar to

Table 36-1. Safe Internal Working Pressure
[in psi (kPa)] for Tubes
(Cold-Drawn Seamless Steel)
Soft Annealed Rockwell B-50

*Yield Point—30,000 psi (206.842.7 kPa). Ultimate—48,000 psi (330,948.3 kPa)

Wall Thickness in. (mm)	Tubing Size			
	¼ in.	5/16 in.	3/8 in.	7/16 in.
	psi (kPa)	psi (kPa)	psi (kPa)	psi (kPa)
0.028 (0.71)	2240 (15,444)	1795 (12,376)	1493 (10,294)	1281 (8832)
0.035 (0.89)	2800 (19,305)	2244 (15,472)	1867 (12,872)	1602 (11,045)
0.042 (1.07)	3360 (23,166)	2692 (18,561)	2240 (15,444)	1922 (13,252)
0.049 (1.24)	3920 (27,027)	3141 (21,656)	2613 (18,016)	2243 (15,465)
0.058 (1.47)	4640 (31,992)	3718 (25,635)	3093 (21,325)	2654 (18,299)
0.065 (1.65)	5200 (35,853)	4167 (28,370)	3467 (23,904)	2975 (20,512)
0.072 (1.83)	5760 (39,714)	4615 (31,819)	3840 (26,476)	3295 (22,718)
0.083 (2.11)	6640 (45,781)	5321 (36,687)	4427 (30,523)	3799 (26,193)
0.095 (2.41)	7600 (52,400)	6090 (41,989)	5067 (34,936)	4348 (29,978)
0.109 (2.77)		6987 (48,174)	5813 (40,079)	4989 (34,398)
0.120 (3.05)		7692 (53,034)	6400 (44,126)	5492 (37,866)
0.134 (3.40)			7147 (49,277)	6133 (42,285)

*1 psi (1b/in²) = 6.894757 kilopascals (kPa)

Tubing Size				
½ in.	5/8 in.	3/4 in.	7/8 in.	1 in.
psi (kPa)	psi (kPa)	psi (kPa)	psi (kPa)	psi (kPa)
1120 (7722)	896 (6178)	747 (5150)	640 (4413)	560 (3861)
1400 (9653)	1120 (7722)	933 (6433)	800 (5516)	700 (4826)
1680 (11,583)	1344 (9266)	1120 (7722)	960 (6619)	840 (5791)
1960 (13,514)	1568 (10,811)	1307 (9011)	1120 (7722)	980 (6757)
2320 (15,996)	1856 (12,797)	1547 (10,666)	1326 (9142)	1160 (7998)
2600 (17,926)	2080 (14,341)	1733 (11,949)	1486 (10,245)	1300 (8963)
2880 (19,857)	2304 (15,885)	1920 (13,238)	1646 (11,349)	1440 (9928)
3320 (22,890)	2656 (18,312)	2213 (15,258)	1897 (13,079)	1660 (11,445)
3800 (26,200)	3040 (20,960)	2533 (17,464)	2171 (14,968)	1900 (13,100)
4360 (30,061)	3488 (24,049)	2907 (20,043)	2491 (17,174)	2180 (15,033)
4800 (33,095)	3840 (26,476)	3200 (22,063)	2743 (18,912)	2400 (16,547)
5360 (36,956)	4288 (29,565)	3573 (24,635)	3063 (21,119)	2680 (18,478)

Fig. 36-1. Thermoplastic tubing finds many uses in industry. (Courtesy Gould Inc., Valve and Fittings Division)

those for rigid pipe, except for the fact that most tubing nuts serve as unions, which makes the individual union fitting unnecessary for that purpose. Since steel tubing is available normally in straight lengths, the tube union usually serves to connect the lengths of tubing, rather than to facilitate the assembly and disassembly of the lines.

Fittings for tubing normally fall into two general groups: (1) flared fittings (Fig. 36-3) and (2) flareless fittings (Figs. 36-4 and 36-5). The general design of fittings readily available from a number of manufacturers is shown in Fig. 36-6.

When flared tubing connections are used, extreme care is required in producing the flare. This can be facilitated by using the proper flaring tools. Two different flare angles are used for flared tubing. The *USASI* standards call for a 37-degree flare. *SAE* standards specify a 45-degree flare. Tube fittings are available with either of the two flare angles. However, the 37-degree flare is becoming more and more popular for hydraulic use in industry.

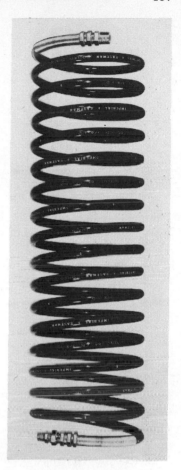

Fig. 36-2. Self-storing nylon tubing is used where shop air is conducted from the main line to portable air-operated tools, air blow guns, air nozzles, etc. where a self-retrieving type line is desirable. Such tube assemblies are available in coils of 12-, 25-, and 50-ft. lengths. (Courtesy Gould Inc., Valve and Fittings Division)

Flexible

Flexible hose is available for a variety of different pressures and classes of work. The hose size is usually specified by its nominal inside diameter. The pressure rating is determined by the numbers of layers and the type of material in the layers in its structure. The so-called *tube* is the lining or inner part that actually contacts the fluid or material being carried. The *carcass* is the supporting structure of the hose. It lies between the tube and the cover (Fig. 36-7). The carcass material can be cotton, synthetic fiber, asbestos, or wire that is woven, braided, lapped, or wound spirally. Or it can be a combination of these materials. The *cover* is the outside covering of

Fig. 36-3. A flared-type tube fitting.
(This illustration is the courtesy of
Parker Hannifin Corporation.)

the hose. Normally, the cover protects the carcass and the hose
assembly from abrasive or other destructive external forces, such as
pulsating pressures, falling objects, weather, oils, greases, acids,
and chemicals. See Table 36-2 for characteristics of hose stock types.

In the classification of hose by pressure ratings, three primary
pressure factors are usually given. The *recommended working
pressure* is that pressure at which a given hose can be operated safely
for satisfactory service. The *test pressure* is the pressure that a hose

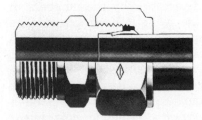

Fig. 36-4. Flareless tube fitting that
provides safe connections which hold
beyond the burst strength of the tube.
(Courtesy Gould Inc., Valve and Fittings Division)

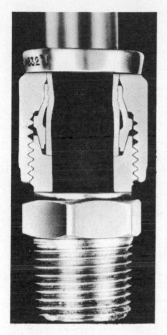 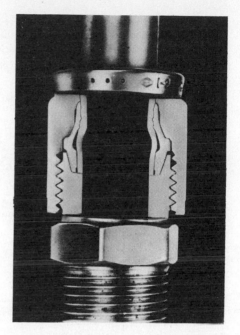

Fig. 36-5. Tube fitting after initial makeup (left). Note that the first tooth is gripping the tube to provide a pressure-tight seal. Same type of tube fitting (right) after repeated remakes. Note that the second tooth is now gripping the tube. Positive seal is maintained, even after repeated disassembly and reassembly, without tubing constriction. (Courtesy Gould Inc., Valve and Fittings Division)

is guaranteed to withstand. The *burst pressure* is the pressure at which the hose is ruined and rendered unfit for further service. A hose should be flexible and easy to handle, although the flexibility usually decreases as the pressure rating of the hose increases. Hose that contains pressurized fluid is usually less flexible than hose at zero internal pressure. This loss of flexibility at the higher pressures is a result of the expanding and interlocking actions in the cross-woven carcass layers (Fig. 36-8).

INSTALLATION

The successful operation of any fluid power installation depends on proper installation. One of the primary requirements is the removal of all dirt particles or foreign matter from the piping being installed. If rigid pipe is used, specific care should be exercised to remove any scale that may be adhering to the inside surface of the pipe. It is good preventive maintenance practice to apply a rotating wire brush to the

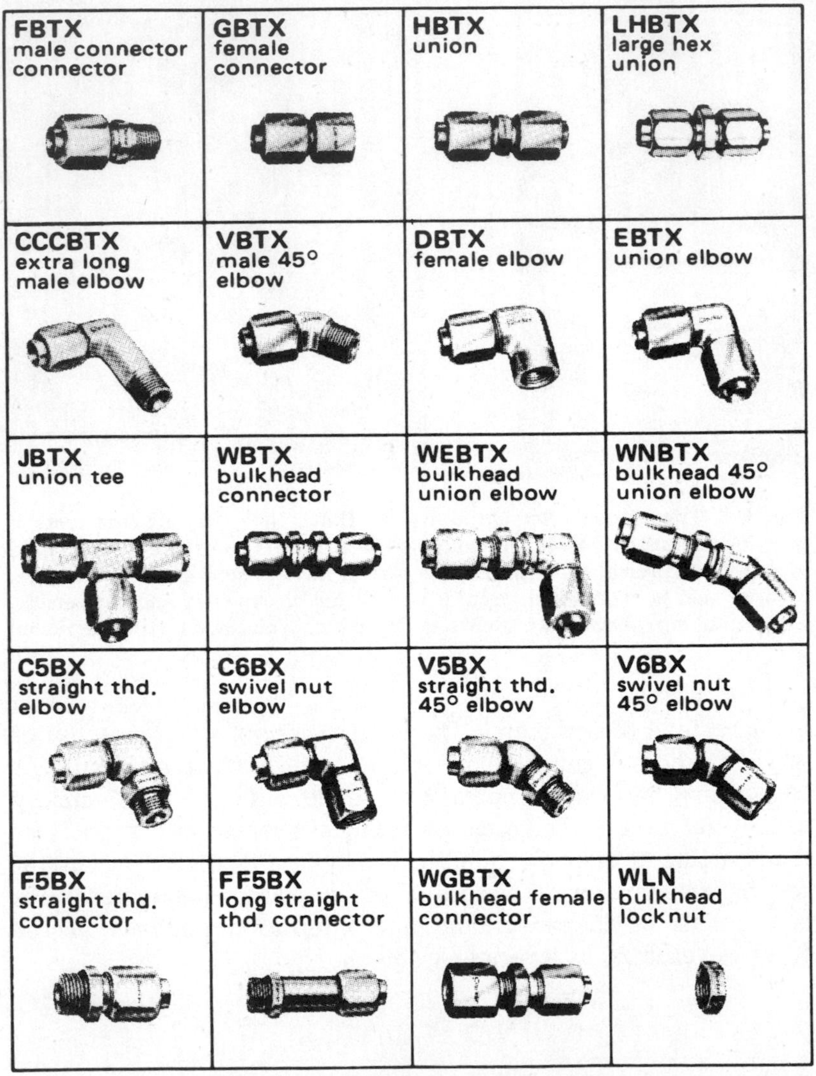

FBTX male connector connector	**GBTX** female connector	**HBTX** union	**LHBTX** large hex union
CCCBTX extra long male elbow	**VBTX** male 45° elbow	**DBTX** female elbow	**EBTX** union elbow
JBTX union tee	**WBTX** bulkhead connector	**WEBTX** bulkhead union elbow	**WNBTX** bulkhead 45° union elbow
C5BX straight thd. elbow	**C6BX** swivel nut elbow	**V5BX** straight thd. 45° elbow	**V6BX** swivel nut 45° elbow
F5BX straight thd. connector	**FF5BX** long straight thd. connector	**WGBTX** bulkhead female connector	**WLN** bulkhead locknut

Fig. 36-6. Tube fittings used in fluid power systems. (This

entire length of the inside diameter of the pipe. This should be followed by flushing the pipe with a good commercial solvent.

The selection of the pipe and tubing diameters is usually determined by the port sizes supplied within the components of the

T22X mountie	**TRBTX** reducer	**CBTX** male elbow	**CCBTX** long male elbow
RBTX male run tee	**SBTX** male branch tee	**MBTX** female run tee	**OBTX** female branch tee
WJBTX bulkhead branch tee	**WJJBTX** bulkhead run tee	**PNTX** plug	**FNTX** cap
R5BX straight thd. run tee	**R6BX** swivel nut run tee	**S5BX** straight thd. branch tee	**S6BX** swivel nut branch tee
BTX nut	**TX-MM** sleeve/ sleeve (metric)	**F4BTX** British parallel pipe connector	**F8BTX** I.S.O. metric connectorc

illustration is the courtesy of Parker Hannifin Corporation.)

Fig. 36-7. Two-wire braid hose (top) recommended for high-pressure oil lines. Meets or exceeds requirements of *SAE* 100R2A. A three-wire braid hose (center) is recommended for extra-high-pressure oil lines. The four-wire spiral hose (bottom) is employed for high-pressure oil lines. The hose has four layers of alternated, spiraled, high-tensile steel wire over a layer of yarn braid. (Courtesy Gates Rubber Co.)

system. If several different components are used, however, care should be exercised to install piping that is adequate in cross-sectional area for carrying the total fluid required with minimum resistance. The flow rate of the hydraulic oil should never exceed 15 feet per second (4.57 m/s) of linear travel.

In both pneumatic and hydraulic systems, the efficient flow of fluid through the fluid conductor is hindered by pipe fittings, bends, or indentations that interrupt the smooth flow. The piping should always be as short as possible. A minimum number of bends should be used. Sharp bends should be eliminated whenever possible. The fluid conductors should be fastened securely to prevent harmonic reactions to internal vibration.

Special attention should be paid to both the piping size and the presence of restrictions when installing an air system within a plant. The *Compressed Air Handbook* (McGraw-Hill, New York, 1954,

Table 36-2. Characteristics of Hose Stock Types

Chemical and physical properties are subject to a considerable amount of control through compounding. Characteristics given below for each stock type are therefore generalized to some degree.

Tube and cover stocks may be occasionally upgraded to take advantage of improved materials and technology.

For detailed information on a specific hose tube or cover stock, contact Denver.

GATES TYPE	A	C	H	M	P
Chemical Name	Poly-chloroprene (Neoprene)	Acryloni-trile and Butadiene (Buna-N)	Isobutylene and Isoprene (Butyl)	Chlorosul-fonated Poly-ethylene (Hypalon)	Ethylene Propylene Diene (EPDM)
ASTM-SAE Designation SAE J14 SAE J200	SC BC	SB BG	R AA	TB CE	R AA
Tensile Strength	Good	Fair to Good	Fair to Good	Good	Good
Tearing Resistance	Good	Fair to Good	Good	Fair	Good
Abrasion Resistance	Good to Excellent	Fair to Good	Fair to Good	Good	Good
Flame Resistance	Very Good	Poor	Poor	Good	Poor
Petroleum Oil and Commercial Gasoline	Good	Good to Excellent	Poor	Good	Poor
Resistance To Gas Permeation	Good	Good	Outstanding	Good to Excellent	Fair to Good
Weathering	Good to Excellent	Poor	Excellent	Very Good	Excellent
Ozone	Good to Excellent	Poor	Excellent	Very Good	Outstanding
Heat	Good	Good	Excellent	Very Good	Excellent
Low Temperature	Fair to Good	Poor to Fair	Very Good	Poor	Good to Excellent
General Chemical Resistance	Good	Fair to Good	Good	Good	Good

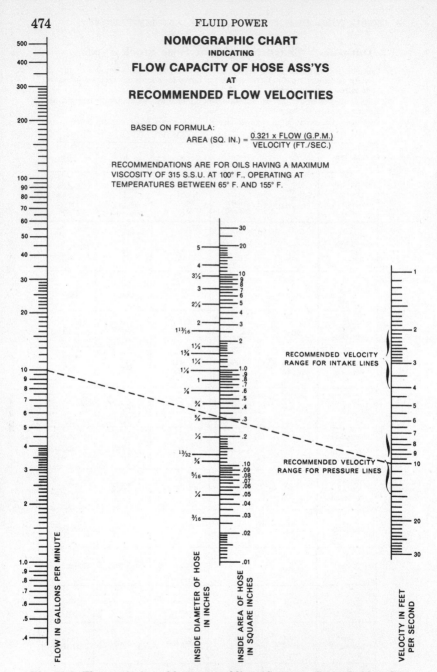

Fig. 36-8. Flow capacity of hose assemblies. (Courtesy Gates Rubber Co.)

pages 3-6) states that the following general rules should be observed in planning a compressed-air distribution system:

1. Pipe sizes should be large enough that the pressure drop between the receiver and the point of use does not exceed 10 percent of the initial pressure. Provision should be made not only for present air capacity but also for reasonable future growth.
2. Where possible, use a loop system around the plant and within each shop and building. This provides a two-way distribution to the point where the air demand is the greatest.
3. The longer distribution lines should be provided with receivers that are adequate in size and located near the far ends, or at points of occasional heavy use. Many peak demands for compressed air are instantaneous, or of short duration, and storage capacity near these points avoids excessive pressure drop.
4. Each header or main should be provided with outlets which are as close as possible to the points of use. This permits shorter hose lengths and avoids large pressure drops through the hose. The outlets should be located at the top of the pipe line to prevent carryover of condensed moisture to the tool.
5. Arrange all piping so that it slopes toward a drop leg or moisture trap in order that condensed moisture may be removed to prevent it from being carried into the air tools or air-operated devices, where it is harmful. The slope of the air lines should always be away from the compressor to prevent condensate from draining backward into the compressor cylinder.

The use of rigid piping requires threading the pipe ends with a pipe die—or using a welded assembly. Thread production on the pipe ends should be performed with extreme care, to effect proper sealing when the pipe is mated to the companion pipe tap. On completion of the threading operation, all shavings, threading compound, and burrs should be removed carefully from both the inside and outside surfaces of the pipe. When welded assemblies are created, all the loose welding beads and scale should be removed from the interior of the pipe.

The installation of semirigid tubing, although it is more flexible than rigid piping, requires handling with extreme care. Proper cutting tools can be most helpful in eliminating distorting, flattening, or damaging the tubing externally during the cutting operation. The final cut should be true, smooth, and free from both external and internal burrs. A typical tube cutter is shown in Fig. 36-9. On the completion of any cutting or forming operation on tubing, the tube should be cleaned thoroughly.

The successful use of flared fittings depends on the proper flaring of the tube ends. The capability of the flared joint for containing the fluid depends entirely on the accuracy of the flare. A typical flaring tool is shown in Fig. 36-10. The manufacturer's instructions for using these tools should be followed closely, especially in gripping and positioning the tubing within the tool.

Nonrigid or flexible tubing, with its capacity for bending, can be used to eliminate many fittings and the resulting restrictions on internal flow. The successful use of tubing bends, however, depends on bending the tubing properly. Again, this operation can be expedited by employing the proper tools. The bends should not be too sharp, and extreme care should be taken that the ID of the tube is neither distorted nor reduced appreciably in creating the bend. The radius of the smallest bend (from centerline to centerline of the tubing) should measure not less than three times the outside

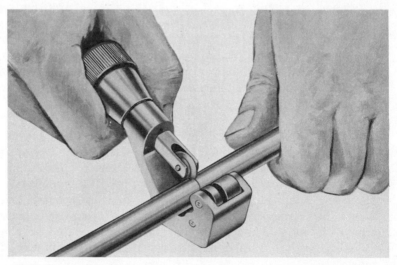

Fig. 36-9. A typical tube cutter. (Courtesy Gould Inc., Valve and Fittings Division)

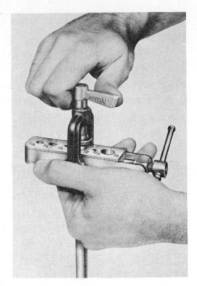

Fig. 36-10. The hand vise and clamp-type, or combination, flaring tool. (Courtesy Gould Inc., Valve and Fittings Division)

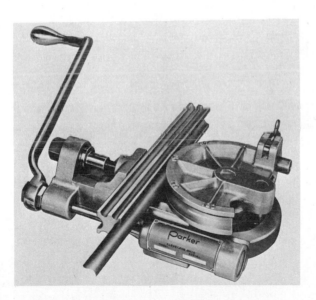

Fig. 36-11. A crank-operated tube bender. (This illustration is the courtesy of Parker Hannifin Corporation.)

diameter of the tubing. Two different styles of tube benders are shown in Figs. 36-11 and 36-12.

The installation of semirigid piping or tubing requires special attention for some details. Straight line connections, especially the shorter lengths, should be avoided, if possible. Care should be exercised to eliminate stresses on tubing. The longer lengths of tubing should be supported by brackets. The lengths of tubing should not be used to support other devices. All parts installed on the tubing lines, such as heavy fittings, valves, etc., should be bolted down to eliminate tubing fatigue. A number of correct and incorrect installations are shown in Fig. 36-13.

In bending the tubing for an assembly, the bends should provide nearly perfect alignment, if possible, between the centerline of the tube and the centerline of the fitting. This avoids distortion or tension in the coupled assembly. Before an installation is made, the ends of the tube section should be examined to determine whether they have been cut true. The inner wall of the tube end should be reamed to

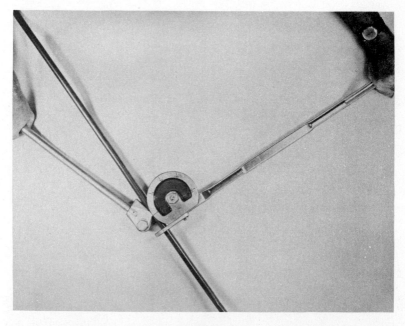

Fig. 36-12. A lever-type hand-operated tube bender. (Courtesy Gould Inc., Valve and Fittings Division)

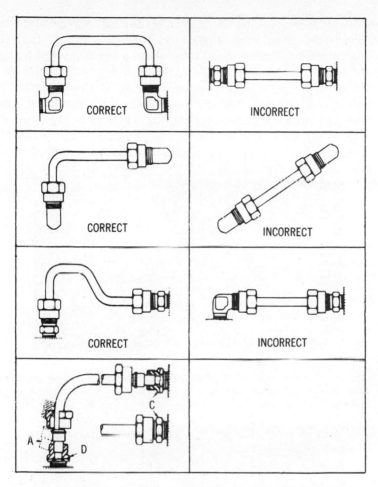

Fig. 36-13. Correct methods (left) and incorrect methods (right) of installing tubing in a fluid system. (Courtesy The Weatherhead Company)

Fig. 36-14. Hydraulic hose assembly with permanent couplings. (Courtesy Gates Rubber Co.)

remove the burrs. Neither the inner nor the outer portion of the cut end should be chamfered excessively, since this can destroy the bearing or sealing surface of the tube where it seats onto the mating fitting.

Flexible piping or hose can be made in any desired length. The connecting ends of the hose are available in a variety of different designs for fitting the various types of tube fittings or pipe threads. The assembly of the tube fittings to the hose ends is usually performed either by permanent assembly at the factory (Fig. 36-14)

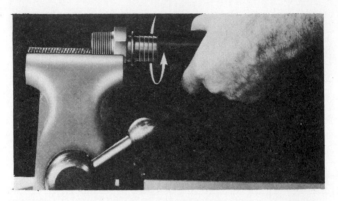

(A) Grip hex of hose stem in vise. (On flanged head couplings, grip shank of stem in vise, using wood spacer blocks to avoid distortion of head.) Lubricate freely stem and inside of hose. Push hose over barbed stem until end abuts the collar.

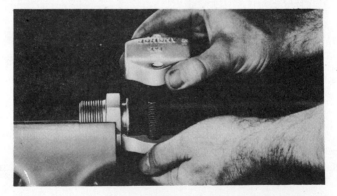

(B) Place lower clamp in position with forward rib between collar and hex. Insert bolts from below and hold in position while placing on upper clamp in line with lower clamp.

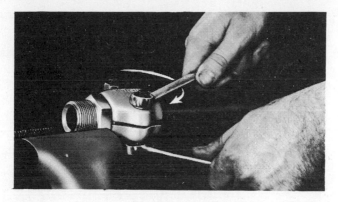

(C) Thread nuts onto bolts and tighten each alternately with automotive wrench on the hose.

Fig. 36-15. Steps in assembling a clamp-type coupling to a hydraulic hose. (Courtesy of Anchor Coupling Co., Inc., Libertyville, Ill.)

Fig. 36-16. A reusable fitting for use with hose. Fitting has a male pipe connection. (Photo courtesy of Aeroquip Corporation)

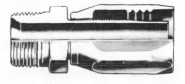

Fig. 36-17. A socketless reusable fitting for use with hose. Fitting has a female swivel on one end. (Photo courtesy of Aeroquip Corporation)

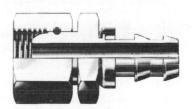

or by assembly at the installation, with the latter fittings referred to as "reusable" fittings. (See Figs. 36-15, 36-16, and 36-17.) The permanently assembled hose fittings are usually crimped onto the hose by machine. This cannot be duplicated easily in the field. The field assembly of one type of reusable hose fitting is shown in Fig. 36-18. Some preparation of the hose end is involved, with the hose fittings being assembled by threading into the ID of the hose and a

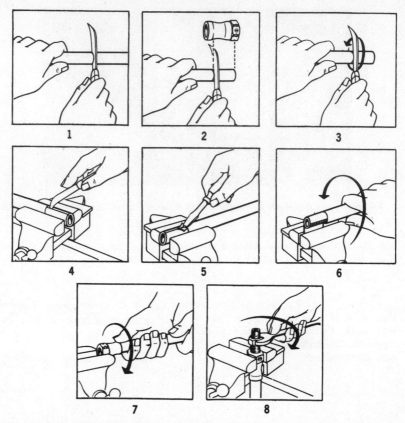

Fig. 36-18. A method in which reusable fittings are utilized in a hose assembly. (Courtesy The Weatherhead Company)

captive end threading onto the OD of the hose. Correct and incorrect ways to install hose are shown in Fig. 36-19.

In making hose-assembly installations, the hose should not be bent or twisted sharply. Also, it should not be placed in torsion during installation or operation. Sharp or excessive bends may cause the hose to kink or rupture. At best, the service life of the hose can be reduced considerably. Most hose decreases in length by approximately 5 percent when rated pressure is applied. A "slack" allowance should be made for this change in length under operating conditions.

Machine installations that use a variety of valves and control components are readily adaptable to manifold mounting. The

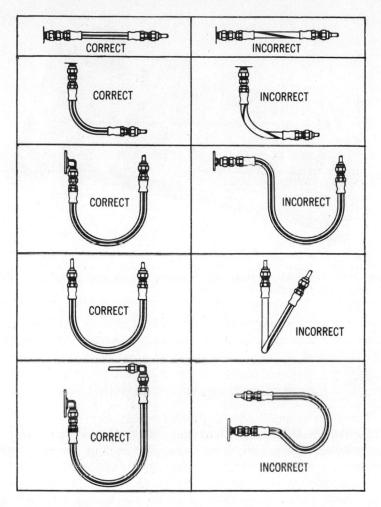

Fig. 36-19. Correct ways (left) and incorrect ways (right) to install hose. (Courtesy The Weatherhead Company)

manifold is a compact unit with a flat mounting surface for the control components (Fig. 36-20). The components are connected by compact passages inside the manifold. This eliminates piping between the valves or control components and reduces the possibility of external leaks. Although the manifolds are relatively expensive in terms of initial cost, they can facilitate the assembly of system components in

Fig. 36-20. Manifolds eliminate much piping and sources of leaks. (Courtesy Double A Products Co.)

their proper working relationships. The hydraulic manifold can be produced either by drilling the internal passages or by assembling the channeled layers of metal plate in a sandwich-type design.

QUICK-CONNECTION COUPLINGS

There are numerous applications for quick-connection couplings not only in manufacturing plants but also in areas of transportation, aircraft, aerospace, agriculture, material handling, and many other areas. It is estimated that a savings can be effected by the use of quick-connection couplings if a line is to be disconnected more than once each week. Think how often an air tool is removed and reassembled to an air line over a short period of time (Fig. 36-21).

All quick-connection couplings have certain elements in common—a plug and a socket. When these elements are connected properly, they seal the joint and lock to resist internal pressures and tensile forces which tend to pull them apart. These couplings are easily disconnected by disengaging the locking mechanism and pulling the components apart without the use of tools.

A number of locking designs are used in quick-connection couplings. Among them are cam-lock, bayonet, roller-lock, ball-lock

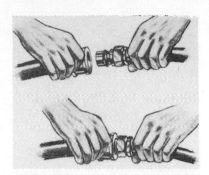

Fig. 36-21. How to connect the two sections of a quick-disconnect coupling. (Courtesy The Hansen Mfg. Co.)

(Fig. 36-22), and pin-lock. The ball-lock is the most common design. A group of individual balls is positioned in the holes in the socket body. The holes are usually tapered or stepped from the largest on the outside diameter to the smallest on the inside diameter, so that the balls will not fall into the center hole of the coupling when the plug is removed; yet, they are free to move.

To obtain the best results, the quick-connection couplings must be matched to the job (Fig. 36-23). In selecting a coupling, a number of factors are to be considered. Among them are the following questions:

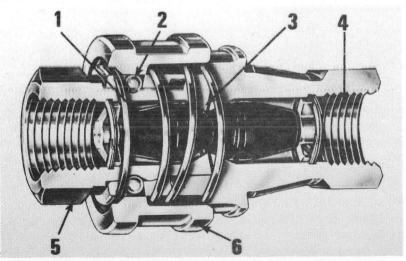

Fig. 36-22. Cross section of a two-way shutoff coupling which provides positive shutoff of both ends of pressurized lines when disconnected. (Courtesy The Hansen Mfg. Co.)

Fig. 36-23. Two-way shutoff coupling (top) provides positive shutoff of both ends of pressurized lines when disconnected. The one-way shutoff coupling (center) requires only automatic shutoff of socket end of the line, and a straight-through coupling (bottom) is recommended for lines where quickly detachable couplings without automatic shutoff valves are needed. (Courtesy The Hansen Mfg. Co.)

1. How much fluid will the coupling be able to pass at the least amount of pressure drop?
2. What type of fluid is to be passed through the coupling and what effect will the fluid have on the seals?
3. Is it necessary to stop all of the flow when the coupling is disconnected; or does it need to be stopped in only one section of the coupling?

4. What will be the maximum pressure that the coupling must withstand?
5. What will be the maximum and minimum temperatures to which the coupling will be subjected?
6. How often will it be necessary to couple and uncouple the quick-connection coupling?
7. Will the coupling be subjected to unusual abuse or extremely poor environmental conditions?
8. What is the size of the hose or pipe to which the coupling will be attached?

Quick-connection couplings are now available in many sizes and for a wide range of pressures and temperatures.

Much thought must be given to the piping layout in a large hydraulic or pneumatic system; otherwise, the efficiency of the system can be greatly reduced and malfunctions can become a regular occurrence. A neat, clean hydraulic piping layout is shown in Fig. 36-24.

Fig. 36-24. Large transfer machine used in the machining of steering gear housing components. Note the clean piping layout for the hydraulic system. (Courtesy The Cross Company, Fraser, Michigan)

Chapter 37

Temperature
Control Devices

In a fluid power system, heat is a problem that may or may not be significant, depending on the application and the circumstances under which the system exists. Although some heat may be present in a system, it may not be a problem unless it exceeds acceptable levels or creates side effects that are detrimental to the proper operation and functioning of the system. Normally, heat results either from friction or from the transformation of mechanical energy into heat energy. The heat problem in a pneumatic system is considerably different from the heat problem in a hydraulic system.

HEAT PROBLEM

The problem caused by the presence of heat in a pneumatic system is usually in the form of side effects or indirect complications. As discussed previously, the heat created normally within a compressed air system is generated within the compressor. Therefore, it is commonly termed the *heat of compression*. Since the compressed air in a pneumatic system is not reused immediately and since the heat

489

induction is not accumulative, some problems result from the presence of the heat and from the subsequent loss of the heat as the air moves through the system.

Pneumatic

The presence of both heat and moisture within a compressed air system creates the ever-present possibility that the heat and moisture combination may reach the critical point at which steam is generated with pressure-increasing or explosive-type forces. If the process remains unchecked, the constant feeding of superheated air into a pneumatic receiver can result in a receiver temperature in excess of safety levels.

A change takes place as the air from the receiver cools—either in the receiver or in moving through the conduit lines. This cooling action results in two undesirable changes in the properties of the compressed air. The first, and less obvious, change is the loss of pressure in the compressed air. The pressure loss results in an efficiency loss between the receiver and the ultimate point of air application.

The second, and more obvious, change in the compressed air is its capacity for holding moisture or water vapor. As the air cools, its capacity for holding moisture in suspension is reduced considerably. Since the moisture holding capacity of the air is reduced, the entrained moisture condenses and begins to accumulate as droplets of water. Of course, the water droplets hinder the proper operation and functioning of the delicate pneumatic instruments. The most effective method of combating or controlling this type of temperature problem in a pneumatic system is to cool the air at some point between the air compressor and the receiver. By cooling, the compressed air is reduced to a workable temperature. Then the condensation of the moisture from the compressed air can be controlled and the condensate can be removed at the air receiver. Since the control signal that governs the pressure and the unloading cycle of the air compressor is received from the pressure within the air receiver, reducing the temperature of the compressed air within the receiver to its operating temperature also can eliminate a further loss of pressure from subsequent cooling downstream.

Hydraulic

In the hydraulic system, the heat generated within the hydraulic fluid remains in the fluid, except for the heat that can be transferred from the oil by means of radiation, convection, or conduction. The hydraulic system is a completely closed system. Therefore, any residual heat that is not transferred from the fluid by one of the aforementioned methods results in cascading or accumulating of heat within the fluid. The undesirable effects of these elevated oil temperatures have been discussed previously in the chapters pertaining to the various components.

To understand the significance of heat properly, heat should be considered in terms of its basic unit of measurement. The most commonly used unit of measurement for heat is the British thermal unit (Btu), which is defined as the quantity of heat required to raise the temperature of 1 pound (453.6 g) of water from 62°F to 63°F (16.7°C to 17.2°C). Btu equivalent values expressed in other units of measurement can be found in Table 37-1.

Table 37-1. Conversion Factors

1 Btu	1.0550×10^{10} ergs
1 Btu	778.3 foot-pounds
1 Btu	3.931×10^{-4} horsepower-hours
1 Btu	2.928×10^{-4} kilowatt-hours
1 Btu/hr	0.2931 watt
1 Btu/min	0.02356 horsepower
1 Btu/min	0.01757 kilowatt

HEAT TRANSFER

As mentioned above, lowering the temperature of the fluid in a fluid power circuit involves transferring the heat within the fluid to another fluid or substance. The three basic methods of transferring heat are: (1) conduction, (2) convection, and (3) radiation.

Conduction is the direct transfer of heat from a material to an adjacent material by heat flow from the warmer material or area to the colder material or area. The rate of heat transfer depends on the area, the temperature difference, and the distance between the two materials. The rate of heat transfer, then, increases as the area and the temperature difference increase. The rate of heat transfer varies inversely with the distance between the two areas.

The various materials display different capacities for transferring heat by means of conduction. The capacity for transfer of heat by conduction is identified by the value K, which is the thermal conductivity of a specific material. The K factor is defined as the number of heat calories transferred per square centimeter per second with a temperature difference of 1°C per centimeter distance from the surface. The conductivity factors for some of the common substances normally encountered in the construction of fluid power components are given in Table 37-2. Although silver is seldom encountered in the construction of a hydraulic component, it is listed in the table as a relative value, because silver is one of the best metallic conductors. It can be noted in the table that the higher value for a material indicates a higher capacity for heat conduction.

The thermal conductivity table is based on the metric system. In most English-speaking countries, however, the K factor is usually expressed in Btu per hour per square foot, with a normal temperature gradient of 1°F per inch, and it is equivalent to 3.4 × 10^{-4} metric units.

Convection is the transfer of heat from one portion of a fluid or material to a contiguous portion of that fluid or material, resulting from the natural heat flow currents within the substance. Normally, the significance of this method of heat transfer is not great in the control of heat, but it defines or explains the lack of isolation of heat to a given area.

Table 37-2. Thermal Conductivity Factors for Common Materials

Material	K Factor
aluminum	0.49
brass	0.25
copper	0.93
iron	0.16
magnesium	0.37
silver	1.00
zinc	0.27
air	0.000057
asbestos paper	0.00019
glass	0.0018
sponge rubber	0.00009

Radiation is the method of heat transfer that most nearly resembles the emanation of light in various directions without a material contact with the source of the heat.

In the study of heat control in fluid power systems, the most important methods of heat transfer are conduction and radiation. For all practical purposes, convection is considered only as a factor that aids in distributing the heat evenly throughout a fluid or component.

COOLING DEVICES

In fluid power systems the device that is used to remove or transfer heat from the fluid is termed a *heat exchanger* in hydraulic systems and an *aftercooler* in pneumatic systems. In the pneumatic unit or aftercooler, heat is transferred from the air to water. In hydraulic systems, the heat transfer is made either from the hydraulic oil to air or from the hydraulic oil to water.

Although the oil-to-air type of unit (Fig. 37-1) is more economical to operate, it is not effective, unless the difference in the air temperature and the hydraulic oil temperature exceeds 10°F ($-12.2°C$). The oil-to-water type of heat exchanger is slightly more

Fig. 37-1. Hydraulic power unit incorporating an oil-to-air cooler to cool the hydraulic fluid. (Courtesy Young Radiator Company)

expensive to operate, depending on the availability of cooling water. However, the capacity of the water to receive the transferred heat permits a more efficient transfer of the heat in a smaller quantity and to a lower temperature. For practical purposes, then, this discussion is aimed more directly toward the oil-to-water heat exchanger, since the principles involved are quite similar to both the oil-to-air and air-to-water units.

The oil-to-water heat exchanger is simple in construction, as shown in Fig. 37-2. Basically, it consists of a series of tubes enclosed within a circular shell and is spaced by a series of locating baffles within the outer shell. The baffles serve the additional purpose of directing the fluid flow along a constantly changing path to ensure maximum exposure of the fluid to the transfer areas. Usually, it is recommended that the more viscous fluids should be channeled through the outer shell area. The less viscous fluids should be channeled through the tube.

This type of heat exchanger is available either as a single-pass unit or as a multiple-pass unit. The single-pass units are usually provided with an inlet port at one end of the assembly and an outlet port at the other end to transfer the less viscous fluids through the unit. In the two-pass units, both the inlet and the outlet ports are on the same end of the heat exchanger. The fluid is channeled to one end of the bundle.

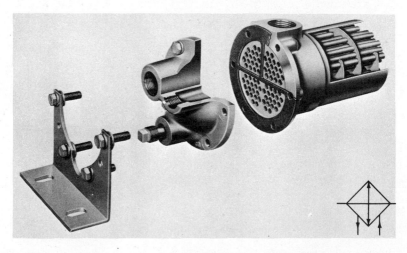

Fig. 37-2. A two-pass type of heat exchanger. (Courtesy Young Radiator Company)

Its direction is then reversed for a return pass through the entire length of the assembly. In the single-pass units, it is recommended that the cooling fluid move in a direction opposite that of the fluid being cooled. This permits a counterflow action of the fluid, which provides maximum heat transfer, since the maximum mean-temperature difference is maintained throughout the entire length of the exchanger unit. In the multiple-pass units, the shell fluid should enter the exchanger at the same end that the tube fluid enters.

The materials used in the construction of heat exchangers vary between units, with special attention provided the type of fluids to be carried and the thermal conductivity of the materials. Although most hydraulic specifications do not recommend the use of copper materials for a hydraulic fluid, the high rate of thermal conductivity of copper makes it an ideal material for the tube bundle (see Table 37-2). The outer shell can be either brass or steel. Normally, the baffles are brass. The end covers containing the ports are usually made from cast iron or bronze.

Usually, this type of heat exchanger is mounted in the circuit in a location where it receives the hydraulic fluid returning to the tank from the directional control valve and the relief valve ports (Fig. 37-3). Occasionally, on the smaller and more simple systems, a bypass arrangement can be used that is similar to the arrangement used by return-line bypass filters.

Temperature-sensing thermocouples can be mounted inside the reservoir to sense the need for more or less cooling water to flow through the coils of the heat exchanger. The thermocouple is used to control a water flow valve. This optional device can be quite valuable, especially when the supply of cooling water is critical. This device also can be quite economical, because the water is turned off automatically when the hydraulic unit is idle (Fig. 37-4).

With petroleum-base hydraulic oils, there is potential danger if the fluid temperature rises above 200°F (93.3°C). The maximum safe temperature in most systems is approximately 180°F (82.2°C). High temperature produces two undesirable effects. One undesirable effect is that the rate of oil oxidation, which creates insoluble gum, sludge and varnish, and insoluble acids is increased. The other undesirable effect is seal deterioration. The materials used in the seals for a piece of equipment are designed to operate within a specified temperature range. When the temperature rises above its rated maximum temperature, a seal may crack, leak, or rupture. A

Fig. 37-3. Temperature control is provided on this packaged hydraulic power system by oil-to-water heat transfer. A circulating pump draws oil directly from the reservoir, passes it through the heat exchanger and back to the reservoir. (Courtesy Young Radiator Company)

combination of faulty seals, oxidized lubricants, and insoluble acids accelerates normal wear and decreases service life drastically.

An overheated hydraulic system is hot to the touch. However, the most positive method for checking a system for over-heating is to use a thermometer. This provides a specific temperature reading. It protects against low-temperature problems in addition to overheating. Careful observations usually reveal the presence of overheating both in localized areas and throughout the entire system.

When the hydraulic oil becomes dark and thick, especially when accompanied by an increase in pressure drop throughout the system and by sluggish machine operation, an overheating condition is indicated. The odor of scorched oil and the appearance of heat-peeled paint on the hydraulic components are also indications of extreme overheating. However, when these conditions appear, other damage is bound to have occurred. An increase in the neutralization number of the oil is a common and fairly reliable symptom of oil overheating

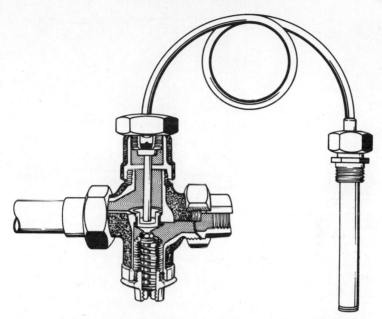

Fig. 37-4. Solenoid-operated on-off valves controlled by thermostatically actuated switches provide automatic control of the flow of cooling water. A modulated flow in response to oil temperature changes is most effective when accomplished with temperature control valves. (Courtesy Sterling, Inc.)

and oxidation. However, special personnel and facilities are required to perform this type of chemical analysis of the hydraulic fluid.

Cleanliness of the component surfaces should be maintained at all times, since a buildup of sludge, scum, or debris on these surfaces dissipates the heat by radiation and inhibits heat transfer from the system. As discussed in Chapter 30, the design of the reservoir size and heat radiation surfaces is important in reducing the need for auxiliary heat transfer equipment.

Chapter 38

Packings and Seals

The tremendous growth in the use of fluid power in industry is a result, in a large measure, of research in the development of packings. Despite the great gains that have been made in the precision fitting of moving metal parts, sufficient clearance is required between these metal parts to make their manufacture and operation practical. This need has given rise to the continued development of flexible and resilient devices for fulfilling the requirement for seals and packings. The development of packings has involved both mechanical design and material development. To understand the success or failure of a given packing application, a study of the different designs and their underlying principles can be most helpful (Table 38-1).

MATERIALS USED

Leather is probably the oldest type of material used in the construction of packings and seals. Although the leather is formed and thereby compacted, it remains a porous material. The

Table 38-1. Various Types of Packing Materials

Material	Compatibility	Temp. Range
Leather	Good for water or oil	(− 54° to 93°C)
Impregnated poromeric material (*Corfam*)	Good for water, oils, fuels, greases; not recommended for phosphate esters.	(− 54° to 121° C)
EP rubber (ethylene-propylene)	Good for water, air, steam, phosphate esters.	(− 54° to 149° C)
Nitrile rubber (*NBR, Buna-N, Hycar*)	Good for water, aliphatic-base petroleum oils, and some synthetics.	(− 54° to 121° C)
Butyl rubber (*11R*)	Good for water and some synthetics; not good with petroleum-base oils.	(− 54° to 121° C)
Polyacrylates (*Hycar 4021*) (*Vyram*)	Good with some synthetics; not good with water-base fluids.	(− 17.8° to 177° C)
Silicone rubbers	Good for water; fair with petroleum oils; good with some synthetics.	(− 84° to 260° C)
Fluoro-silicon	Good with water, oils, and some synthetics.	(− 84° to 260° C)
Urethane	Good for air, oil, or water.	(− 40° to 100° C)
Fluoro-elastomers (*Viton-A*)	Good for water, petroleum-base fluids, and most synthetics.	(− 29° to 260° C)
Fluoro-plastics (*Teflon*) (*Kel-F*)	Good for all fluids.	(− 195° to 260° C)

commercially available leather packings, therefore, are impregnated with various compounds. Until recent years, wax has been the most common material used for impregnation or filling the cellular voids. Since the melting point of wax is low, it has been replaced in recent years by synthetic rubber compounds. The use of rubber compounds for leather impregnation has increased the working temperature of the packing. It is the leather itself that establishes the maximum temperature specification, which is approximately 180° (82.2° C) to 200° F (93.3° C).

The leather packings have advantages or disadvantages, depending on the application. They should not be used for steam-pressure applications or for applications where either strongly alkaline or strongly acid conditions exist. Leather packings are quite satisfactory at temperatures as low as -65° F (-54° C). These packings possess a low coefficient of friction. Leather packings provide high tensile strength. Therefore, they resist extrusion into the clearances between the metal parts. This latter property permits their use in high-pressure applications and where the tolerances between the fitted metal parts are not close.

Packings made from leather usually do not accumulate abrasive materials. Therefore, they do not create a detrimental abrading action on the metal surfaces. The continual movement of the leather against a metal surface can produce the same polishing result obtained in stropping a razor. The fiber structure of the leather surface tends to retain lubrication, thereby resisting the tendency of some packings to wipe themselves dry. The antiextrusion property of leather is desirable, because the material resists the extrusion of rubber-type packings which may be forced into a cold-flow penetration into the clearances or openings.

Another type of material often used in the construction of packings is the fabricated rubber material. This is a combination of synthetic rubber and a contained fabric. The fabric, in this instance, is contained within the structure of the molded rubber compound to impart more strength to the packing and to resist the extrusion or cold-flow tendency of the rubber under high pressure. The fabric used in fabricated rubber construction is usually cotton, asbestos, or nylon. The type of rubber used depends on the operating conditions, such as the temperature and the type of fluid. The most commonly used types of rubber are polychloroprene and *Buna N*, for hydraulic oil service; *Buna S*, for water service; butyl, for phosphate esters; and *Viton*, for temperatures above 250°F (121°C).

Although the fabricated rubber packings are not as effective in retaining lubricating fluid, their temperature range is wider than the range for leather. Also, they can be compounded to resist alkaline and acid conditions. Fabricated rubber packings are slightly more vulnerable to rough surfaces. They require finer finishes for the mating metal parts than do the leather packings.

In general, the hardness of the packing material contributes both to its coefficient of friction and to its resistance to extrusion. Assuming a given pressure and a given clearance between the metal parts, the harder material can withstand the extruding forces better, but at the expense of its frictional properties.

The homogeneous packings are made entirely from the rubber compounds, with no fabric reinforcement. They consist of the same basic rubber compounds used in the fabricated rubber packings. Also, the same temperature ranges apply, with their resistance to extrusion depending entirely on the hardness or softness of the rubber compound.

The homogeneous packings require a finer surface finish on the matching metal surfaces than any other type of packing. Most packing manufacturers recommend that the metal finishes in contact with the dynamic homogeneous packings should not exceed 16 rms. The reason for this recommendation is, of course, that the packing does not possess the ability to trap and carry the lubrication fluid. A metal surface finish of less than approximately 8 rms increases the total frictional drag of a compound moving across it.

Another resilient material that has been widely accepted in recent years is available under its trade name *Teflon*. It is a plastic material or a *tetrafluoroethylene resin*. *Teflon* possesses toughness and resistance to heat and nearly all the chemicals, except the molten alkaline metals. Its lubricity produces a low coefficient of friction against nearly any surface. The compatibility of *Teflon* with a wide range of fluids and compounds makes it ideal for matching unusual temperature ranges with unusual fluids. The successful use of *Teflon* depends on its proper application and the mechanical design of the sealing component. *Teflon* tends to "cold flow" under the stress of pressure. Its plastic properties are not entirely adequate in pressure sealing of the lip-type seals. *Teflon* has proved most satisfactory when it is used in conjunction with a rubber material. The rubber material supplies the resilient elasticity required for a dynamic seal.

The pistons in hydraulic cylinders often use a type of seal similar to the metal piston rings found on automotive-type pistons. This type of metal ring seal is usually made from a cast-iron material when it is used inside a steel tube and from a steel material when it is used inside a cast cylinder tube. The bronze rings are used in both types of cylinders. The metal rings are usually open-end, notched-type construction, with a spring-like feature that causes them to expand forcedly against the inner surface of the cylinder wall. These rings do not create a leakproof seal. They permit a varying volume of fluid to pass, depending on the pressure of the operating fluid. Normally, they are not recommended for pneumatic applications.

TYPES OF PACKINGS AND SEALS

In terms of their physical design and their characteristics, two general types of packings are available. The mechanical seals rely on fluid pressure for expansion against the mating surfaces. The compression packings rely on the physical compression of a mechanical member to force the sealing material against the mating surfaces. Although most of the mechanical packings are preloaded to some extent, they should not be confused with the compression-type packings which require actual mechanical contact or pressure to effect the sealing function.

The *mechanical*, or lip-type, packing is available in a variety of designs, depending on the design of the component into which the seal is incorporated. They are either single-lip or multiple-lip packings. A single-lip packing is commonly referred to as a cup-type packing (Fig. 38-1). This packing is a typical design that can be found on many cylinder pistons. The true cup-type packing is invariably a single-lip type. A variation, sometimes referred to erroneously as a cup packing, is shown in Fig. 38-2. The packing shown is correctly called a "U" packing when it is formed from leather. It is called a "U"-cup packing when it is constructed from a homogeneous material. Another variation of the "U"-cup packing is diagrammed in

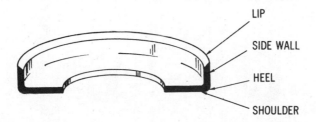

LIP

SIDE WALL

HEEL

SHOULDER

Fig. 38-1. A cup-type packing.

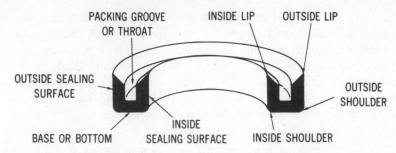

Fig. 38-2. A "U" packing.

Fig. 38-3. It is known as a block-vee packing. It resembles the "U" cup on the closed side (heel). It more nearly resembles a "vee" on the open or pressure side.

A *multiple-lip* mechanical packing, often referred to as a "V" packing or chevron packing, is shown in Fig. 38-4. In many installations, the vee packing is pressure loaded by a mechanical means, such as a lightweight spring load. Packings of this type often use alternating rings of different materials, sometimes alternating a rubber ring with a formed leather ring. This arrangement tends to satisfy a variety of operating requirements, since the leather rings provide the solid antiextrusion properties required in high-pressure applications.

The *cup-type* packing should not be installed with excessive pressure exerted on the flat center portion of the cup. Excessive pressure in this area can cause the lips of the cup to "toe in," thereby losing their ability to seal against the surface of the mating member. The piston cup actually seals by the pressure of the fluid against the inside areas of the cup, which forces both the lip and the heel of the cup against the cylinder wall. Whether the cup is sealing properly can often be determined by examining the worn cylinder cups. Wear

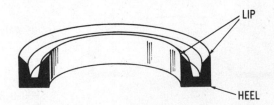

Fig. 38-3. A block-vee packing.

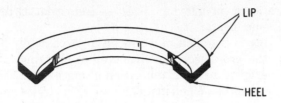

Fig. 38-4. A multiple-lip packing, often called a "vee" or chevron packing.

usually appears to be excessive at the heel of the cup, with little wear in evidence at the edge of the lip. If excessive wear is evident at the heel of the cup, especially on the homogeneous type of cup, a tendency of the cup to expand and to extrude into the mechanical clearances is indicated. The extruding tendency can be eliminated effectively by installing a flat leather backup ring which tends to expand under pressure and cover the mechanical clearance. The leather backup ring or washer is used for a number of different packing designs, as will be indicated later in the discussion of "O"-ring seals.

Another common packing design seals against the outside diameter of a moving member. It is known as the *flange-type* or hat packing (Fig. 38-5). The "hat" packing is used for rotary motion, as well as for reciprocating motion. However, it is usually unsatisfactory for high pressures.

The *"O" ring* is probably the most popular and the most widely used packing. As its name implies, the "O" ring is a molded rubber packing. It is symmetrical in design. It is usually made from a synthetic rubber material. These packings are molded to standard sizes that are accepted in industry. They are, undoubtedly, the most economical packing available. "O" rings are used primarily for reciprocating motion at the slower speeds. They are most commonly

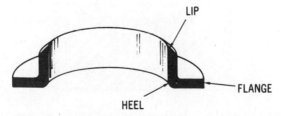

Fig. 38-5. The flange-type or "hat" packing.

used as static or nonmoving seals in industrial hydraulic and pneumatic components.

The "O" ring is installed in a groove approximately 10 percent wider than the cross-sectional diameter of the ring. Its sealing function is performed by a combination of mechanical compression and fluid compression. When installing an "O" ring, its cross section should be subjected to a compression which is referred to as its *preload*. In actual operation, the distortion of the "O" ring increases as the operating pressure increases.

The "O" ring is usually not successful, unless it is used against hard metal surfaces. When it is used on the softer metals, such as aluminum, brass, or bronze, its effectiveness depends on the surface hardness and smoothness of the material. The moving members to be sealed should be manufactured with an absolute minimum clearance, to avoid pressure extrusion of the "O" ring into the clearance (Fig. 38-6). If this type of extrusion occurs between two moving members, nibbling or peeling of the "O" ring usually results.

The metal surfaces which an "O" ring is to move against should be quite smooth—not exceeding a finish of 16 rms. The surfaces should be smooth overall, that is, without nicks, burrs, or scratches.

The tendency of the "O" ring to extrude into the metal clearances can be avoided by using the aforementioned leather backup rings (Fig. 38-7). The diagram shows an application in which an "O" ring and leather backup rings to counteract the extrusion tendencies of the seal are used. In the diagram, the "O" ring is under pressure. It is distorted, as described previously.

The "O"-ring manufacturers recommend that backup rings be used when fluid pressures exceed 1500 psi (10,342.1 kPa). Below 1500 psi, extrusion is usually not a problem, unless the minimum clearances are exceeded. The "O" rings can be made in various degrees of

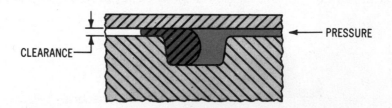

Fig. 38-6. Diagram showing pressure extrusion of an "O" ring into the clearance gap.

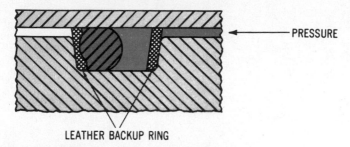

LEATHER BACKUP RING

Fig. 38-7. An application in which an "O" ring and leather backup rings are used to counteract the extrusion tendencies of the seal.

hardness. The optimum hardness is a Shore durometer reading (hardness) of 70 on the A scale. Shore durometer readings of 80 to 90 also can be produced, which increase the resistance to extrusion. However, the frictional drag is also increased.

The preload squeeze imparted to an "O" ring varies, depending on the operating pressure. Increasing the preload squeeze also increases the frictional drag of the seal against the mating metal part. The preload squeeze also varies for dynamic seals and static seals. The supplier of the "O"-ring seals should be consulted to determine the recommended preload squeeze for the various installation conditions.

Extreme care should be exercised when installing "O" rings, especially in passing them over sharp edges, such as the crests of threads. In actual operation, the "O" ring is seldom used successfully where it is required to pass over ports or drilled holes, especially under pressure.

The successful use of "O" rings for dynamic seals in hydraulic valves depends on flow characteristics that permit frictional drag to oppose the pressure distortion of the seal. When these two forces are in opposition, they tend to cancel out. Ignoring this principle in the application of the "O" ring inevitably results in nibbling or peeling, even at comparatively moderate pressures.

Special designs and special materials for specific applications are often used in fluid power components. Although the foregoing discussion has not included all the different packing designs, the student should be reasonably familiar with packings in general.

Chapter 39

Design of Fluid Power Safety Circuits

In designing fluid power circuits, the designer should consider, first, safety for the operator. Then he should consider safety for the equipment on which the fluid power circuit is used and for the workpiece on which the various operations are performed. Since the operator, in many instances, is exposed to fast moving components and to components that impart great force, provision should be made in the circuit design for protecting the operator. For example, the circuit should be designed to prevent any possibility of the operator placing his hands beneath a fast moving ram, either through carelessness or through a malfunction in the circuit. Also, if a hose should burst, provision should be made for shutting off the fluid supply. In a complicated circuit, this can be an extremely difficult problem to solve.

If a circuit malfunction could cause considerable damage to a machine, the circuit should be designed to be fail-safe, if possible. For example, in a circuit in which a sequence of operations is set up by means of sequence valves, either a sudden surge or a sudden requirement for more pressure can change the sequence of operations, thereby causing considerable damage to the slides, connectors, piston rods, and other parts of the machine.

Since many workpieces are quite expensive, they should be afforded ample protection. A loss of several expensive pieces as a result of a circuit malfunction or improper circuit design can mean the difference between profit and loss on a production run.

CIRCUITS FOR OPERATOR SAFETY

The demands for increased safety by safety committees, insurance groups, and workers have resulted in safer plant operations, including the safeguards for fluid power equipment. Two suggested circuits which have been designed to provide operator protection are discussed here (Figs. 39-1 and 39-2). However, some operators attempt to circumvent safety designs in an effort to speed up some machine processes, especially if they are working on a piece-rate schedule.

In the pneumatic safety circuit diagrammed in Fig. 39-1, the operator is required to keep his hands on the two palm buttons while the piston of the cylinder is advancing. It should be remembered that the location of the controls is also important. The controls should be located where they cannot be actuated or tripped accidentally by a hand and a knee or some other part of the body. To operate the circuit, the workpiece is placed on the slide table and locked in place. The operator depresses the palm buttons of both three-way control valves A and B. Air pressure is directed to the safety control valve C and onward to port X of the four-way control valve D. The flow director of valve D is shifted. Air pressure is directed to the blind end of the cylinder E. The piston of the cylinder advances at a rapid rate, moving the work slide forward. When the operation has been completed, the operator removes his hands from the palm buttons on the control valves A and B. The air pressure is released from port X of the control valve D. The flow director of the control valve D returns to its normal position. The air pressure is then directed to the rod end of cylinder E. The piston retracts to its starting position. If the operator should remove his hands from the palm button of either valve A or valve B during the advance stroke of the piston in cylinder E, the flow director in the control valve D returns to its normal position. Then the air pressure is directed to the rod end of cylinder

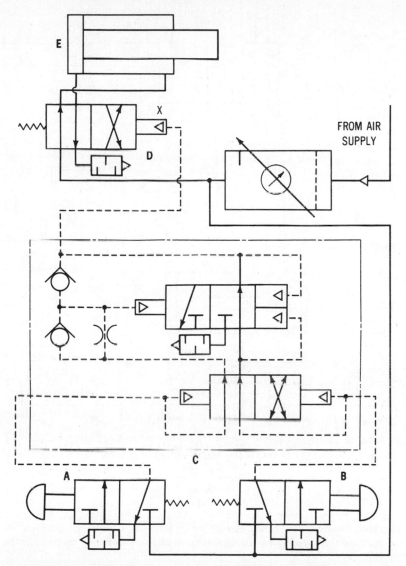

Fig. 39-1. Diagram of a pneumatic circuit in which the operator is required to keep his hands on the two palm buttons while the piston in the cylinder is advancing.

E. The piston retracts. If the operator should tie down the actuator of either valve *A* or valve *B*, the cycle cannot be repeated.

In some applications, it is necessary to protect the operator from

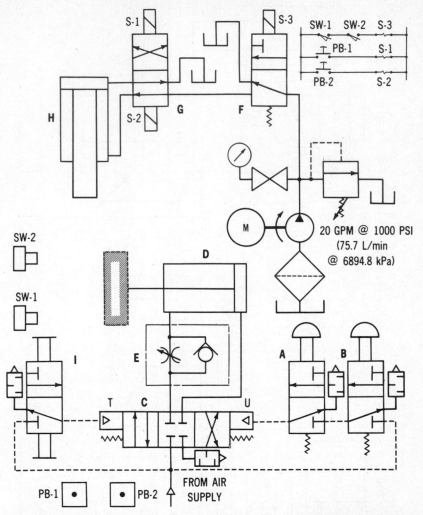

Fig. 39-2. Diagram of a safety circuit which provides a barricade between the operator and the workpiece while work is being performed on the workpiece by the machine.

flying particles. Therefore, some type of barricade is required between the operator and the workpiece. Numerous types of barricades are used. Some barricades use a wire-mesh enclosure, and other types use solid steel panels with either shatterproof-glass peepholes or small windows. All barricades should be designed to protect the operator while work is being performed on the workpiece.

In the diagram in Fig. 39-2, a hydraulic circuit is used in conjunction with a pneumatic circuit. The pneumatic circuit closes and opens the barricade. The operator depresses the buttons on the three-way directional control valves A and B. The air pressure is directed through valve B to valve A and onward to the pilot connection U of the four-way directional control valve C. The flow director of the control valve C which is a spring-centered three-position valve is shifted. Air pressure is directed to the blind end of the barricade closer cylinder D. The piston of cylinder D advances at a speed determined by the orifice in the speed control valve E. If the operator removes either hand from the button control of valve A or B, the flow director of the control valve C centers, and the piston of cylinder D stops. When the barricade is fully closed, the operator releases the buttons of valves A and B. When the barricade is completely closed, the pins on the electrical switches SW-1 and SW-2 are depressed. This causes the solenoid of the three-way valve F to shift, directing the oil flow into the hydraulic circuit. The operator momentarily depresses the push button PB-1. This momentarily energizes the solenoid S-1. The flow director of the control valve G is shifted, directing oil pressure to the blind end of cylinder H. The piston of cylinder H advances. The work is performed. The operator then momentarily depresses the push button PB-2. This momentarily energizes the solenoid S-2. The flow director of the control valve G is shifted to its original position. The oil pressure is directed to the rod end of the working cylinder H. The piston retracts. The operator then shifts the handle of the control valve I. Air pressure is directed to the pilot connection T of the control valve C. The flow director control of valve C shifts. Air pressure is directed to the rod end of cylinder D. The piston retracts to complete the cycle.

CIRCUITS FOR MACHINE SAFETY

Many safety circuits have been devised to reduce tool breakage and to reduce damage to the parts of the machines. A pneumatic circuit that causes a reversal of the cylinder if an excessive force is encountered is shown in the diagram in Fig. 39-3. This feature provides relief for the tool and for the machine.

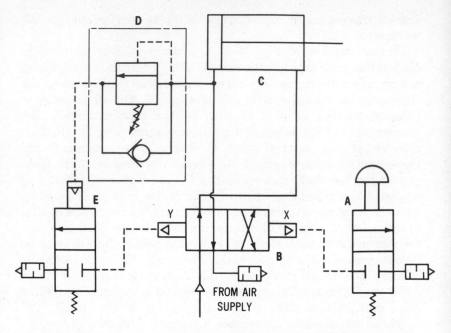

Fig. 39-3. Diagram of a pneumatic circuit which automatically reverses the cylinder movement when an excessive load or force is encountered.

To operate the circuit (see Fig. 39-3), the operator momentarily trips the actuator of the two-way control valve A. The air is bled from the pilot chamber X of the four-way control valve B. The spool of the control valve B shifts as a result of the unbalanced condition. The air pressure is directed from the inlet of the control valve B to the blind end of the cylinder C. A takeoff to the sequence valve D is placed in the line between the control valve B and the cylinder C. If either the full line pressure or a predetermined set pressure is suddenly applied (as the result of a jam or other malfunction) against the blind end of the piston of cylinder C while the piston is advancing, the sequence valve D opens. The two-way control valve E is actuated. This permits the air pressure to be bled from the pilot chamber Y of valve B. The spool shifts. The air pressure is directed immediately to the rod end of the cylinder. The piston retracts.

Normally, when the piston in cylinder C (see Fig. 39-3) reaches the end of its forward stroke, the pressure in the line between the control valve B and the cylinder builds up to full line pressure. The sequence valve D opens. This causes the spool in the control valve B to shift, eliminating the need for the two-way control valve E. In applications where an exceptionally high external force may contact the piston rod during the advance stroke, it may be desirable to set the pressure required to open the sequence valve D higher than the full line pressure. In this instance, the two-way control valve E would be required in the circuit.

A hydraulic circuit which provides safety to a machine by means of positive sequencing is shown in the diagram in Fig. 39-4. When dealing with hydraulics, it should be remembered that large forces which are not handled properly can cause considerable damage to a machine. In the circuit diagram, two pumping systems are used—one system to control each cylinder. The transfer cylinder A can be operated at a much lower pressure than the working cylinder B. To operate the circuit, the work load is placed on the table which is supported by the piston rod of cylinder B. A substantial tail rod is provided to afford additional bearing in the cylinder. The operator momentarily depresses the push button $PB-1$. This causes the solenoid $S-1$ of the control valve C to be energized momentarily. The flow director of the control valve C shifts. Oil pressure is directed to port X of cylinder B. As the piston reaches the end of its forward movement, a swing-type cam trips the limit switch $LS-1$. This momentarily energizes solenoid $S-3$ of the control valve D. Oil pressure is directed to the blind end of the transfer cylinder A. The piston advances and removes the work load from the table. At the end of the stroke, the limit switch $LS-2$ is tripped momentarily. This momentarily energizes solenoid $S-4$ of the control valve D. The flow director of the valve is shifted to its original position. The oil pressure is directed to the rod end of cylinder A. The piston retracts. Just before the piston reaches its retracted position, the swing-type cam on the transfer slide momentarily trips the limit switch $LS-3$. This causes the solenoid $S-2$ of control valve C to be energized momentarily. The flow director in valve C is shifted. Oil pressure is directed to port Y of cylinder B. The piston retracts to complete the cycle. The relief valves E and F are installed to dump the oil to the reservoir at low pressure during the loading period. Each of these relief valves should be set to maintain only enough pressure in the system for retracting the pistons of the two cylinders.

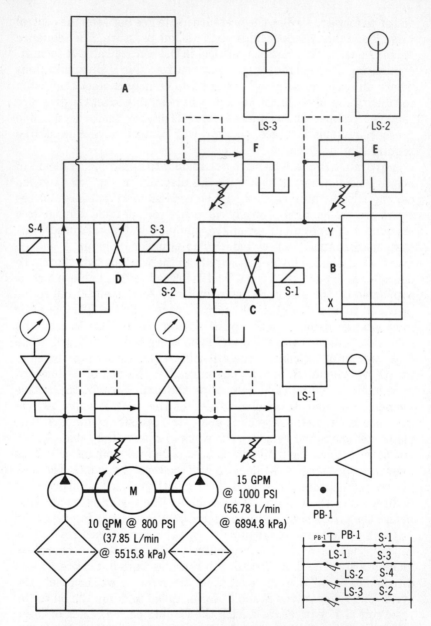

Fig. 39-4. Diagram of a hydraulic circuit which provides safety for a machine by means of positive sequencing.

CIRCUITS FOR SAFETY TO WORKPIECES

Workpieces can become quite expensive after several different operations have been performed on them. If an error occurs during any operation, considerable loss can be experienced. When fluid power equipment is to be used on expensive workpieces, built-in safety features are essential. This is shown in the diagrams of the hydraulic circuit (Fig. 39-5) and the pneumatic circuit (Fig. 39-6). A hydraulic circuit that can be used for straightening a workpiece, for press fitting a subassembly into a workpiece, for disassembling two

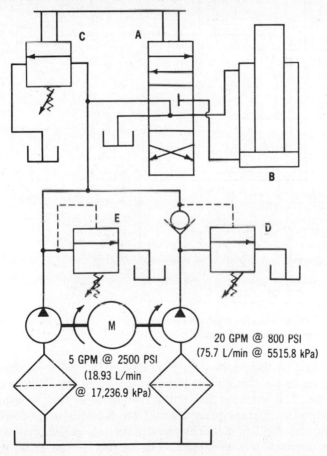

Fig. 39-5. Diagram of a hydraulic press circuit which is used to protect the workpiece.

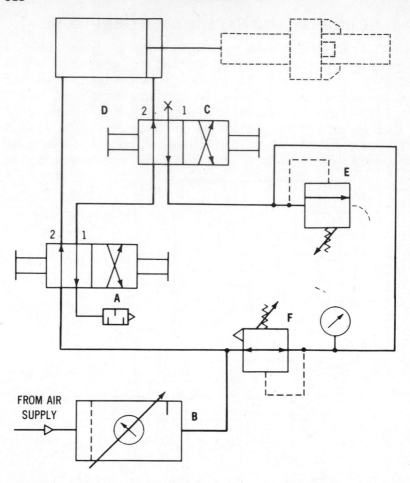

Fig. 39-6. Diagram of a pneumatic circuit that provides both high and low pressure for a power chucking application.

parts, etc., is diagrammed in Fig. 39-5. The operator places the workpiece to be straightened on vee-blocks directly beneath the press ram. The operator then shifts the handle of the three-position four-way valve A. Low-pressure oil from both pumps is directed to the blind end of cylinder B. The piston advances at low pressure until the ram contacts the workpiece. The setting on the relief valve C controls the circuit pressure. The unloading valve D unloads the low-pressure pump. When the operator shifts the valve handle farther, external pressure is placed on the spring in the relief valve C. The pressure increases. The greater the distance that the operator shifts the handle, the higher the operating pressure becomes, until the pressure setting on the relief valve E is exceeded. Then the excessive pressure is dumped. After gauging the workpiece during the straightening process and after reaching the apparent correct setting, the operator can shift the flow director in the control valve A to the neutral position, thereby dumping the pumps while he is inspecting the workpiece. If the work is satisfactory, the operator then shifts the flow director of the control valve A to the opposite position. The oil pressure is directed to the rod end of cylinder B. The piston retracts. The operator can gauge the force applied quite accurately by reading the pressure gauge. He can either increase or decrease the force on the workpiece by moving the handle on the control in small increments.

A pneumatic circuit (see Fig. 39-6) can be applied on power chucking applications in which it is desirable to take a heavy cut on the workpiece while the workpiece is being gripped under high pressure. Then a light finishing cut can be made on the workpiece, after the gripping pressure has been reduced to eliminate distortion. This action can be accomplished by several different methods without removing the workpiece from the chuck. In one example, the workpiece is placed into the power chuck. The operator shifts the handle of the four-way directional control valve A. The flow director in the control valve A is shifted, so that the air pressure set by the pressure regulator in the combination B is directed to the inlet port of the control valve A and from port 1 to the inlet port of the control valve C. Air pressure leaves port 2 of the control valve C, and is directed to the rod end of the rotating cylinder D. The jaws of the power chuck close under full pressure. After the roughing cut has been completed on the workpiece, the operator shifts the handle of the control valve C. The air pressure is reduced on the rod end of the cylinder as a result of the lower pressure setting on the sequence valve E and the regulating valve F. After the finishing cut has been completed, the operator shifts the flow directors of both valves to their original positions. Full line pressure is directed to the blind end of the rotating cylinder. The piston of the rotating cylinder advances, causing the chuck jaws to open. This action usually occurs after the machine spindle has been stopped. To effect a proper working condition for the finishing operation, the sequence valve E should be set to open at a pressure value that is slightly higher than the pressure setting of the regulating valve F.

Many pneumatic and hydraulic circuits can be designed to provide safety. The control system should be kept as simple as possible.

Chapter 40

Air and Oil in
A Single System

The use of both compressed air and hydraulic oil or hydraulic fluid in a single fluid power circuit permits considerable flexibility in circuit design. Several different combinations should be considered. It should be remembered when using these two fluids in a single system that they should never be mixed, which means that the two fluids should always be separated by a barrier, such as a seal, a diaphragm, a packing, a deflector, etc.

ADVANTAGES

Many advantages are derived from the basic circuits that use both fluids. In a number of basic feed circuits, it is desirable to provide a smooth, even feed without involving the expense for a complete hydraulic system. If the force required to provide a smooth, even feed is no greater than the force that can be exerted by an air cylinder, an air-hydraulic cylinder (Fig. 40-1) is satisfactory. An air-hydraulic cylinder not only produces an extremely fine feed but

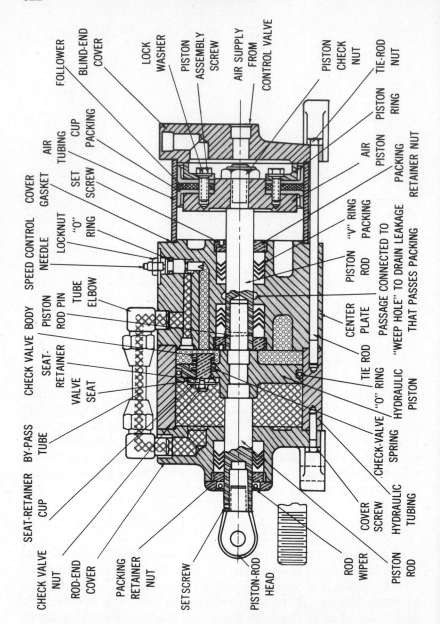

Fig. 40-1. Parts diagram of an air-hydraulic cylinder. (Courtesy Logansport Machine Co., Inc.)

also provides a quick action in those portions of the stroke in which feed is not required. Also, the heat problem does not exist when an air-hydraulic cylinder is used for fine-feeding operations. A feed rate of 1/8 inch (3.18 mm) per minute is quite common. This is a lower rate than can usually be obtained with a hydraulic system using a cylinder having the same bore size.

To obtain rates of feed in the higher ranges, the air-hydraulic cylinder can be constructed with a smaller bore size for the oil section than for the air section. For example, the bore sizes for the air section can be 8 inches (20.3 cm) and the bore size for the hydraulic section can be 3 inches (7.6 cm).

As shown in Fig. 40-1, provision can be made to prevent mixing of the air and oil. A weep hole is provided between the two sets of packings in the center cover of the cylinder for any leakage (either oil or air) from either set of packings to drain. It is insignificant if a small volume of oil leaks into the air portion of the system. Only a small air leak into the oil portion causes erratic feeds.

Controlled Feed

An air-hydraulic cylinder can be installed to provide a controlled feed as the piston in the cylinder advances and to provide a high-speed return stroke as the piston retracts (Fig. 40-2). To operate the circuit, the operator places the workpiece on a slide which is connected to an air-hydraulic cylinder A. The operator then momentarily depresses push button PB-1. This, in turn, momentarily energizes the solenoid S-1 of the two-position four-way valve B. The spool in the control valve B is shifted. Air pressure is directed from port 2 to the blind end of cylinder A. The piston advances at a rate controlled by the flow control valve C. At the end of the stroke, the limit switch LS-1 is contacted. The solenoid S-2 is energized. The spool in the control valve is shifted. The air pressure is directed to the rod end of the air section. The piston returns rapidly.

In many applications, a controlled feed is required in both directions of the piston movement in the cylinders. However, a faster feed may be required in one direction (Fig. 40-3). This circuit does not require electrical equipment. The cylinder reciprocates until the stop-start valve is actuated. To operate the circuit, the operator loads the workpiece onto the machine table, and then shifts the actuator of the "normally closed" two-position two-way valve A. This

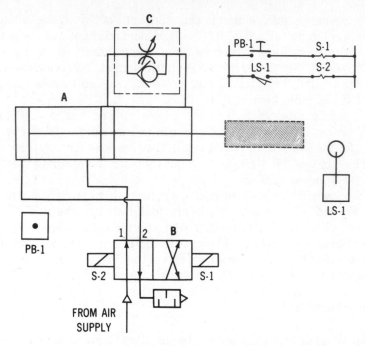

Fig. 40-2. Diagram of a circuit in which an air-hydraulic cylinder provides a controlled feed as the piston advances and a quick-return as the piston retracts.

permits the air pressure to bleed from the pilot chamber X of the bleeder-operated two-position four-way valve B through the control valve A and through the "normally closed" two-position two-way control valve C, since the actuator of valve C is in the actuated position when it is riding on the cam of the machine table. The spool of the four-way control valve B shifts, directing air pressure from port 2 of valve B to the blind end of the air-hydraulic cylinder D. The pistons of cylinder D advance at the speed set by the flow control valve E. As the pistons in cylinder D reach the end of their outward travel, the cam on the machine table contacts the actuator of the "normally closed" two-position two-way valve F. As the passage through the valve F opens, air pressure bleeds from the chamber Y of the four-way control valve B. The spool shifts to its original position. The air pressure is directed from port 1 of the valve to the center section of the cylinder D. The pistons in the cylinder retract at a feed rate

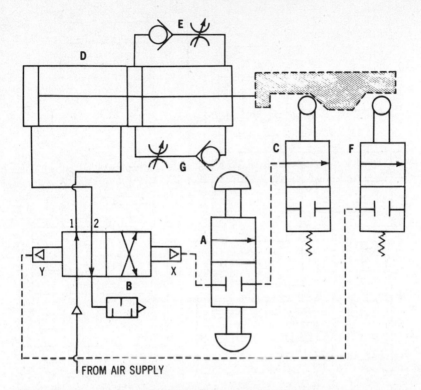

Fig. 40-3. Diagram of an air reciprocating circuit that provides controlled feeds in both directions.

controlled by the flow control valve G. When the pistons in the cylinder retract, the two-way valve C is contacted. The cycle begins again and continues the reciprocation. To stop the cycle at the starting point, the operator shifts the actuator of the control valve A to close the passage through the valve. The stroke of the cylinder can be shortened by moving the control valve F closer to the control valve C.

Skip-Feed Applications

An air-hydraulic cylinder can be used for skip-feed arrangements (Fig. 40-4). This arrangement is often used when interrupted cuts are required on the workpieces. In the circuit shown in the diagram, solenoid-operated control valves are used. However, bleeder or

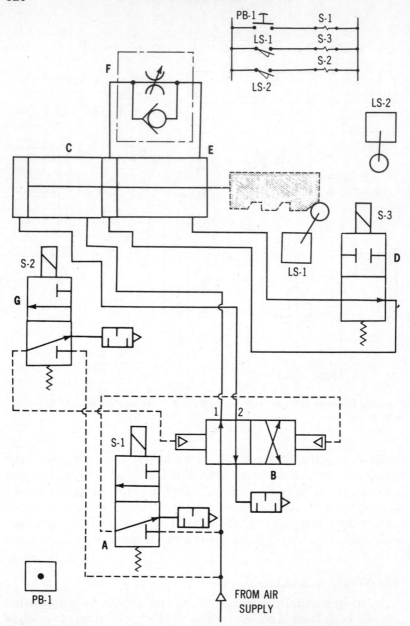

Fig. 40-4. Diagram of a circuit that utilizes an air-hydraulic cylinder to provide a skip-feed arrangement.

pilot-operated pressure controls also can be used to accomplish the same results. The operator places the workpiece on the table, and clamps it in place. Then the operator momentarily depresses the push button *PB-1*. This, in turn, causes momentary energization of the solenoid *S-1* of the control valve *A*. The spool of valve *A* shifts, directing air pressure to valve *B* and to the blind end of the cylinder *C*. The pistons in the cylinder begin to advance rapidly as the passage in the "normally open" two-way valve *D* is opened in the normal position. Also, some oil is passed through the line *E*. When the work position is approached, the cam on the table depresses the roller on the limit switch *LS-1*. This causes the solenoid *S-3* of the control valve *D* to become energized. The passage through valve *D* is closed. The oil is then required to flow through line *E*. The volume of oil passed is controlled by the flow control valve *F*. The speed of the pistons of cylinder *C* are controlled by the flow control valve *F*, until the cam rides off the limit switch *LS-1*. The number of cams located on the table depends on the number of work positions desired. When the pistons of the cylinder approach the end of the out-stroke limit, the limit switch *LS-2* is contacted. Current is directed to the solenoid *S-2* of the control valve *G*. The spool of the control valve *B* is shifted to its original position. The air pressure is directed to port *1* of valve *B* and then to the center cover of the cylinder *C*. The pistons of the cylinder retract rapidly as the oil flows through the check in the flow control *F*. The pistons return to their starting position to complete the cycle.

Positioning Applications

Air-hydraulic cylinders are well adapted to positioning applications, because the oil flow can be shut off quickly to stop the piston movement. A circuit in which an elevating mechanism is operated by an air-hydraulic cylinder is diagrammed in Fig. 40-5. The electrical controls can be used in several different ways to effect the desired results. In this circuit, the electrical controls are tied together. Therefore, when the spool in the three-position four-way control valve *A* is centered, the shutoff valve *B* closes. To operate the circuit, the operator momentarily depresses the push button *PB-1*. This energizes a holding relay. The solenoid *S-1* of the control valve *A* and the solenoid *S-3* of the shutoff valve *B* are energized. Air pressure is directed to port *2* of valve *A*. The line *H* in the oil section of cylinder *C* is opened. The pistons advance at a speed set by the orifice in the flow control *D*. When the limit switch *LS-1* is contacted, the relay drops

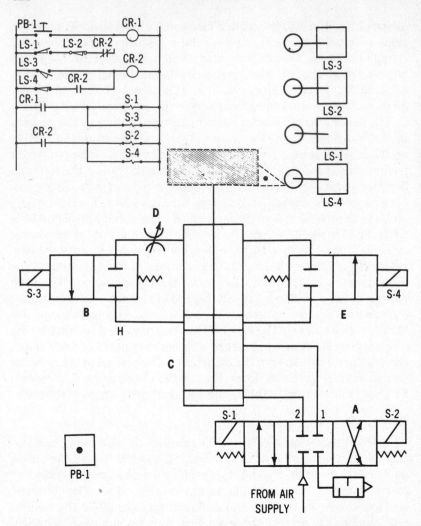

Fig. 40-5. Circuit diagram in which an air-hydraulic cylinder is used to operate an elevating mechanism.

out and solenoids *S-1* and *S-3* are de-energized, causing the spool in valve *A* to move to the center position. This blocks all the ports. Also, the spool in valve *B* closes the passage through the valve, which stops piston movement immediately. To start the piston movement again, the push button *PB-1* is contacted. The valve spools shift, permitting the pistons of the cylinder C to advance until the limit switch LS-2 is contacted, etc. When the limit switch LS-3 is contacted, the solenoids S-1 and S-3 are de-energized. The solenoids S-2 of valve A and S-4 of valve E are energized. This permits the pistons to retract at a rapid rate. As the pistons approach the retracted position, the limit switch LS-4 is contacted. The solenoids S-2 and S-4 are de-energized to complete the cycle. The cam which trips the limit switches LS-1, LS-2, and LS-3 should be a swing-type cam. This prevents the limit switches from being tripped during the retracting stroke.

Synchronized Motions

The air-hydraulic cylinders are especially well adapted for circuits in which the motions are synchronized. Synchronization is extremely difficult in fluid power circuits. However, air-hydraulic cylinders have solved many of these problems. In the circuit shown in Fig. 40-6, three air-hydraulic cylinders are synchronized to raise a long beam. The cylinders A, B, and C are identical in bore and stroke. The workmen load the beam onto the platform which is attached to the ends of the piston rods of cylinders A, B, and C. The operator then shifts the handle of three-position four-way valve D to position X. Air pressure is directed to port 2, and then to the blind ends of cylinders A, B, and C. As the pistons of the cylinders advance, the oil moves from port 2 of cylinder A to port 1 of cylinder B. At the same time, the oil moves from port 2 of cylinder B to port 1 of cylinder C and from port 2 of cylinder C to port 1 of cylinder A. This keeps the pistons of all the cylinders moving at the same rate of speed, regardless of the evenness of their loadings. When the operator desires to stop the cylinder pistons, the spool of the control valve D is shifted to the center position. To retract the pistons, the handle is shifted to position Y. The air pressure is directed to port 1 of the control valve and than to the center sections of cylinders A, B, and C to retract the pistons in unison.

In the circuit diagram in Fig. 40-7, an air cylinder is used in conjunction with a closed hydraulic system to effect a drill press feed. The operator places the workpiece in a vise on the drill press table, and momentarily depresses the button on the "normally closed" pilot-operated three-way control valve A. The air pressure is directed to the pilot chamber Y of the four-way control valve B, shifting the spool. Pressure is then directed to port 2 of valve B and to

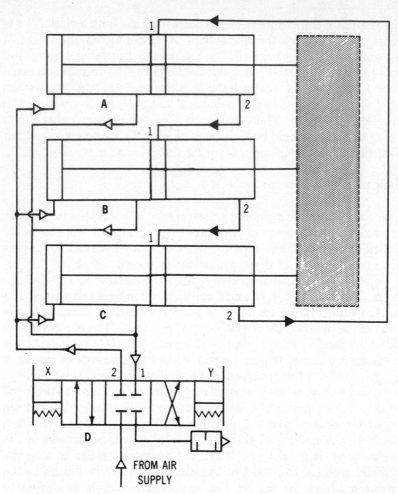

Fig. 40-6. Circuit diagram in which three air-hydraulic cylinders are snychronized to elevate a long beam.

the blind end of the vise cylinder C. The piston of the cylinder advances, clamping the workpiece securely. Air pressure builds up in the line to the sequence valve D. Its orifice opens, permitting air pressure to flow to the blind end of the air cylinder E. The piston of cylinder E begins to move the drill press quill at a rapid rate, until the arm on the piston rod of cylinder E contacts the stop nut on the rod of the hydraulic cylinder F. The piston of the cylinder advances the drill

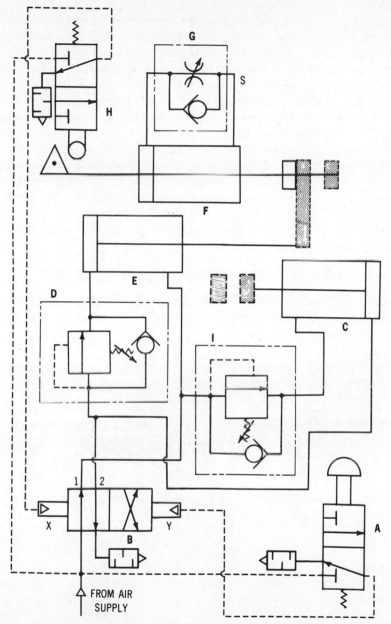

Fig. 40-7. Diagram of a circuit in which an air cylinder is used in conjunction with a closed hydraulic system to effect a feed for a drill press.

into the work at a rate of speed set by the orifice of the flow control G in the line S of the cylinder F. When the drilled hole has been completed, the cam on the end of the piston rod of cylinder F contacts the roller of the three-way control valve H. Air pressure is directed to the pilot chamber X of the four-way control valve B. The spool of valve B is shifted to its original position. Air pressure is directed to port 1 of valve B. Then it is directed to the rod end of the air cylinder E. The piston of cylinder E retracts at a rapid rate, since the built-in check valve in the flow control G permits a free-flow return. When the piston of cylinder E reaches its retracted position, air pressure builds up in the line to the sequence valve I. The orifice opens, permitting air to flow to the rod end of the vise cylinder C. The piston of cylinder C retracts, releasing the workpiece in the vise, and the cycle is completed. A drill press equipped with an air-hydraulic feed mechanism is shown in Fig. 40-8.

AIR-ACTUATED HYDRAULIC CONTROLS

Many interlocking circuits can be devised and used to advantage by using hydraulic controls that are actuated with compressed air. This

Fig. 40-8. A drill press equipped with an air-hydraulic feed mechanism.

eliminates the need for the more expensive actuating mechanisms. A typical air-actuated hydraulic control valve is diagrammed in Fig. 40-9. Low-pressure air is used in the pilot chamber of the hydraulic valve. High-pressure oil up to 3000 psi (20,684.3 kPa) and higher is used in the oil section. Note that the air pressure cannot enter the oil section of the valve. A small air valve with a 1/8 inch pipe port can satisfactorily actuate a large hydraulic valve with a 2- to 4-inch (5.1 to 10.2 cm), or even larger, pipe port.

A simple circuit that uses air pressure to actuate a hydraulic control valve is shown in Fig. 40-10. The four-way control valve A can be located at some distance from the four-way control valve B if necessary. The air provides fast, snappy action in moving the spool of valve B. It is often advisable to install speed control valves in the pilot lines to the hydraulic valve to prevent the spool from moving too fast, which can create a shock in the system. This is especially true when large-volume pumps are employed. The present trend is to install pilot chambers with larger diameters on the air pilot-operated hydraulic valves, so that low-pressure air (approximately 2 psi or 13.8 kPa) can be used.

Air can be used, along with hydraulic fluid, to produce intensified pressure on the hydraulic fluid. A basic circuit is shown in the

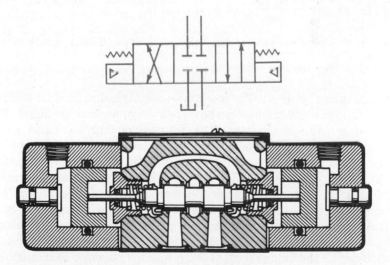

Fig. 40-9. Cutaway diagram of an air-operated hydraulic control valve with manual operators. (Courtesy Double A Products Co.)

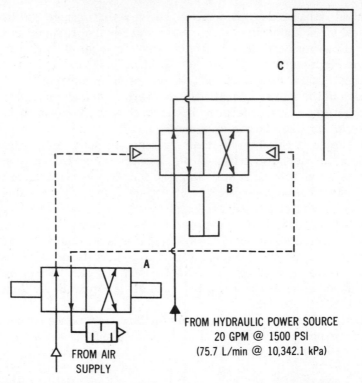

Fig. 40-10. Diagram of a simple circuit that uses air to actuate a hydraulic control valve.

diagram in Fig. 40-11. The workpiece is attached at the test station. The fluid to be intensified can be oil, water, or some other liquid. The operator shifts the actuator of the two-position four-way valve A. Air pressure is directed to port 2 of valve A and then to the blind end of the air cylinder B. The air cylinder is usually larger in bore size than the hydraulic cylinder C. The piston of cylinder B begins to advance, moving the piston of cylinder C which is smaller in diameter. The piston of the hydraulic cylinder C should be leakproof—or as leakproof as possible. Since the area of the piston of the air cylinder B is several times larger than the area of the piston of the hydraulic cylinder C, the fluid in front of the piston in cylinder C is intensified. The fluid ahead of the piston in cylinder C moves into the workpiece for the testing operation. When the testing operation has been

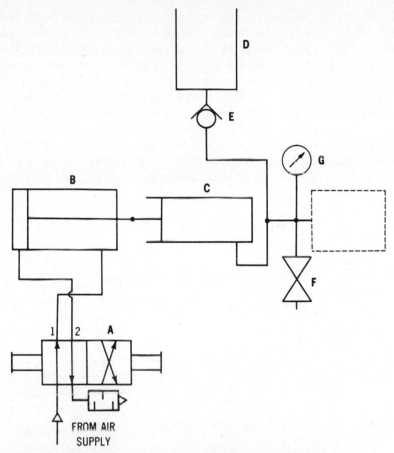

Fig. 40-11. Diagram of a basic circuit in which an air cylinder is used in conjunction with a hydraulic cylinder to produce an intensified pressure.

completed, the operator moves the actuator of control valve A to its original position. Air pressure is directed to port 1 of valve A and then to the rod end of cylinder B. The pistons of cylinders B and C retract. The oil reservoir D is used as a storage tank. The check valve E prevents the high-pressure fluid from entering the storage tank. The shutoff valve F is used as a drain valve. The gauge G is used to record the test pressure.

Air (gas) can be used successfully in conjunction with hydraulic fluids in hydraulic accumulators (see Chapters 33 and 34 for a discussion of the design and application of these devices). The diagram (Fig. 40-12) shows how an accumulator can be applied to a press circuit to provide lengthy holding periods at high pressure. The operator loads the workpiece into a molding press, and then moves the actuator handle on the four-way control valve *A*. Oil pressure is directed to port *2* of valve *A* and then to the blind cover of cylinder *B*. The piston of cylinder *B* advances to the workpiece. Pressure builds up as the accumulator *C* is filled and the gas is compressed. When the peak pressure is reached, the operator permits the handle of the control valve *A* to return to the center position. This blocks the two cylinder ports of valve *A*, and connects the inlet port to the tank port.

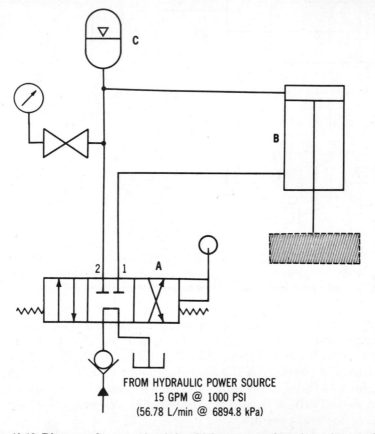

FROM HYDRAULIC POWER SOURCE
15 GPM @ 1000 PSI
(56.78 L/min @ 6894.8 kPa)

Fig. 40-12. Diagram of a press circuit in which an accumulator is used to provide lengthy holding periods at high pressure.

Therefore, a heat condition is eliminated, because the oil is returned to the reservoir at nearly zero pressure.

When the press cycle (see Fig. 40-12) has been completed, the operator shifts the actuator handle of the control valve A to the extreme or opposite position. Oil pressure is directed to port 1 of valve A and then to the rod end of cylinder B. The piston of cylinder B retracts, completing the cycle. The operator releases the actuator handle. The valve spool moves to the center position. It should be remembered that compressed air and hydraulic oil, if they are applied properly, can be used in a single system to solve many complex fluid power problems.

Chapter 41

Design of Basic
Press Circuits

The applications of the various air and hydraulic components to the operation of industrial presses are discussed in this chapter. This comprises one of the larger categories of applications for fluid power equipment. Both air-actuated and hydraulic-actuated components are used widely in all types of press applications. Fluid power is used to eliminate the need for brute strength and to provide safety for the operator.

When air pressure is used to actuate a component, it affords the rapid movement that is ideal for press operations, such as high-speed punching, staking, assembling, etc. By the very nature of hydraulics, a large force can be applied by a relatively small package. Minor adjustments can be made to obtain a wide range of forces. These characteristics make hydraulic pressure ideal for press operations, such as forming, drawing, riveting, molding, coining, etc.

PRESS CIRCUITS

The circuits studied here are basic press circuits. The designer can add the necessary components to fulfill the requirements of the application. For example, the basic press circuit can be made automatic by the addition of infeed and outfeed mechanisms.

Hydraulic Press

The basic hydraulic press circuit in Fig. 41-1 uses a hydraulic power unit A, a three-position four-way valve B, and the press cylinder C. In this type of circuit, the power unit should be ample in size for providing adequate fluid volume and pressure for operating the press cylinder at the required speed and pressure. The reservoir

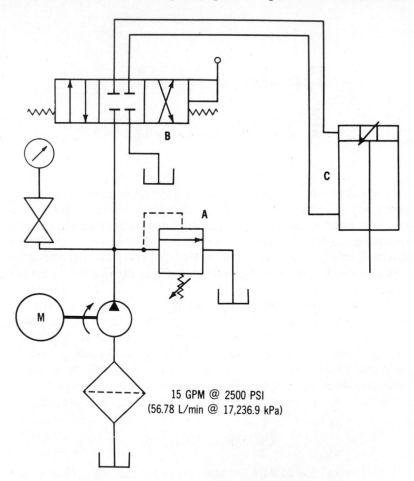

15 GPM @ 2500 PSI
(56.78 L/min @ 17,236.9 kPa)

Fig. 41-1. Diagram of a basic circuit for a hydraulic press.

of the power unit should be large enough to maintain the proper oil level in the system and to provide ample cooling facilities for the oil, unless a heat exchanger is used. The power unit should be located near the press, so that the operator can adjust the relief valve without leaving the press area. In some instances, remote relief valves are located on the press. Then the power unit can be located at a short distance from the press.

The press cylinder C should be rugged enough to resist all shock loads that are likely to be met. The rod bearing should be of proper length to compensate for side thrust. Press cylinders with a 2-to-1 return-advance ratio are advantageous, since the piston returns at twice the speed at which it advances, provided the porting in the blind cover, the control valve, and the piping are large enough in size for handling the added volume of oil. In some applications, a guide on the end of the piston rod or ram is recommended for press work.

In regard to valving, a directional control valve (B in Fig. 40-1) is common in press circuits, since there is usually considerable idle time involved during loading, unloading, and other operations. An open-center control valve permits the oil to circulate or return to the reservoir at nearly zero pressure during the idle periods.

To operate the circuit (see Fig. 41-1), the workpiece is placed on the press platen beneath the fixture on the press ram. The flow director of the spring-centered three-position four-way valve, with the inlet port connected to the exhaust when in the center position, is shifted to direct the oil flow to the blind end of the cylinder C. The piston begins to advance. If the flow director is placed in the neutral position at any time, all movement stops. By using a three-position control valve, the piston of the cylinder can be moved in extremely small increments. If the workpieces vary in height, the piston in the cylinder needs to be retracted only to the point where the workpiece can be removed from the press, which saves much time.

If an exceptionally heavy load, such as a work fixture or a tool, is attached to the end of the ram, a pilot-operated check valve D (Fig. 41-2) can be added to the basic circuit. The check valve closes the line between the rod end of the press cylinder and the control valve. The line is opened when the external pilot pressure is directed to the pilot connection of the check valve. The only internal leakage present is the volume of oil that leaks past the piston. This leakage can be prevented by using a leakproof packing, such as an "O" ring, cups, or block-vee packings, on the piston.

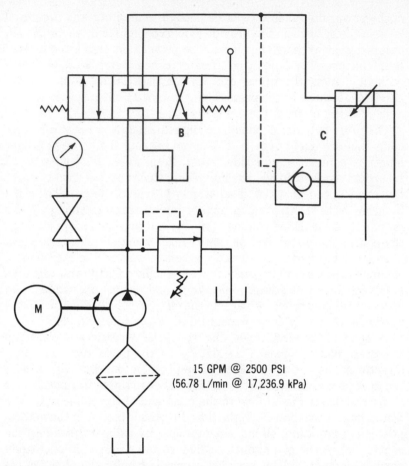

15 GPM @ 2500 PSI
(56.78 L/min @ 17,236.9 kPa)

Fig. 41-2. Diagram of the basic circuit with a check valve placed between the press cylinder and the control valve.

The piston can be held in the "UP" position during a long stand-by period, as shown in the circuit diagram in Fig. 41-3. The relief valve *A* is set at a pressure that is barely high enough to return the press ram *B*. The oil at low pressure is returned to the reservoir during the stand-by period. In this application, a two-position four-way control valve *C* is employed.

Pneumatic Circuit

In a pneumatic circuit for a press application (Fig. 41-4), a

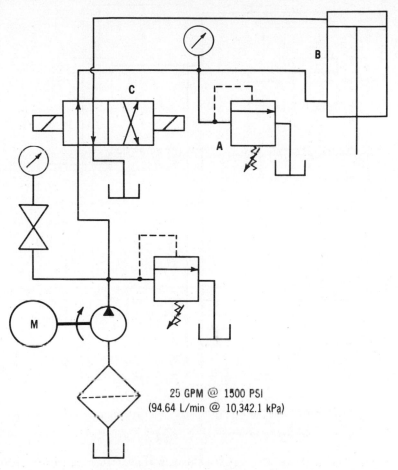

Fig. 41-3. Circuit diagram in which a secondary relief valve is utilized to hold the piston in the "up" position during a long standby period.

sequence valve is used to return the press ram automatically when a predetermined pressure is reached. To operate the circuit, the workpiece is placed on the platen of the press. The flow director of the two-way valve *A* is shifted momentarily. This momentarily opens the orifice through the valve. The air bleeds from the pilot section *T* of the two-position four-way control valve *B*. This causes the flow director of valve *B* to shift, directing the air pressure to the blind end of the cylinder *C*. The piston advances. Work is performed on the

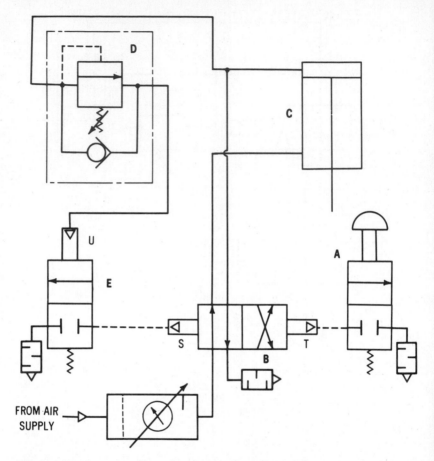

Fig. 41-4. Diagram of a pneumatic press circuit in which a sequence valve is used to return the press ram automatically when a predetermined pressure setting is reached.

workpiece. The pressure builds up. The orifice through the sequence valve D is opened, directing air pressure to the pilot chamber U of the two-way valve E. The flow director in the two-way valve E is shifted. The air is bled from the pilot chamber S of the control valve B. The flow director is shifted to its original position. Air pressure is directed to the rod end of cylinder C. The piston is retracted to complete the cycle.

ELECTRICAL CONTROLS IN CIRCUIT

A hydraulic press circuit diagram in which electrical controls are used is shown in Fig. 41-5. In operating the circuit, the workpiece is placed on the press platen. The operator depresses both electrical push buttons *PB-1* and *PB-2*. This, in turn, energizes the solenoid *S-1* of the closed-center three-position four-way control valve *A*. The flow director of valve *A* is shifted. Oil pressure is directed to the blind end of the cylinder *B*. The piston begins to advance. If the operator

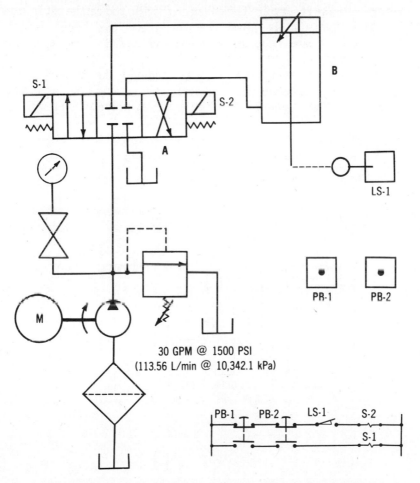

Fig. 41-5. A hydraulic press circuit controlled by electrical push buttons.

releases one of the two push buttons, the piston of the cylinder stops. After the ram has completed its full forward travel, the operator removes his hands from the push buttons *PB-1* and *PB-2*. The flow director in valve *A* is shifted to the opposite end of the valve by the solenoid *S-2*. The oil pressure is directed to the rod end of cylinder *C*. The piston retracts, until the collar on the trip rod attached to the end of the piston rod contacts the limit switch *LS-1*. This causes the solenoid *S-2* to be de-energized. The flow director in valve *A* is centered. This piston movement is stopped. The oil from the pump is returned to the reservoir. The collar on the trip rod can be set to stop the ram immediately after the workpiece has been cleared. Therefore, the ram is not required to travel the full return stroke for short workpieces.

After the workpiece has been placed beneath the press ram, the operator then depresses the electrical push buttons *PB-1* and *PB-2* (Fig. 41-6). The solenoid *S-1* on the four-way valve *A* is energized.

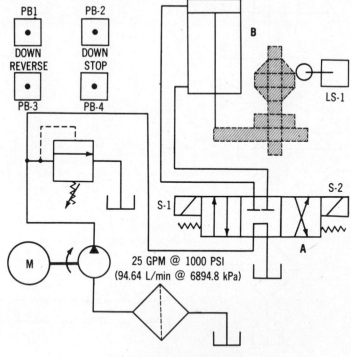

(A) Hydraulic press circuit.

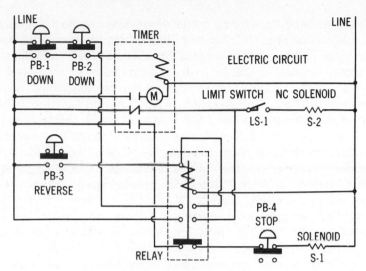

(B) Electrical press circuit.

Fig. 41-6. Diagrams of typical press circuits which are used for a molding press application.

The flow director is shifted to direct oil pressure to the blind end of the press cylinder B. The piston advances. This exerts the maximum pressure on the workpiece (set by the relief valve on the power unit). When the preset time is ended, the flow director of valve A is shifted to the extreme opposite end of the valve body by the solenoid S-2. The oil pressure is directed to the rod end of the cylinder. The piston retracts to a point where the cam on the trip rod depresses the roller on the limit switch LS-1. This de-energizes the solenoid S-2. The flow director of valve A shifts to neutral. The oil is directed to the reservoir at 65 psi (448.2 kPa), which is enough back pressure for operating the pilot-operated control valve A.

If an emergency should occur on the downward stroke of the press ram, the operator depresses the reverse push button PB-3. The piston of the cylinder retracts. If the operator desires to stop the press ram during its downward stroke, he depresses the push button PB-4. When the push button PB-4 is released, the ram continues on the downward stroke. This feature facilitates moving the ram in small increments. The electrical circuit, which includes a relay, timer, limit switch, and the four electrical push buttons, is also shown in Fig. 41-6. This is a typical circuit for a molding press application.

A suggested circuit for an automatic cycle for a press is shown in the diagram in Fig. 41-7. Several different types of controls can be used in the circuit to set up a press for automatic cycling. But, in this instance, solenoid-operated hydraulic controls are employed. In operating the circuit, the workpiece moves downward on the conveyor to the transfer station, where the limit switch LS-1 is contacted. The limit switch LS-1 is in series with the limit switch LS-2. The solenoid S-1 of the four-way control valve A is energized. The flow director is shifted, directing the oil flow to the X end of the transfer cylinder B. The piston advances. The workpiece is transferred to the work station. As the workpiece approaches the work station, the limit switch LS-3 is tripped momentarily. This momentarily energizes the solenoid R-1 of the two-position four-way control valve C. The flow director is shifted, directing oil pressure to the X end of the press cylinder D. The piston advances. As it reaches the end of its stroke, the limit switch LS-4 is tripped momentarily. This causes the solenoid R-2 of valve C to be energized momentarily. The flow director is returned to its normal position. Oil pressure is directed to the Y end of cylinder D. The piston retracts. As the piston of cylinder D approaches the retracted position, the limit switch LS-5 is tripped momentarily. The solenoid T-1 of the four-way control valve E and the solenoid S-2 of valve A are energized momentarily. The flow directors in both valves are shifted. Oil pressure is directed to the X end of the ejection cylinder F and to the Y end of the transfer cylinder B. The piston of cylinder F advances as the piston of cylinder B retracts, and the workpiece is ejected. The limit switch LS-6 is tripped. The solenoid T-2 of valve E is energized momentarily. The flow director is shifted. The oil pressure is directed to the Y end of the ejection cylinder F. Its piston is retracted. When the ejection cylinder contacts the limit switch LS-2, the cycle is completed. The next cycle is ready to begin. Emergency stop buttons and other safety controls can be added to the basic circuit.

A pneumatic press circuit is diagrammed in Fig. 41-8. Many applications require the piston of a cylinder to move at an extremely rapid rate to produce a hammer-like blow on the workpiece. The circuit shown in the diagram not only produces a hammer-like blow from the cylinder piston but also provides safety to the operator from the fast-moving piston rod.

To operate the circuit (see Fig. 41-8), the operator places the workpiece in an air-operated collet chuck, and momentarily

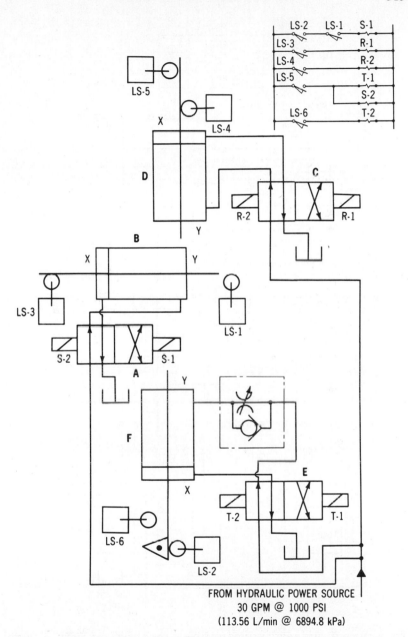

Fig. 41-7. Diagram of a circuit that can be used for an automatic cycle for a press.

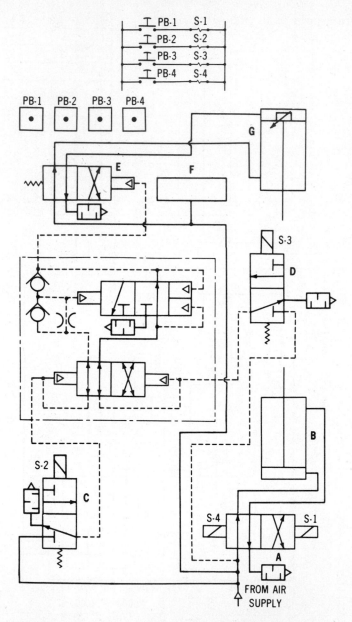

Fig. 41-8. Diagram of a pneumatic press circuit that produces a hammer-like blow on the workpiece and provides safety for the operator.

depresses the push button *PB-1*. This energizes the solenoid *S-1* of the four-way control valve *A*. The piston shifts, directing air pressure to the rod end of cylinder *B*. The jaws of the collet chuck close to hold the workpiece securely. The operator then depresses the push buttons *PB-2* and *PB-3*. The solenoids *S-2* of the control valve *C* and *S-3* of the control valve *D* are energized. Air pressure is directed to the pilot connection of the control valve *E*. The spool of valve *E* shifts. The large volume of air under pressure from the surge tank *F* and the air from the source are directed through the valve *E* to the blind end of the cylinder *G*. The piston of cylinder *G* advances at an extremely rapid rate. Work is performed on the workpiece. The operator releases the push buttons *PB-2* and *PB-3*. This, in turn, causes the air to be released from the pilot section of the control valve *E*. When the spool of valve *E* is shifted by spring pressure, the air pressure is directed to the rod end of cylinder *G*. The piston retracts. The operator momentarily depresses push button *PB-4*, which energizes the solenoid *S-4* of the control valve *A*. The spool of valve *A* shifts. Air pressure is directed to the blind end of cylinder *B*. The jaws of the collet chuck open. The operator removes the completed workpiece.

A regenerative-type circuit for hydraulic press applications is shown in the diagram in Fig. 41-9. This circuit provides a quick approach to the workpiece, and then full pressure is applied on the workpiece. Here is the operation: When the four-way directional control valve *A* is in the neutral position and the power-unit motor is running, the oil is exhausted to the reservoir of the power unit. To start the cycle, the operator shifts the handle of the control valve *A*. Oil flows to both the blind end and the rod end of the cylinder *B*. Since the cylinder has a 2:1 advance-return ratio piston, the force on the blind end is greater. The oil is forced from the rod end of the cylinder through the control valve *C* and then through the control valve *A* to the blind end of the cylinder *B*. The piston advances at a rapid rate. When the resistance is encountered, the passage through the valve *D* is opened. The oil pressure shifts the spool of the four-way valve *C*. The piston completes the work stroke at full pressure.

When the work has been completed, the operator shifts the handle of the directional control valve *A* to the opposite position. Oil under pressure is directed through valve *C* to the rod end of the cylinder *B*. The oil is then exhausted from the blind end of the cylinder through the control valve *A*. If the operator desires to stop the piston of the

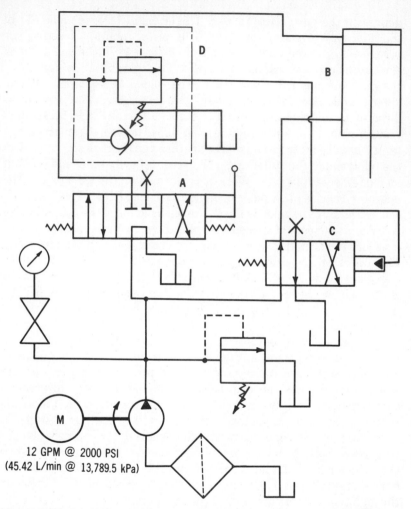

Fig. 41-9. Diagram of a regenerative-type circuit for a hydraulic press application. This circuit provides a quick approach to the workpiece, and then applies full pressure on the workpiece.

cylinder at any time, he merely releases the handle of the control valve *A*. The piston centers, blocking the cylinder port.

A press circuit in which several different pressures are readily available is shown in the diagram in Fig. 41-10. In operating the circuit, the operator loads the workpiece into the press, and shifts the

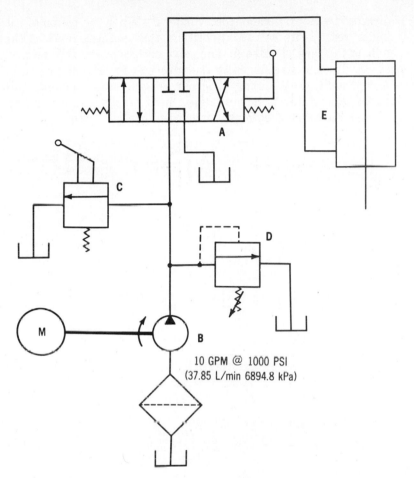

Fig. 41-10. Diagram of a press circuit in which several different pressures are made available.

handle of the three-position four-way valve A. Oil from the pump of the power unit B is directed to the inlet port of the control valve A at low pressure. The pressure is controlled by the relief valve C. The press ram descends until the workpiece is contacted. Then the operator shifts the handle of the relief valve C farther. The pressure in the system cannot exceed the pressure setting of valve D. After the work has been completed, the operator shifts the handles of valves A and C to their extreme opposite positions. Low-pressure oil is directed to the rod end of the cylinder E. When the piston of the cylinder reaches its retracted position, the operator releases the handle of the control valve A. The valve spool moves to the neutral position. The oil is returned to the reservoir under low pressure.

A hydraulic press circuit diagram in which the movement is initiated by a limit switch being closed momentarily is shown in Fig. 41-11. As the workpiece moves inward to the work station on the

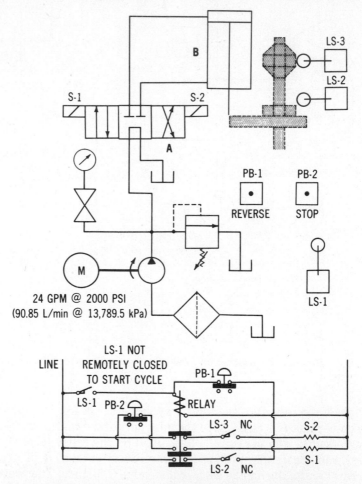

Fig. 41-11. Schematic diagram of a hydraulic press circuit (upper) in which an automatic cycle is initiated by a limit switch being closed momentarily. The electrical circuit diagram is shown (lower).

press platen, the "normally open" limit switch *LS-1* is closed momentarily. This energizes the relay (which closes). Current is directed to the solenoid *S-1* of the solenoid-operated three-position directional control valve *A*. The flow director in the control valve *A* is shifted, directing oil pressure to the blind end of the press cylinder *B*. The piston of the cylinder advances. When the press ram reaches the predetermined downward point, the cam on the trip rod opens the "normally closed" limit switch *LS-2*. This breaks the contact to the solenoid *S-1*. Both the solenoid and the relay are de-energized. The contacts on the relay, which cause the solenoid *S-2* to be energized, are closed. The flow director in control valve *A* is shifted. Oil pressure is directed to the rod end of the cylinder *B*. The piston retracts until the cam on the trip rod opens the limit switch *LS-3* to stop the piston movement. To reverse the downward movement of the press ram, the reverse button *PB-1* is depressed. To stop the downward movement of the press ram, the stop button *PB-2* is depressed. When the stop button is released, the ram continues its downward movement and completes the automatic cycle.

Chapter 42

Pneumatic Logic Control Systems

A dictionary definition describes *LOGIC* as being "Connection, as of facts or events in a rational way." *Pneumatic Logic* is precisely that. *Pneumatic Logic*, then, is simply the connecting or combining of control signals to obtain a desired resulting output action.

To meet the ever-increasing demands of industry and to make easier the packaging and functioning of such control systems, a whole new generation of control components has been developed. While most of these components have been with us for some time, they are, nevertheless, comparatively new in their modern form. They have also been given new names based on their functional role within a control system, and they are designed for compactness.

Throughout this chapter, an attempt will be made to identify each logic component in terms of both the *Pneumatic Logic* concept and the more traditional control valve nomenclature. It is not the goal of the authors to present an all-inclusive and detailed story of *Pneumatic Logic*, but rather to lay the groundwork for an understanding of the basic logic concepts and nomenclature employed in the industry.

557

A logic system and its various components can be most readily understood by first identifying the components in terms of their function within the system. A logic system is best described by the following diagram:

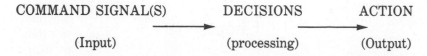

COMMAND SIGNAL(S)   DECISIONS   ACTION

(Input)   (processing)   (Output)

All signals passing through the logic system are binary in nature. That is, there are only two conditions that may exist with a signal anywhere within the system. The signal is either present or not present; either on or off; either passing or not passing; and either yes or no. In the language of Boolean Algebra, these signals are either Logic 1 (on) or Logic 0 (off).

COMMAND SIGNALS

Command or input signals are originated by small three-way valves that may be manually or mechanically actuated. Within a logic circuit, they are usually referred to as "push buttons" or "limit valves." They are to the Air Logic System what push-button switches and limit switches are to the electrical or relay logic circuits. A common variation of the limit valve is the jet sensor. This is the approximate equivalent of the electrical proximity switch.

DECISIONS

The processing portion of the logic circuit is where the input signals or combinations of the input signals are received, processed, and turned into logic-type decisions. This portion of the circuit, in effect, reads the input signals and signal combinations. Then, it sends forth an output signal or signals to create action or perform work as dictated by the input signal or signal combinations. The basic logic elements used in pneumatic logic systems have equivalents in other disciplines. Comparisons can readily be made by referring to such equivalents to facilitate a clearer understanding of the various logic functions. For reasons of simplicity, electric logic comparisons have been chosen for this discussion, since they are the most common and widely known.

A logic system, whether it is pneumatic, electric, electronic, or any combination of these, can best be understood by first analyzing the various functions involved in the processing of command signals. The basic logic functions are AND, OR, NOT, MEMORY, and TIME. Combinations of these basic functions or inversions can produce secondary logic functions such as NAND, NOR, and FLIP-FLOP. However, this presentation will be limited to the basic functions.

When examined individually, and particularly within a discipline with which one is familiar, the various logic functions are easily understood and aptly named.

The circuits involved in the various basic logic functions are:

1. AND (series circuit)—*two or more* signals; *both* are required for output signal (Figs. 42-1 and 42-2).

2. OR (parallel circuit)—*two or more* signals; *one* is required for output signal (Figs. 42-3 and 42-4).

3. NOT (series circuit)—*one* signal opens switch (if input signal *not* present, there is output), cutting off output signal.

4. MEMORY—holding circuit.

5. TIME—on-delay (delayed input, or emerging); off-delay (delayed cutoff of output).

AND Function

The AND function is accomplished by combining two or more input signals to obtain a single output signal. The output signal is realized only if all inputs are present. The removal of one or more of the input signals will result in the loss of the output signal. The AND function, as it is used electrically with limit switches in series or with control relay contacts in series, is shown in Fig. 42-1. This same AND function can be accomplished with a single pneumatic unit that is commonly identified as a three-way, pilot-operated, normally closed valve. Such a valve is symbolically diagrammed in Fig. 42-2. An input air signal must be present at both A and B, in order for an output signal to exist at C. Removal of either of the two inputs will result in discontinuation of the output signal. In the diagram, the B port is

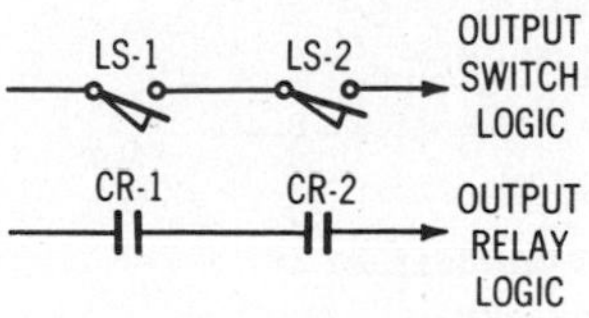

Fig. 42-1. Diagram of electric *AND* functions.

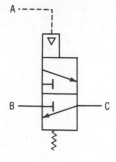

Fig. 42-2. Diagram of a pneumatic *AND* function utilizing a three-way pilot-operated air valve.

normally the valve "in" port. The *A* port is normally the pilot signal port.

OR Function

The OR function is accomplished by two or more input signals, any one of which alone can effect an output. The electrical OR function is diagrammed in Fig. 42-3. It is commonly referred to as a "parallel" circuit. Limit switch *LS-1* alone or limit switch *LS-2* alone can produce an output signal. This same OR function can be accomplished by a single pneumatic element known as the "shuttle valve," as represented in Fig. 42-4. An input signal at either *A* or *B* will produce an output at *C*. If neither input signal is present, there will be no output signal present at *C*. This simple pneumatic device thus permits the use of an output air signal that may arrive from either of two alternate sources.

NOT Function

The NOT function is accomplished electrically by the use of a single normally-closed limit switch (Fig. 42-5). If there is *not* an actuation of the limit switch, an output signal is present. If the limit

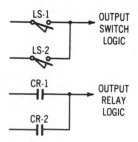

Fig. 42-3. Diagram of electric *OR* functions.

Fig. 42-4. Pneumatic *OR* function using shuttle valve.

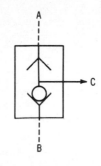

Fig. 42-5. *NOT* function using electric switching.

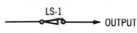

switch is actuated, the output signal ceases to exist. The NOT function is accomplished pneumatically by using a single three-way, pilot-operated, normally open valve (Fig. 42-6). In this instance, the *B* port is not considered to be a signal port. A constant supply of pressurized air is available to it. The *A* port remains an input signal port. If an input signal is *not* present at the *A* port, an output signal is present at port *C*. Conversely, the presence of an input air signal at port *A* will eliminate the output signal at *C*.

MEMORY Function

The MEMORY function is easily recognized in electric relay logic as a simple lock-in by means of employing a set of the subject relay's contacts (Fig. 42-7). Momentary actuation of manual switch *PB 1* energizes the coil of *R-1* relay. A sustaining or "holding" signal is then supplied to the *R-1* relay coil through *(Y)* set of *R-1* contacts. The "reset" function is accomplished by momentary actuation of manual switch *PB-2*. This interrupts the "holding" or "lock-in" signal and

Fig. 42-6. Pneumatic *NOT* function employing a three-way, normally open air valve.

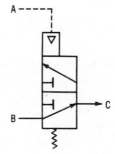

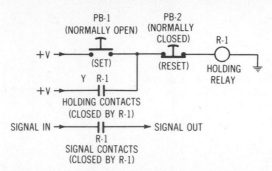

Fig. 42-7. *OFF-RETURN MEMORY* function utilizing switches and relays.

eliminates the output signal. Interruption of the main power supply will also "reset" the relay and return it to its de-energized state. This latter characteristic gives rise to the term "OFF-RETURN MEMORY" as a more accurate title of the described memory function. This function is the standard system for employing push-button control of magnetic motor starters.

OFF-RETURN MEMORY Function

This same OFF-RETURN MEMORY function is accomplished pneumatically by employing the same three-way, pilot-operated, normally closed valve used to create the AND function, but with variations (Fig. 42-8). The air supply at B is constant. Therefore, it is not considered to be a signal port. An input air signal at A passes through the shuttle valve S to shift valve (L), resulting in an output at C. The output air signal also passes upward through orifice (V) to supply the "holding" or "lock-in" signal through the shuttle valve (S). The memory is now accomplished, since the output signal will sustain itself until a "reset" function causes it to cease. This reset function is represented by valve No. *2* in the diagram, a two-way, manually operated, normally closed valve. Manual (or other) operation of valve No. *2* will bleed off the "lock-in" air signal. Valve No. *1* will return to its original "de-energized" state. As in the relay logic diagram (Fig. 42-7), interruption of the main air supply to port B will reset the unit and return it to its "de-energized" state.

TIME Function

TIME functions, as they are considered in logic systems, are merely the delaying of an input signal or delaying the loss of an output

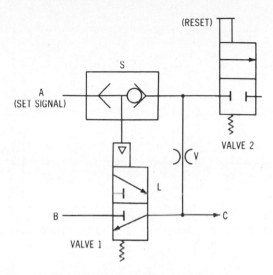

Fig. 42-8. Pneumatic *OFF-RETURN MEMORY* function.

signal. The delayed input is known as an "on-delay." The delayed loss of output signal is known as an "off-delay." In its simplest form, as diagrammed in Fig. 42-9, the delay is accomplished by the use of a flow-control valve that retards the rate of passage of the input signal, thereby delaying its effectiveness. An "on-delay" timing function is represented in Fig. 42-9. The "on-delay" function can be changed to the "off-delay" function diagrammed in Fig. 42-10 merely by reversing the effective direction of the flow control. In the "off-delay" function, the input signal is effective immediately. However, upon interruption of the input signal, its effectiveness is dissipated slowly through the flow control, resulting in delaying the loss of output signal.

Within the DECISION portion of pneumatic logic systems, the functional elements might be designed to operate at full line pressures. Some systems, however, employ units that are designed to operate at very low pressures in the range of 1 ½ to 3 psi (10.3 to 20.7 kPa). Supplying these low pressures is accomplished by using a good low-pressure air regulator capable of stable resolution at such pressures. Conversion of output signals to working pressures is simply a matter of interfacing the low-pressure logic signal with standard pressure valves. Such an interface device can best be

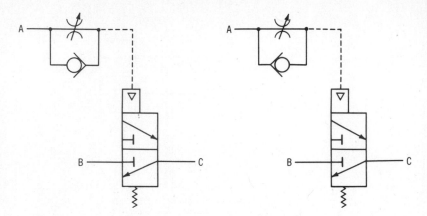

Fig. 42-9. *ON-DELAY* **timing function.**   **Fig. 42-10.** *OFF-DELAY* **timing function.**

described simply as a pilot-operated air valve capable of responding to very low pilot signals.

The pneumatic logic field is a rapidly expanding field. Quite sophisticated componentry is readily available. This presentation has been greatly simplified. It is intended to be only a very fundamental and brief introduction to the philosophy of pneumatic logics.

Chapter 43

Troubleshooting of Pneumatic Circuits

To be capable of tracing the cause of a malfunction in a circuit—pneumatic, hydraulic, or electric—is important. Quite often, a malfunction is caused by a minor problem within a component. Malfunctions also can be attributed to faulty installation of components or piping and to the entry of foreign matter into the system.

A systematic check of the circuit should be conducted whenever failure occurs. The symptoms, causes, and remedies of the various components are listed here.

CYLINDERS

Excessive Wear on One Side of Piston Rod

Probable Cause	*Remedy*
1. Misalignment or side thrust.	1. Provide a better guide for the piston rod. A rod-end cover with a larger rod diameter, a longer rod bearing, or a stop tube inside the cylinder can be

2. Incorrect mounting style.

used. A double-end rod (Fig. 43-1) may solve the problem.

2. Change the mounting style of the cylinder. Foot-mounted and center-line mounted cylinders are extremely rigid, and their alignment is critical. A pivot-mounted cylinder (Fig. 43-2) may be substituted to compensate for misalignment.

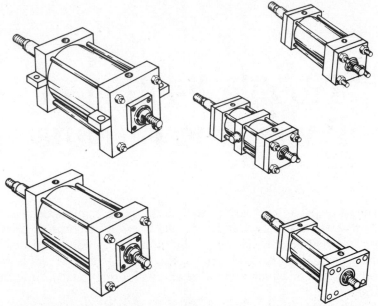

Fig. 43-1. Typical double-end pneumatic cylinders. (This illustration is the courtesy of Parker Hannifin Corporation.)

Fig. 43-2. A heavy-duty pivot-mounted nonrotating cylinder.

Bent Piston Rod

Probable Cause	*Remedy*
1. Excessive side load on end of the piston rod.	1. Utilize cylinders with oversized piston rods.
2. Cylinder stroke too long in relation to diameter of piston rod, resulting in column failure.	2. Same as No. 1.
3. Cylinder used for pressures higher than the pressure for which it was designed.	3. Keep working pressures for cylinder within safe limits.

Jerking Movement of Cylinder Piston

Probable Cause	*Remedy*
1. Insufficient lubrication.	1. Check the lubricator unit for proper operation and proper lubricant.
2. Leaks within or at the cover gaskets.	2. New packings should be installed if the piston packing is leaking. Then the piston should be stopped at intermediate points in its travel, rather than at the ends. If the cylinder tube is worn excessively, the tube and the tube gaskets should be replaced. A worn cylinder tube sometimes can be chrome plated on the ID (if not worn too much), honed, and then returned to service.
3. Score marks inside the cylinder tube or on the piston rod.	3. When score marks occur either on the piston rod or inside the tube, the parts should be replaced if the score marks cannot be polished out. The cylinder should not be returned to operation until the cause of the problem has been determined and corrected.
4. Rod packing gland too tight.	4. The rod packing gland can be too tight on some cylinders; this not only causes a jerking action of the cylinder, but also can stall the movement of the piston. Generally, a cylinder that utilizes a packing gland which can be "taken up" as the packing wears is loose enough for the packing to "breathe."
5. If the piston rod is attached to a moving member, the member may be binding at some point.	5. Light lubrication added to a way or slide sometimes eliminates the jerking action caused by tight spots or binding in the path of a moving member.

6. If the load that the piston is moving requires the full force that the piston can develop, friction at any point in the length of its travel can cause a jerking action.

7. Intermittent air flow or a restriction to the parts in the cylinder.

6. A cylinder with a bore size that is larger than the size required to do the job should be employed.

7. Check the air lines from the source for restrictions or intermittent air flow. Restrictions usually can be eliminated, but intermittent air flow can be a problem. A surge tank (Fig. 43-3) can be used to provide a quick supply of air for the circuit.

Cylinder Piston does not Move

Probable Cause

1. On a cushion-type cylinder in which the cushion nose or collar and the orifice are made from metal, the required close tolerances cause the nose or collar to stick in the orifice on the long-stroke, small-bore cylinders as a result of sag in the piston rod.

Remedy

1. In some instances, the diameter of the cushion nose or collar can be decreased slightly. Sometimes, either a cylinder with a larger bore size or a piston rod with a larger diameter can be installed; or a cylinder with a floating cushion (Fig. 43-4) or a

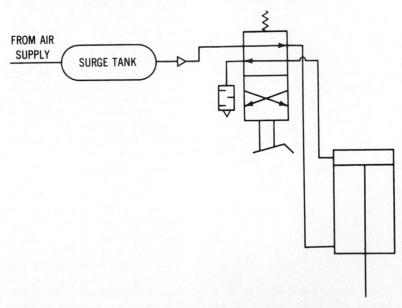

FROM AIR SUPPLY

SURGE TANK

Fig. 43-3. Diagram of a pneumatic circuit that includes a surge tank for supplying air quickly to the cylinder.

2. The load may be too large for the cylinder to move.

3. The speed control valve can be closed to the point where the cylinder cannot exhaust the air. In the three-position valves, the control may have been moved accidentally to the center position, blocking the passages to the cylinder.

4. The piston may have broken, with the pieces lodged between the piston and the cylinder wall.

5. The ball check in the cushion assembly may be stuck; therefore, air pressure on only the small area of the cushion nose is not sufficient to force the piston from the cushion. This occurs more often with nonadjustable cushions.

nonmetallic orifice can be used.

2. The loads should be calculated carefully, and provision should be made to compensate for a drop in line pressure.

3. Make sure that all valves are functioning properly.

4. Check the action of the cylinder. If it seems to be jammed, the cylinder should be disassembled, checked, and repaired.

5. Clean the ball check in the cushion assembly, and check the parts to determine whether the cylinder is operative.

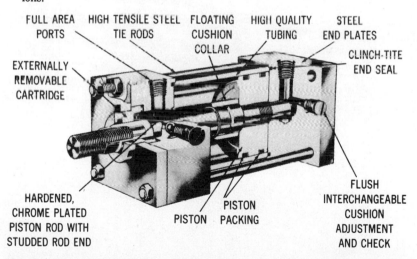

FULL AREA PORTS   HIGH TENSILE STEEL TIE RODS   FLOATING CUSHION COLLAR   HIGH QUALITY TUBING   STEEL END PLATES

CLINCH-TITE END SEAL

EXTERNALLY REMOVABLE CARTRIDGE

HARDENED, CHROME PLATED PISTON ROD WITH STUDDED ROD END

PISTON   PISTON PACKING

FLUSH INTERCHANGEABLE CUSHION ADJUSTMENT AND CHECK

Fig. 43-4. Cross section of air cylinder which meets *NFPA* standards. Note the studded rod end. These studs are made from 125,000 psi (861,844.6 kPa) tensile point material. The extra-long rod bearing is in the form of a cartridge which houses the rod wiper and the rod seal. (Courtesy The S-P Manufacturing Corporation)

DIRECTIONAL CONTROL VALVES
(SOLENOID-OPERATED)
(Fig. 43-5)

Flow Director does not Shift

Probable Cause

1. Solenoid coil is burned out.

Remedy

1. Before making the electrical connections, check for correct voltage, and check the markings on the solenoid coils. Determine whether the solenoid plunger is bottoming; a loud humming noise indicates that the plunger is not bottoming, and the coils usually burn out quickly. Failure to bottom can be the result of:

    a. Foreign matter lodged in the path of the solenoid plunger; this may be

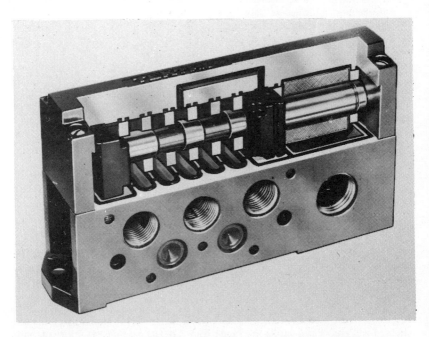

Fig. 43-5. Single-solenoid, lapped-spool stainless steel four-way air control valve. Valve is designed to provide high flow capacity and fast response. (Courtesy Logansport Machine Co., Inc.)

caused when the valve covers are removed and when the valves are operated with the covers removed. In the newer designs, the solenoids do not remain in place if the covers are removed.

b. The end of the flow director can be damaged by continual hammering of the solenoid plunger, causing the plunger to stick in a guide bushing. This peening action is prevented in newer valve designs by using a synthetic shock-absorber material on the end of the flow director.

c. In a double-solenoid valve, both solenoid plungers may be energized at the same time, resulting in neither plunger bottoming. This is usually caused by a malfunction in an electrical control circuit. If complicated electrical circuits are involved, check the electrical system and the control valves by manual push-pin operation, if possible.

2. Actuating or limit switch for the valve is malfunctioning.

2. When a solenoid-operated control valve malfunctions, the control valve is often suspected; however, the electrical control switches or relays are often at fault. The mechanism which trips the limit switch is also often at fault.

3. Flow director is cramped by the mounting of the valve body.

3. Faulty mounting is the usual cause of a binding action within the valve after it has been mounted. The valve should be mounted on a flat surface. On valves that contain threaded ports in the valve body, a binding action can result if the pipes are turned in too tightly.

4. A loose electrical connection.

4. Loose electrical connections resulting from vibration or other causes fail to direct current to the solenoid at the proper time.

5. Dirt particles may cause the flow director to stick.

5. Filters should be utilized in the circuit, and the piping should be clean when it is installed.

6. Lack of lubrication may cause sticking.

6. Control valves should be lubricated properly and with the proper lubricant. The lubricator should be installed at a short distance upstream from the control valve. An improper lubricant can cause a residue of grime, causing the flow director to stick.

DIRECTIONAL CONTROL VALVES (PILOT-OPERATED)

Flow Director does not Shift on either a Pressure Pilot-Operated or a Bleeder Pilot-Operated Control Valve

Probable Cause

1. Inadequate pressure for shifting the flow director.
2. Lack of lubrication in the valve.

3. Dirt particles causing a binding action in the valve.
4. Faulty mounting causing a cramped condition in the valve.
5. Worn seals on the mechanism that shifts the flow director.

6. Failure in the pilot valve that directs pressure to the pilot chamber of the control valve.

Remedy

1. Check for adequate pressure to the pilot chambers.
2. Lubrication should be checked both in the flow-director section and in the pilot-operating section of the control valve.
3. Check the line filters for contaminants.
4. Check for a faulty mounting condition that results in a binding action.
5. Check for leaks in the seals of the pilot operators, and replace any worn seals. Scratches in the bore of the pilot section of the valve should be eliminated.
6. Check the flow through the pilot controls, and repair or replace them if unsatisfactory. The pilot valves should function properly.

SEQUENCE CONTROL VALVES

Valve does not open Properly to Permit Proper Sequence Action of a Cylinder or Other Device

Probable Cause

1. Valve spring setting results in a higher tension on the flow director than the tension provided by the operating pressure of the system.

Remedy

1. Set the sequence valve to open at a pressure level that is higher than the pressure requirement in the primary circuit.

2. Valve spring may be broken.

2. Broken springs should be replaced immediately by springs that are recommended for the valve.

3. Valve packing may be worn or broken.

3. Defective packing should be replaced immediately by a recommended packing.

4. Dirt particles may be lodged between the flow director and the valve body.

4. Clean the valve thoroughly to remove dirt particles from the flow director or valve bore.

5. Dirt particles lodged on the poppet seat may permit leakage.

5. If dirt particles are lodged in the poppet seal, it may be necessary to replace the seal. If the poppet seat is scored, it should be refinished; or the body may be reworked and a bushing with a built-in poppet seat pressed in, if there is sufficient material in the valve body to receive the bushing.

6. Leakage may be caused by improper seating of the ball check.

6. If either the ball-check seat or the ball is damaged, it should be repaired to effect a tight seal.

SPEED CONTROL

Component (Cylinder, Actuator, etc.) Being Controlled is not Permitted to Repeat Accurately

Probable Cause

Remedy

1. Dirt particles in the air line may alter the size of the orifice through which the metered air passes.

1. Air should be filtered well.

2. Fluctuations in line pressure alter the operating pressure of the device and, in most instances, the operating speed of the device.

2. Use regulators near the controlled device to reduce fluctuation in air pressure on the downstream side of the regulator.

3. Dirt particles in the check assembly may cause the check to remain open when it should be closed.

3. Check the check-valve assembly, and replace the worn parts.

4. Leakage in the seals or seats may permit the control to be accurate, if the leakage is constant; however, a small particle of dirt, lubricant, or water may stop the leakage temporarily.

4. Check the seals to make sure that they are airtight.

Forward Stroke of Cylinder cannot be Controlled When Speed Control is located Between Rod End of Cylinder and the Four-Way Control Valve

Probable Cause

1. Speed control is reversed in the line.

Remedy

1. Turn the speed control 180 degrees, and return it to the line. The arrow that indicates controlled flow should be pointed *away from* the cylinder. These valves are often installed incorrectly.

System Responds Sluggishly

Probable Cause

1. Inadequate piping size, including the hose and fittings.
2. Oil-laden exhaust mufflers.

3. Valving restrictions.

4. Collapsed piping, resulting from damage by shop trucks, etc.
5. Clogged intake filters.

Remedy

1. Check the ID size of hose to be sure that full-size piping is being used.
2. Exhaust mufflers should be checked regularly, since they often become clogged with lubricants, which slows the action of the system.
3. Valving restrictions can be eliminated by replacing them with valves that permit a full flow.
4. Protect the pipe lines, and replace the damaged lines.
5. Clean the intake filters at regular intervals (Fig. 43-6).

NOTE: When troubleshooting a pneumatic system—before removing any components, make absolutely certain that the air pressure is shut off to the system and that all air pressure is released in the system.

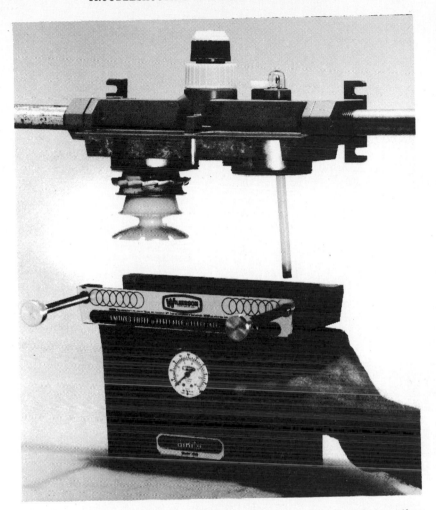

Fig. 43-6. This *F-R-L* unit is easily serviced. By using universal mounting brackets, the entire unit can be removed from the air line without disturbing the piping. This unit is designed with a 5-micron reusable filter element, a self-relieving regulator, and a lubricator which can be filled while the line pressure is on. (Courtesy Wilkerson Corp.)

Chapter 44

Troubleshooting of Hydraulic Circuits

Diagnosing the ills of a hydraulic circuit can be quite frustrating if some plan of analysis is not followed. Troubleshooting is usually more complicated in a hydraulic system than in an air system, because of more complicated valving, the fluid involved, and the pumping unit. Either faulty installation or misapplication of components causes most hydraulic circuit failures. It should be remembered in dealing with hydraulics that high pressures are usually involved and these are important factors when installing and using the system.

The symptoms, causes, and remedies are listed here in regard to failures in the hydraulic systems. Of course, as the student delves further into the study of hydraulics, other conditions that cause failures can be found.

577

HYDRAULIC POWER DEVICE
(Fig. 44-1)

Hydraulic Pump does not Produce Liquid, or does not Produce Rated Volume

Probable Cause	*Remedy*
1. The shaft that drives the pump (usually the shaft of an electric motor) rotates in the wrong direction, causing the shaft of the pump to rotate backward.	1. When the pump is first placed in operation, the pump shaft should be checked carefully to determine whether the driving means is rotating the shaft in the correct direction. The pump manufacturers usually place a directional arrow on the pump to indicate the correct direction of rotation.
2. The shaft that drives the pump is not rotating at the recommended speed.	2. If the pump shaft is not rotating at the correct speed, check the driving means. If an electric motor is the driving means, check its speed and its voltage. An incorrect voltage for the motor can reduce its speed.
3. The internal parts of the pump are worn badly, or they may be broken.	3. If the internal parts are worn badly, the pump should be repaired or replaced. Occasionally, a shaft key is sheared inside the pump; then the pumping mechanism does not rotate, and the pump does not deliver fluid. Broken springs inside the pump also can cause a pumping problem.
4. The coupling between the driving means and the pump is worn; then slippage occurs between the driving means and the pump.	4. If the coupling between the motor and pump is worn, it should be repaired or replaced.
5. The intake volume to the pump can be restricted by an intake line that is inadequate in size, an intake filter that is too small, or an intake filter that is clogged.	5. The intake of the pump should be free of restrictions. Only filters that are adequate in size should be employed.
6. The viscosity of the oil is too high.	6. The oil that is used should permit the pump to perform at its top operating efficiency.
7. An oil level that is too low does not permit the pump to deliver its full capacity, especially when the intake filter is partially covered.	7. The oil level in the pump should be maintained at the proper level.

Hydraulic Pump does not Produce Rated Pressure

Probable Cause	*Remedy*
1. The relief valve is not set at the correct pressure.	1. To set the operating pressure, close the downstream side of the relief

Fig. 44-1. Hydraulic power unit equipped with a variable-volume pressure-compensated tandem-vane pump unit. Note the large clean-out plate on the end of the reservoir which facilitates cleaning. The legs on the reservoir allow for circulation of air underneath the reservoir to help in the cooling of the oil. The large eye-bolts at each corner of the cover must be used to lift or transport the unit. (Courtesy Continental Hydraulics, Inc.)

valve; then gradually increase the tension on the regulating spring in the relief valve. A pressure gauge placed in the gauge connection in the relief valve permits the operating pressure to be read visually. Do not increase the pressure on the regulating spring to a value higher than the rated pressure.

2. An open-center valve in the system may not permit the pressure to build up.

2. Open-center valves should be employed only when they do not affect the operation of the system.

3. The internal parts of the pump, relief valve, and other components may be worn so badly that excessive leakage does not permit a build-up in pressure, especiallay when small-volume pumps are used.

3. Repair or replace the defective parts. New seals in the cylinder or a new spool in a control valve may remedy the problem.

4. Pump is rotating in the wrong direction.

4. Carefully check the direction of rotation of the pump shaft when installing a power unit.

5. Broken spring in relief valve or a stuck poppet.

5. The springs in direct-operated relief valves may be a problem. Remove the spring, and determine whether it is broken or set permanently. In some instances, the valve springs break and maintenance personnel attempt to substitute another spring for a broken spring, if the recommended replacement spring is not readily available. This often results in an early malfunction of the spring. It is often difficult for personnel to realize that the springs are designed for a specific job and that only the recommended spring is suitable. Spring breakage in relief valves often results from sudden shocks in the system. The poppet may be stuck in the open position, if dirt particles become lodged between the poppet and the bore in the valve body.

Hydraulic Pump Excessively Noisy

Probable Cause

1. A leak in the inlet line to the pump or in the pump-shaft seal permits air to

Remedy

1. Check for leakages in the inlet lines to the pump and at the pump-shaft seal

enter the pump, causing cavitation.

2. The oil viscosity is not compatible with the pump. This also causes cavitation inside the pump. Starting the pump in extremely low temperatures, even though the correct oil for normal operation is employed, may cause a problem.

3. If the oil level in the power unit is too low, air is permitted to enter the suction strainer, causing cavitation.

4. Excessive heat may cause a pump to become noisy and to fail quickly.

5. Dirt particles in the system, which are permitted to enter the pump, usually cause a noisy pump. Sometimes, an intake filter is removed for cleaning, and is not replaced before operation of the system is resumed.

by squirting a small volume of oil on each connection. If the connection is leaking, the oil is pulled into the connection, and the pump becomes quieter immediately. The connection should be sealed, and, if necessary, the pump-shaft seal should be replaced.

2. When installing the pumping system, the oil recommended by the pump manufacturer should be used.

3. Keep the oil level in the power unit at a level which covers the intake filter amply during all types of operating conditions. On systems that utilize long piping lines and large cylinders, either an auxiliary reservoir or a single large reservoir which more than satisfies the requirements of the pump may be necessary.

4. If excessive internal heat is involved, a heat exchanger usually solves the problem. A change in the design of a circuit often overcomes the heat problem.

5. Never permit a system to operate without some type of filtering protection for the pump.

Excessive Heat Created Within Hydraulic System

Probable Cause

1. A system operating under high pressure for lengthy periods of time.

Remedy

1. Check the operation thoroughly to determine whether the application actually requires lengthy periods of high-pressure operation. Perhaps an open-center control valve can be used in conjunction with an accumulator to reduce the need for the pump to operate continuously at high

pressure. A variable-volume pumping system may be employed. In some instances, a high-low pumping system can solve the problem. If high pressures are required for lengthy periods of time, the possibilities for utilizing accumulators, intensifiers, or heat exchangers should be explored. The oil temperature level should be kept between 120° (49°C) and 140°F (60°C). To conserve the packings and seals and to prolong the service of the system components, the oil should be kept in good condition.

CYLINDER
(Fig. 44-2)

Jumping Action of Piston in a Cylinder

Probable Cause	*Remedy*
1. Air in the system.	1. When air is permitted to remain in a hydraulic system, the voids cause a jumping or jerking action of the pistons in the hydraulic cylinders. When a hydraulic system is first placed in operation, the air should be bled from the entire system. Air bleeders can be installed in the high places in the system. Considerable time may be required to bleed all the air from the system.
2. Defective seats in the flow controls.	2. Although defective seats in the flow controls are not always the cause of a jumping or jerking action, the repeated jumping actions often become a problem, and the seats should be replaced.
3. A worn or scored cylinder tube or defective piston seals can cause a jumping action.	3. Replace the worn tube. If the tube is scored, it may be possible either to polish out the scored marks or to hone and chrome plate the ID of the tube. If work is performed on the tube, the piston seals should be replaced.
4. Excessive side loading of the piston rod of the cylinder.	4. Provide adequate guide bearings for eliminating the excessive side loads.

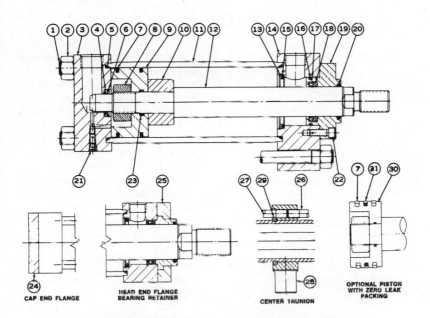

Cylinder Parts List

NO	PART NAME		NO.	PART NAME
1	TIE ROD		18	ROD PACKING
2	TIE ROD NUT		19	GLAND

Fig. 44-2. In troubleshooting hydraulic equipment, it is well to have a parts list drawing of each component in the system. Here is a parts list drawing of a heavy-duty hydraulic cylinder showing clearly the construction. (Courtesy Rexnord Inc., Cyl. Div.—HANNA)

5. The power produced by the cylinder cannot overcome the external friction caused by the tight spots between the piston rod and the guides.

5. Provide a cylinder that is adequate for the job, and keep the guides aligned and lubricated properly.

Movement of Cylinder Piston Sluggish

Probable Cause

1. Restriction in the piping.

Remedy

1. The piping size should be adequate for accommodating the full flow from the pump. The recommended size for the pipe bends should be used, and the number of pipe bends should be kept at a minimum. The oil velocity in the piping should be as low as possible; for practical purposes, it should not exceed 15 feet per second.

2. A control valve in the system may remain partially open.

2. Check all control valves in the system to be sure that their drain lines are not restricted.

3. The cylinder may be working at its full capacity.

3. Use a cylinder with a margin of safety for overcoming conditions that are not ideal.

4. Misalignment of the device and the piston rod of the cylinder.

4. To maintain alignment, it is often necessary to change the mounting style of the cylinder.

DIRECTIONAL CONTROL VALVES (SOLENOID-OPERATED)
Flow Director does not Shift

Probable Cause

1. Solenoid coil is burned out.

Remedy

1. Before making the electrical connections, check for correct voltage, and check the markings on the solenoid coils. Determine whether the solenoid plunger is bottoming; a loud humming noise indicates that the plunger is not bottoming, and the coils usually burn out quickly. Failure to bottom can be the result of:

   a. Foreign matter lodged in the path of the solenoid plunger; this may be caused when the valve covers are removed and when the valves are operated with the covers removed. In the newer designs, the solenoids do not remain in place if the covers are removed.

b. The end of the flow director can be damaged by continual hammering of the solenoid plunger, causing the plunger to stick in a guide bushing. This peening action is prevented in newer valve designs by using a synthetic shock-absorber material on the end of the flow director.

c. In a double-solenoid valve, both solenoid plungers may be energized at the same time, resulting in neither plunger bottoming. This is usually caused by a malfunction in an electrical control circuit. If complicated electrical circuits are involved, check the electrical system and the control valves by manual push-pin operation, if possible.

2. Actuating or limit switch for the valve is malfunctioning.

2. When a solenoid-operated control valve malfunctions, the control valve is often suspected; however, the electrical control switches or relays are often at fault. The mechanism which trips the limit switch is also often at fault.

3. Flow director is cramped by the mounting of the valve body.

3. Faulty mounting is the usual cause of a binding action within the valve after it has been mounted. The valve should be mounted on a flat surface on valves that contain threaded ports in the valve body, a binding action can result if the pipes are turned in too tightly.

4. A loose electrical connection.

4. Loose electrical connections resulting from vibration or other causes fail to direct current to the solenoid at the proper time.

5. Dirt partices may cause the flow director to stick.

5. Filters should be utilized in the circuit, and the piping should be clean when it is installed.

DIRECTIONAL CONTROL VALVES
(PILOT-OPERATED)
(Fig. 44-3)

Flow Director does not Shift

Probable Cause	*Remedy*
1. Inadequate pressure for shifting the flow director.	1. Check for adequate pressure to the pilot chambers.
2. Dirt particles causing a binding action in the valve.	2. Check the line filters for ·contaminants.
3. Faulty mounting causing a cramped condition in the valve.	3. Check for a faulty mounting condition that results in a binding action.
4. Worn seals on the mechanism that shifts the flow director.	4. Check for leaks in the seals of the pilot operators, and replace any worn seals. Scratches in the bore of the pilot section of the valve should be eliminated.
5. Failure in the pilot valve that directs pressure to the pilot chamber of the control valve.	5. Check the flow through the pilot controls, and repair or replace them if unsatisfactory. The pilot valves should function properly.

DIRECTIONAL CONTROL VALVES
(MANUALLY OPERATED)

Flow Director does not Shift or Difficult to Shift

Probable Cause	*Remedy*
1. Broken internal part	1. Disassemble the control. Check for broken parts.
2. Operating pressure too high.	2. Check valve manual for correct operating pressure and reduce

Fig. 44-3. An air pilot-operated four-way hydraulic control valve. This control affords a rapid response and is suitable for machines requiring fast cycling. (Courtesy Continental Hydraulics, Inc.)

3. Stem seals may be swollen.

4. Valve body distorted because of improper mounting surface.

pressure. Too high pressure may cause valve cores to collapse.

3. Check to see whether the stem seals are compatible with the fluid being used. Change seals if necessary.

4. Make certain the valve is mounted on a flat surface.

DIRECTIONAL CONTROL VALVES (ALL TYPES)
Internal and External Leakage

Probable Cause

1. Excessive leakage between the flow director and the valve body.

2. Leakage of an external nature.

Remedy

1. Worn parts. If the valve is expensive, the flow director may be built up by plating to fit properly into the valve body.
   a. If the flow director is scratched because of dirt, it may be replaced or repaired by plating.
   b. Operating pressure may be too high. Check the recommended operating pressure and stay within that range.
   c. Oil may be of the wrong viscosity—either the wrong oil is being used or the oil is too hot.

2. Leakage at valve covers caused by deterioration of cover seals. Make certain that this is not caused by heat or the use of an oil that is not compatible with seals.
   a. Leakage at stem seals. This could be caused by dirt, heat, or oil that is not compatible with the stem seal. A burr on the stem can quickly cause the stem seal to leak. Repair as soon as possible.
   b. Valve covers that are loose can allow external leakage. Make certain the covers are tight.
   c. Porous valve body will allow the valve to leak externally. Repair or replace the body.

d. If the connections at the ports are not tight, leakage will occur. Make certain all pipe connections are tight.

PRESSURE CONTROL VALVES
Cannot Control Pressure Setting

Probable Cause	*Remedy*
1. Pressure control spring becomes broken or takes a "set."	1. Disassemble the valve and check the spring. Replace if necessary.
2. Wrong spring in valve.	2. Check the spring to see whether the correct spring is being used for the operating pressure. Some valves require different springs for different pressures.
3. Faulty valve seat in control head.	3. Inspect for score marks in the valve seat. Remove any imperfections. It may be necessary to replace the seat.
4. Control orifice plugged with dirt.	4. Check the orifice. Remove any dirt or gummy substances.
5. Spool in main valve body is sticking.	5. Remove the spool. Check for imperfections and deposits. Clean the spool thoroughly with a fine emery cloth.
6. Back pressure on exhaust or drain connection.	6. Eliminate any back pressure.
7. Gasket shutting off passage between body and cover.	7. Remove cover. Make certain the passage is clear.
8. Control is worn out.	8. Check clearances in valve. Replace if necessary.

Excessive Internal Leakage

Probable Cause	*Remedy*
1. Fluid viscosity too low.	1. Even though pressure control valves have close fits between the spool and body, a low-viscosity oil causes undesirable internal leakage. Either use a higher viscosity oil or change to a control that will provide the desired results.
2. Operating pressure too high.	2. Check the operating pressure of the system. Reduce it to the maximum recommended for the control, or

3. Excessive temperatures, either ambient or system.

3. If the ambient temperature is causing the problem, shield the control. If the system temperature is causing the problem, install a cooler, since that temperature will likely cause problems in other parts of the system.

Sluggish Action of Controls

Probable Cause

1. Extremely low temperatures.

2. Viscosity of fluid too high.

3. Inadequate pressure or flow to cause controls to properly function.

4. Dirty fluid which clogs orifices and affects the operation of the controls.

Remedy

1. The viscosity of the oil will be greatly increased, thereby reducing the oil flow through small orifices. It may be necessary to install heaters in the system.

2. Either use an oil of a lower viscosity or use valves with larger orifices.

3. Reduce the size of controls.

4. Check the fluid and install better filters.

Leaks Caused by Seal Problems

Probable Cause

1. Chemicals and other fluids splashed on the controls.

2. Fluid not compatible with seal and gasket material.

3. Excessive ambient or operating temperatures.

Remedy

1. Protect the controls from materials that will cause seals to deteriorate or install seals that will withstand chemicals and such fluids.

2. Check the type of fluid being used in the system. It may be necessary to change the seals in all of the components. Some fire-resistant fluids have adverse affects on seals made for petroleum-base fluids.

3. Reduce the temperature by providing adequate cooling of the fluid, or by protecting the controls from any hot blasts.

FLOW CONTROLS
Will Not Control Movement of Component

Probable Cause	*Remedy*
1. Broken internal parts in control.	1. Disassemble control and inspect parts. Replace defective parts.
2. Flow control installed backward in the system.	1. Check to see whether the control arrow on the valve is pointed in the right direction. On a nonpressure-compensated control, the fluid should be controlled as it leaves the cylinder or motor.
3. Dirt in check section.	2. Dirt lodged under the seat of the check section will keep the free-flow passage partially open. Clean the check section.
4. Orifices clogged.	3. Clean the valve thoroughly. Remove any gummy substances.

Flow Restricted in Free-Flow Section

Probable Cause	*Remedy*
1. Valve size inadequate for application.	1. Install larger ported valve.
2. Broken internal parts.	2. Broken parts will not allow check to fully open. Inspect valve and replace defective parts.
3. Parts in free flow section binding.	3. Inspect the check section of the valve. Eliminate any binding conditions. Deposits of dirt, varnish, or gum in the check section may be the cause of trouble.

HYDRAULIC MOTORS
Motor Shaft Rotating in Wrong Direction

Probable Cause	*Remedy*
1. Piping from control to motor in error.	1. Check the ports on the control and switch piping at control or at motor.

Motor Shaft not Rotating

Probable Cause	*Remedy*
1. Fluid not getting to the motor.	1. Check to see whether fluid is getting

2. System operating pressure too low to cause motor shaft to rotate when attached to heavy load.

to the motor. There may be an open control upstream. Correct the condition.

2. Calculate the theoretical torque requirements. Determine whether the motor operating at the system pressure will meet those requirements. It may be necessary to install a larger motor or to increase the operating pressure of the system.

3. Broken internal parts.

3. Dismantle the motor. Replace any defective parts.

4. Binding action in the part to be rotated.

4. Before dismantling the motor, make certain the problem does not lie with the object to be rotated. Check for any binding action.

5. Blocked exhaust port on the motor.

5. Check the circuit to make certain a control is not blocking the exhaust port.

Excessive Shock in System Causing Breakage in Piping and Components

Probable Cause

Remedy

1. Sudden stoppage of high-pressure, high-velocity fluid.

1. Install small accumulators which function as shock alleviators in the strategic points in the system. Short lengths of hose installed in the piping sometimes reduce pipe breakage. The piping should be anchored in the problem spots.

2. If stopped suddenly, heavy loads moving at high speeds may damage a hydraulic system.

2. The piston moving at high speeds should be provided with adequate cushioning in the cylinder. Flow controls also should be used, if possible.

3. Breakage of springs in control valves.

3. In some instances, a pilot-operated control, rather than a direct spring-operated control, can be used to reduce spring tension and spring travel.

4. The sudden shifting of a directional control valve, such as a direct solenoid-operated or an air pilot-operated control valve.

4. Equip the flow director with shock grooves. Replace the direct solenoid-operated control valves with pilot solenoid-operated control valves. On air pilot-operated valves, install snubbers or choke blocks.

Many symptoms, causes, and remedies have not been discussed here. Actual or practical experience with hydraulic circuits is most helpful in recognizing the various problems in hydraulic systems.

In troubleshooting a hydraulic circuit, the use of a portable hydraulic circuit tester can be very helpful. It helps to eliminate down time and to minimize service time. With the tester shown in Fig. 44-4, it is not necessary to remove the individual components from the system for visual or mechanical checking. The tester is connected into the pressure line of the pump (Figs. 44-5 and 44-6).

Before removing any component from a hydraulic system (Fig. 44-7) always make certain that the power unit is shut off and that the pressure is released from the system.

In a plant or an operation where there is a great deal of hydraulic equipment, a hydraulic test bench is a big assist to a troubleshooter.

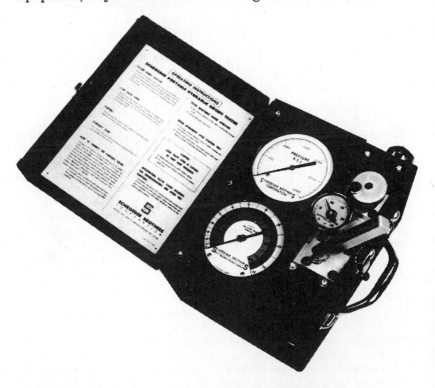

Fig. 44-4. Hydraulic circuit tester for troubleshooting faulty circuit components. (Courtesy Schroeder Brothers Corp. an Alco Standard Co.)

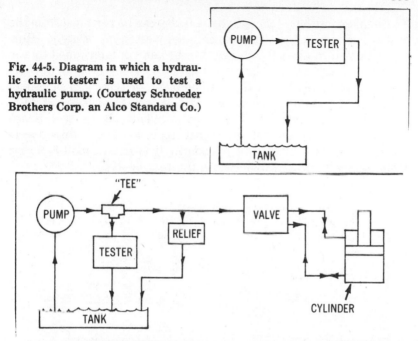

Fig. 44-5. Diagram in which a hydraulic circuit tester is used to test a hydraulic pump. (Courtesy Schroeder Brothers Corp. an Alco Standard Co.)

Fig. 44-6. Diagram in which a hydraulic circuit tester is used to analyze performance of hydraulic system. (Courtesy Schroeder Brothers Corp. an Alco Standard Co.)

Fig. 44-7. Hydraulic systems that must withstand all types of environmental conditions, yet give peak performance for long periods, are employed on mobile-type machines. Note the number of large hydraulic cylinders on this backhoe loader. (Courtesy International Harvester Co., Pay Line Group)

The component that is thought to be faulty can be removed from the system, placed on a test bench, and tested accurately. How sophisticated the test bench needs to be can be determined by the projected usage after it is installed.

A hydraulic test bench that can be used for testing a variety of hydraulic pumps, motors, valves, and cylinders is shown in Fig. 44-8. The operating controls are located conveniently at workbench height. The instrumentation is easy to read. Quick disconnects facilitate hookups and testing procedures. The electric motor and the pump unit are located underneath the test bench.

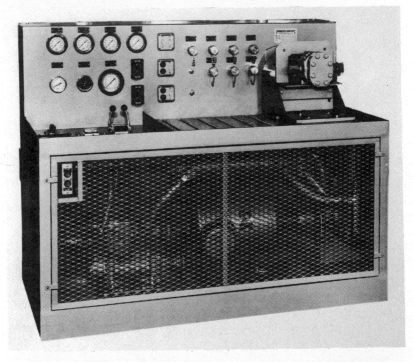

Fig. 44-8. A hydraulic test bench for efficient and accurate testing of pumps, cylinders, valves, and motors. (Courtesy Schroeder Brothers Corporation, an Alco Standard Co.)

Appendix A

Engineering Information

Metric System of Measurements

The principal units are the meter (m) for length, the liter (L) for volume or capacity, and the kilogram (kg) for weight. The following prefixes are used for subdivisions and multiples:

milli = 1/1000; centi = 1/100; deci = 1/10
deca = 10; hecto = 100; kilo = 1000

Measures of Length

10 millimeters (mm)	= 1 centimeter (cm)
10 centimeters	= 1 decimeter (dm)
10 decimeters	= 1 meter (m)
1000 meters	= 1 kilometer (km)

Measures of Weight

10 milligrams (mg)	= 1 centigram (cg)

10 centigrams	= 1 decigram (dg)
10 decigrams	= 1 gram (g)
10 grams	= 1 decagram (dag)
10 decagrams	= 1 hectogram (hg)
10 hectograms	= 1 kilogram (kg)
1000 kilograms	= 1 (metric) ton (t)

Square Measures

100 square millimeters (mm²)	= 1 square centimeter (cm²)
100 square centimeters	= 1 square decimeter (dm²)
100 square decimeters	= 1 square meter (m²)

Cubic Measures

1000 cubic millimeters (mm³)	= 1 cubic centimeter (cm³)
1000 cubic centimeters	= 1 cubic decimeter (dm³)
1000 cubic decimeters	= 1 cubic meter (m³)

Dry and Liquid Measures*

10 milliliters (mL)	= 1 centiliter (cL)
10 centiliters	= 1 deciliter (dL)
10 deciliters	= 1 liter (L)
100 liters	= 1 hectoliter (hL)

1 liter (L) = 1 cubic decimeter (dm³) = the volume of 1 kilogram (kg) of pure water at a temperature of 39.2° F

*National Metric Council prefers that dry and liquid measures be expressed in liters (L) and milliliters (mL).

Appendix B

Useful Conversion Multipliers

Atmospheres × 33.9   = feet of water (at 4°C)
          × 29.92 = inches of mercury (Hg) (at 0°C)
          × 1.0332 = kg/cm^2
          × 14.7   = pounds/in² (psi)

bar × 1.018 = kg/cm^2
   × 14.5 = pounds/in² (psi)
   × 10^5 = pascals (Pa)
   × 10^2 = kilopascals (kPa)

Btu/hr   × 0.0003929 = horsepower
Btu/min × 0.02356   = horsepower
Btu   × 0.00029282= kilowatt hrs

centimeters × 0.03281 = feet
           × 0.3937 = inches
centimeter-grams × 0.00007233 = pound/ft

centimeters of mercury × 101316 = atmospheres
$\qquad$ × 0.1934 = psi
centimeters/sec $\quad$ × 1.9685 $\qquad$ = ft/min
$\qquad$ × 0.03281 $\quad$ = ft/sec
cubic centimeters × 0.061023 $\quad$ = in^3
$\qquad$ × 0.00003531 = ft^3
$\qquad$ × 0.0002642 $\quad$ = gallons (U.S.)
cubic feet × 28,320 $\quad$ = cm^3
$\qquad$ × 1728 $\qquad$ = in^3
$\qquad$ × 7.48052 $\quad$ = gallons (U.S.)
$\qquad$ × 28.316 $\quad$ = liters
cubic feet/min × 472 $\qquad$ = cm^3/sec
cubic inches $\quad$ × 16.387 $\qquad$ = cm^3
$\qquad$ × 0.0005787= ft^3
$\qquad$ × 0.016387 $\quad$ = liters
cubic meters $\quad$ × 1,000,000 = cm^3
$\qquad$ × 35.317 $\qquad$ = ft^3
$\qquad$ × 61,023 $\qquad$ = in^3
$\qquad$ × 264.17 $\qquad$ = gallons (U.S.)
cubic meters/hr × 4.403 = gpm

degrees (arc) × 0.01745 = radians
degrees/sec × 0.1667 = rpm
dynes × 0.000002248 = pounds
dynes/cm^2 × 0.0000145038 = psi

feet × 30.48 $\qquad$ = centimeters
$\quad$ × 0.3048 $\qquad$ = meters
$\quad$ × 0.0003048 = kilometers
feet of water × 0.02950 = atmospheres
$\qquad$ × 0.8826 $\quad$ = in. of mercury
$\qquad$ × 0.03048 = kg/cm^2
$\qquad$ × 0.4335 $\quad$ = psi
feet/min × 0.508 $\quad$ = cm/sec
$\qquad$ × 0.01667 = ft/sec
$\qquad$ × 0.3048 $\quad$ = meters/min
$\qquad$ × 0.01136 = miles/hr (mph)
feet/sec × 30.48 $\quad$ = cm/sec
$\qquad$ × 18.29 $\quad$ = meters/min
$\qquad$ × 0.6818 = miles/hr (mph)

foot-pounds × 0.001286   = Btu
           × 0.000000505 = hp-hr
           × 0.1383   = kg-meters
foot-pounds/min × 0.0000303 = hp

gallons (U.S.) × 3785   = cm^3
          × 0.1337 = ft^3
          × 231   = in^3
          × 3.78543 = liters
          × 0.83267 = gallons (Brit. Imp.)
gallons (Brit. Imp.) × 1.20095 = gallons (U.S.)
gallons/min (gpm)   × 0.002228 = ft^3/sec
                × 0.13368 = ft^3/min
grams × 0.002205 = pounds
grams/cm   × 0.0056 = pounds/inch
grams/cm^3 × 62.43   = pounds/ft^3
grams/cm^2 × 2.0481 = pounds/ft^2
horsepower × 42.44   = Btu/min
         × 33000 = ft-pounds/min
         × 550   = ft-pounds/sec
         × 0.7457 = kilowatts
         × 745.7 = watts
horsepower-hrs × 2547   = Btu
            × 1,908,000 = ft-pounds

inches × 2.540 = centimeters
     × 0.0254 = meters
     × 25.40 = millimeters
     × 1000   = mils
inches of mercury × 0.03342 = atmospheres
               × 0.03453 = kg/cm^2
               × 70.73   = pounds ft^2
               × 0.4912 = pounds/in^2 (psi)
inches of water (at 4°C) × 0.002458 = atmospheres
                     × 0.07355 = inches of mercury
                     × 0.00254 = kg/cm^2
                     × 0.03613 = pounds/in^2 (psi)

kilograms × 2.2046 = pounds
kg/cm^2 × 0.9678 = atmospheres

kg/cm²   × 28.96   = inches of mercury
        × 14.22   = pounds/in² (psi)
        × 0.982   = bars
kilowatts   × 0.0004426 = ft-pounds/min
       × 737.6   = ft-pounds/sec
       × 1.341   = horsepower
kilometers × 3280.8   = feet
        × 0.62137 = miles (statute)

kilopascals (kPa) × 0.1450377 = lb/in² (psi)
kilowatt-hrs × 1.341 = horsepower-hrs
        × 3415 = Btu

liters × 1000   = cm³
    × 0.035317 = ft³
    × 61.02   = in³
    × 0.26417 = gallons (U.S.)
liters/min   × 0.0005886 = ft³/sec

meters/min × 1.667   = cm/sec
        × 3.281   = ft/min
        × 0.05468 = ft/sec
meters/sec × 196.8   = ft/min
        × 3.281 = ft/sec
meters × 1.09361 = yards
    × 3.2808 = feet
    × 39.37   = inches
millimeters × 0.03937 = inches
miles (statute) × 1.60935 = kilometers
mils × 0.00254   = centimeters
   × 0.00008333 = feet
   × 0.001   = inches
minutes (arc) × 0.01667   = degrees
         × 0.0002909 = radians
         × 60   = seconds

pascals (Pa) × 0.000145 = lb/in² (psi)
poise   × 1 = gram/cm-sec
pounds × 453.5924 = grams
    × 0.4536 = kilograms

pounds of water $\times$ 0.01602 = ft^3
$\qquad\qquad\qquad \times$ 27.68 = in^3
$\qquad\qquad\qquad \times$ 0.1198 = gallons
pounds/ft^3 $\times$ 0.01602 = grams/cm^3
pounds/ft^2 $\times$ 0.0004725 = atmospheres
$\qquad\qquad \times$ 0.01414 = in. of mercury
$\qquad\qquad \times$ 4.882 = kg/m^2
$\qquad\qquad \times$ 0.006944 = pounds/in^2 (psi)
pounds/in^2 (psi) $\times$ 0.06804 = atmospheres
$\qquad\qquad\qquad \times$ 2.036 = in. of mercury (Hg)
$\qquad\qquad\qquad \times$ 703.1 = kg/m^2
$\qquad\qquad\qquad \times$ 6894.757 = pascals (Pa)
$\qquad\qquad\qquad \times$ 144 = pounds/ft^2
$\qquad\qquad\qquad \times$ 0.0703 = kg/cm^2

radians $\times$ 57.3 = degrees (arc)
$\qquad \times$ 3438 = minutes (arc)
$\qquad \times$ 206,300 = seconds (arc)
radians/sec $\times$ 9.549 = rpm
$\qquad\qquad \times$ 0.1592 = revolutions/sec
revolutions/min (rpm) $\times$ 6 = degrees/sec
$\qquad\qquad\qquad\qquad \times$ 0.01667 = revolutions/sec
revolutions/sec $\times$ 60 = rpm

slug $\times$ 14.59 = kilograms
$\qquad \times$ 32.17 = pounds
square centimeters $\times$ 0.0155 = in^2
square feet $\times$ 929 = cm^2
$\qquad\qquad \times$ 144 = in^2
square inches $\times$ 6.452 = cm^2
$\qquad\qquad\quad \times$ 0.06944 = ft^2
$\qquad\qquad\quad \times$ 645.2 = mm^2
square millimeters $\times$ 0.01 = cm^2
$\qquad\qquad\qquad \times$ 0.01550 = in^2

temperature ($^\circ$C) $\times$ 1.8 + 32 = ($^\circ$F)
$\qquad\qquad$ ($^\circ$F) $-$ 32 $\times$ 0.555 = ($^\circ$C)
$\qquad\qquad$ ($^\circ$C) + 273 = absolute temperature ($^\circ$C)
$\qquad\qquad$ ($^\circ$F) + 460 = absolute temperature ($^\circ$F)
tons (long) $\times$ 1016 = kg

$$\times \ 2240 \ = \ \text{pounds}$$
$$\times \ 1.120 \ = \ \text{short tons}$$
tons (metric) $\times$ 1000 = kg
$$\times \ 2205 \ = \ \text{pounds}$$
tons (short) $\times$ 907.1848 = kg
$$\times \ 2000 \quad = \ \text{pounds}$$
$$\times \ 0.89287 \ = \ \text{long tons}$$
$$\times \ 0.9078 \quad = \ \text{metric tons}$$

watts $\times$ 3.413   = Btu/hr
$$\times \ 0.05688 \ = \ \text{Btu/min}$$
$$\times \ 0.001341 = \text{horsepower}$$
$$\times \ 0.001 \quad = \ \text{kilowatts}$$
watt-hrs $\times$ 3.413   = Btu
$$\times \ 2656 \quad = \ \text{ft-pounds}$$
$$\times \ 0.001341 = \text{horsepower-hrs}$$
$$\times \ 367.2 \quad = \ \text{kg-meters}$$
$$\times \ 0.001 \quad = \ \text{kw-hrs}$$

yards $\times$ 0.9144 = meters

Index